MODERN
PRACTICAL BOTANY

MODERN PRACTICAL BOTANY

Volume - 2

(PTERIDOPHYTES, GYMNOSPERMS, ANATOMY AND EMBRYOLOGY OF ANGIOSPERMS)

Dr. S. Sundara Rajan
Professor and Head
Dept. of Post Graduate Studies and Research in Botany
St. Joseph's College
Bangalore - 560 025

ANMOL PUBLICATIONS PVT. LTD.
NEW DELHI - 110 002 (INDIA)

ANMOL PUBLICATIONS PVT. LTD.
4374/4B, Ansari Road, Daryaganj
New Delhi - 110 002

Modern Practical Botany

First Edition, 2000

ISBN 81-261-0690-5 (Set)

PRINTED IN INDIA

Published by J.L. Kumar for Anmol Publications Pvt. Ltd., New Delhi -110 002 and Printed at Mehra Offset Press, Delhi.

CONTENTS

Preface

Laboratory study is an integral part of the knowledge that a student acquires in the college. This is true not only for botany but for all science subjects. Theoretical study of a subject has to be supplemented with laboratory experiments to provide a comprehensive understanding.

One of the chief advantages of laboratory or practical study is the first hand knowledge that a student acquires about the material of study.

A systematic, structured and disciplined method of work is necessary in the laboratory to obtain maximum benefit. In order to help the students realize their goal practical manuals indeed play an important role.

Modern practical botany aims to provide the students, a useful hand book to guide them in the proper study of the subject material. Theoretical details are kept at the minimum and more emphasis is laid on understanding the structural details to enable the student a proper identification. A large number of properly labelled original diagrams are incorporated in the book as they are integral to the understanding of the subject.

Modern Practical Botany is a series of three volumes.

Volume One deals with Algae, Fungi, Bacteria, Lichens, Viruses and Bryophytes.

Volume Two deals with Pteridophytes, Gymnosperms and Anatomy and Embryology of Angiosperms.

Volume Three deals with Taxonomy, Economic Botany, Ecology, Physiology and Cytogenetics.

I have taken utmost care in the presentation of matter, style of writing and choice of illustrations so that the volumes are user friendly. The examples chosen for study are such that the books should cater to the needs of all Indian Universities.

Constructive criticisms and suggestions are most welcome.

Dr. S. SUNDARA RAJAN
Professor and Head

Dept of Botany
St. Joseph's College
Bangalore - 560 025.

Chapter 1

INTRODUCTION

The study of science subjects is carried out at two levels,viz., *Theoretical* and *Practical*. Theoretical study is carried out in lecture classes when the students are imparted all the details regarding a particular topic. This has to be supplemented by practical or Laboratory studies, when the students come into active contact with the objects about which they would have heard in lecture classes. Therefore it may be safely said that laboratory studies form an integral part of a comprehensive study of any topic in science subjects.

In the study of plants, laboratory assumes a special significance as the students acquire a meaningful and firsthand knowledge about the plants which they study in lecture classes. In fact as far as botany is considered any study of a plant without being supplemented by laboratory observation is futile. It is only in the laboratory study that the student can dissect, observe and analyze the various features of the plant and it is only a study of this type that makes the student acquire a meaningful knowledge about the pant. Besides observation and study in the laboratory plants are also to be studied in field to obtain an account of their habitat, habit, association and such other details.

In this chapter an attempt is made to explain some of the essentials that a student should understand before he begins a practical work.

Cleanliness: This is the most important point that a student should keep in mind while in the laboratory. The working table and the surroundings should be kept dry and spotlessly clean.. All the requirements needed by the students such as petri dishes, slides, reagent bottles, stains etc. must be properly labeled and arranged neatly. The microscope and particularly the optical parts must be cleaned with a white lint free cloth. After the completion of the assigned work, the student should take care to clean the place, remove any debris of the plant material and put it into the dust bin.

The students should work in silence and independently unless the assigned experiment is to be carried out by a group. When a particular work is assigned to a student, he must carefully plan and divide the available time for experimentation, observation and recording of the results. Whenever a doubt arises, it is always better to consult the teacher in charge, than a classmate. It is only by a proper and systematic work, that a student can derive maximum benefit from the practical classes.

Laboratory requirements

The department provides certain amenities, instruments etc to each student for the purpose of practical work. But a student also should carry some of the requirements in order to be effective in the practical class.

One cardinal principle that every student should remember in a practical class is to have one's own requirements and not to depend on anyone. Never lend or borrow any requirements. Borrowing or lending a pencil, needle or blade will disturb the work of both.

In general every student should individually possess the following requirements

1. *Two sharp pointed needles* mounted on stainless steel or plastic handle. This is essential for the purpose of separating and teasing the microscopic material.

2. *A pair of forceps:* This is essential to remove the material either from the bottle or from petri dish.

3. A sharp properly honed razor to cut sections of the plant material.

4. A new safety razor blade. This will help either to cut section or to trim the material.

5. Two fine hair paint brushes. This will be of use in lifting the sections of plant (with out damage) from the petri dish on to the slide.

6. Two sharpened pencils (H, Hb or 2b).

7. A pencil eraser.

8. A pencil sharpener.

9. Glass droppers.

10. A piece of thin, white, lint free cloth to clean the optical parts of the microscope.

11. An observation book to note down immediately the points of interest.

12. A practical record.

13. Glass slides and cover slips(if they are not provided by the department).

Scheme of Practical Work

Practical work in the laboratory will be most useful and effective if the student keeps in mind the following:

1. Maintain silence during working. All the requirements for the practical work must be absolutely clean.

2. Every student should work independently, unless the experiment involves the participation of others.

3. Before beginning the practical work, listen to the instructions of the teacher attentively and work accordingly.

4. Follow a systematic procedure in a study.

5. If the work involved is the morphological study of an organism, follow the procedure given below:

(a) Study of external characters - for example whether a plant is microscopic or macroscopic, unicellular or multicellular, filamentous or colonial etc.

(b) Study of internal structure like details of cell, tissues etc. If the material requires sectioning, sections are to be cut and internal details should be observed.

(c) Study the reproductive details like asexual and sexual reproductive organs.

6. Mount the material after properly teasing and separating the filaments, mycelia etc. Stain the material wherever necessary and cover the preparation with a cover glass.

7. Observe the slide carefully under the microscope, first under low power and then under high power.

8. Make sure that you have observed all the details as instructed by the teacher.

9. After all the observations are over, record your findings, structural details etc. in the practical records in the form of diagrams. Draw proportionate diagrams in the proper place on the drawing sheet. Diagrams should never be drawn to the very edge of the drawing sheet. Write suitable explanatory notes pertaining to the diagrams on the interleaf.

Laboratory Equipments

In addition to the individual requirements carried by the student, the department generally provides the following equipments/instruments etc.to the students. Sometimes these equipments may be given to each student .

These are :

(1) Students compound microscope .
(2) Dissection microscope.
(3) Binocular Research microscope.
(4) Watch glasses.
(5) Cavity blocks.
(6) Slides and cover glasses.
(7) Petri dishes.
(8) Enamel tray.
(9) Polythene/glass wash bottle.
(10) Spirit lamps.
(11) Glass marking pencil.
(12) Microtome.
(13) Staining /bottles.
(14) Stone/ for honing the razor.
(15) Chemicals and stains.
(16) Culture media(to be prepared by students).

In the list of requirements mentioned above, we will discuss some in detail.

Microscope

Of all the instruments used by students in the botany laboratory, the one that is most indispensable to the students of biology is the 'Microscope'. It is not an exaggeration to say that no biological study in the laboratory is possible without the microscope.

The function of the microscope is to enlarge the small objects, so that they become visible to the human eye. In reality the magnification power of the microscope lies in its resolving capacity between one particle and another. A microscope distinguishes one particle from another which are situated very closely. A human eye cannot resolve particles that are lying closer than 0.025 mm. In its function a microscope can be compared to a human eye, but much more effective than that.

There are different types of microscopes, but all of them are based on the same principle with variations in their construction. Some of the types of microscopes used in the laboratory are:

(1) Compound microscope.
(2) Binocular Research microscope.
(3) Dissection microscope.
(4) Stereoscopic dissection microscope.
(5) Phase contrast microscope.
(6) Electron microscope etc.

Of the six types of microscopes mentioned above the compound microscope and the dissection microscope are the ones that are most frequently used by students. The Electron microscope is very costly and a very highly sophisticated instrument.

In this book the compound microscope and the dissection microscope are explained in detail as the students should be familiar with them.

Compound Microscope

The microscope has a strong 'U' shaped or 'V' shaped base. It is also called the horse shoe shaped base. Attached to the base is a bent limb. This limb can be tilted in any direction as it is attached to the base with an inclination joint.

Attached to the limb is a stage at its lower end and a body tube at its upper end. The body tube is attached to the limb with the help of a rack and pinion so that it can move up and down. To help in the movement of the tube during focusing, there are 2 knobs viz.,*Coarse adjustment* knob and *Fine adjustment knob*.

The body tube at its top end consists of the ocular or eye piece lens. The eye pieces usually have three magnifications 5x,10x,and 15x, and can be used to get a low or high magnification. Attached to the lower end of the tube is a revolving nose piece with three objective lenses low (10x) medium (40x) and high (60x). Some times there will be a fourth objective of the magnification of 100x. This is also called the oil immersion lens (while viewing through an oil immersion lens a drop of oil-liquid paraffin has to be put on the viewing slide).

Below the objectives and attached to the limb is the *stage* of the microscope where the slides are kept for viewing. The stage may be circular, square or rectangular in shape. In order to keep the slides properly placed on the stage, there are two stage clips. In some of the costly microscopes, the stage has movable sliding system operated by knobs to move the slides. This is called the mechanical stage. The mechanical stage permits the movement of the slide backwards or forwards, left or right with the turning of the knobs provided for the purpose. A hole is provided in the stage to receive the light from below and illuminate the object. Attached to the limb, below the stage is a condenser (fixed or movable up and down) lens. Below the condenser lens is an Iris diaphragm to regulate the light passing through. Iris diaphragm is a metallic disc with an aperture. The aperture can be narrowed or widened by turning the lever. The condenser which is a system of lenses condenses the light rays passing through and converges them on the object.

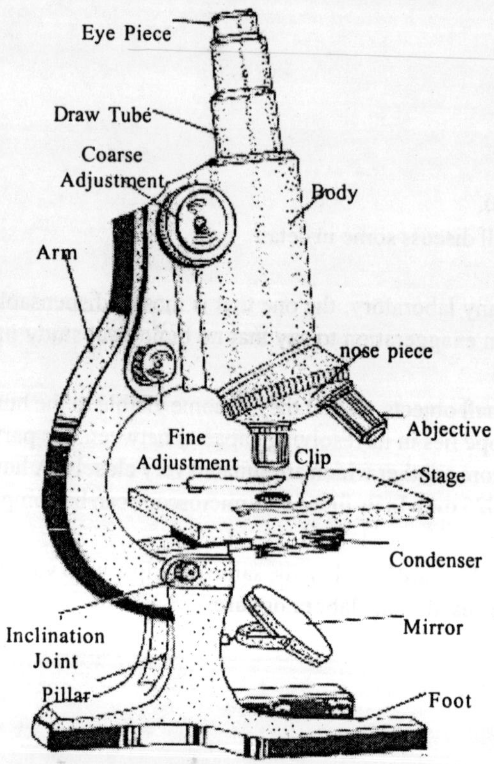

Fig.1.1 A compound microscope

The source of illumination in a student's microscope is a mirror which reflects the natural light or electric light. The mirror has two faces - plain and concave and is attached to the limb below the condenser. The mirror is adjustable and can be turned to the direction of light. When day light is the source of illumination, the concave side of the mirror is usually used to converge the rays of light. In some of the costly microscopes, instead of the mirror there is a built in illumination ,This is a tube with a lamp (9v or 12v) attached to a transformer, which can be plugged to any AC source.

Operation of the Microscope

Microscope should be placed in a well illuminated area but avoiding the direct light as it is harmful to the eyes. Generally northern light is preferable. When ' built in illumination' is employed, it is always better to use a blue or green filter to subdue the light.

The light has to be first focused on the condenser by adjusting the mirror towards the source of light. The amount of light entering the condenser may be regulated by closing or opening the iris diaphragm. The slide to be viewed is kept on the stage and fixed securely with the help of the stage clips. The object is first observed with a low power objective and then turned on to medium or high power, the specimen or section has to be covered with a cover slip. The slide should be observed with both the eyes open even while observing with monocular microscope.

Care of the Microscope

1. Clean all the parts of the microscope before and after use. The lenses and the mirror are to be cleaned with a lint free cloth or tissue paper.

2. Do not pull the microscope to the very edge of the table.

3. While lifting the microscope, use both hands. With one hand hold the base and with the other hold the arm.

4. The slide to be observed must be clean and dry (specially the under surface of the slide).

5. Do not remove the eyes piece or objectives from the microscope.

6. The microscope should be covered when not in use.

Binocular Research Microscope

This is like the ordinary compound microscope except that, it has binocular head for providing a natural stereoscopic view.

Dissection Microscope

This is a very simple microscope with only one set of lenses. Microscopes of this type will not give a very high magnification, but are very useful in dissecting flowers, plant parts etc.

The microscope has a 'V' shaped base and a limb. The stage is made up of a transparent glass plate. Below the stage is a mirror to focus the light. Attached to the limb is a mechanical arm which can be moved up and down with the help of a rack and pinion movement. The arm can also be moved horizontally. Attached to the free end of the arm is a holder to accommodate the lens (10x,15x etc.).

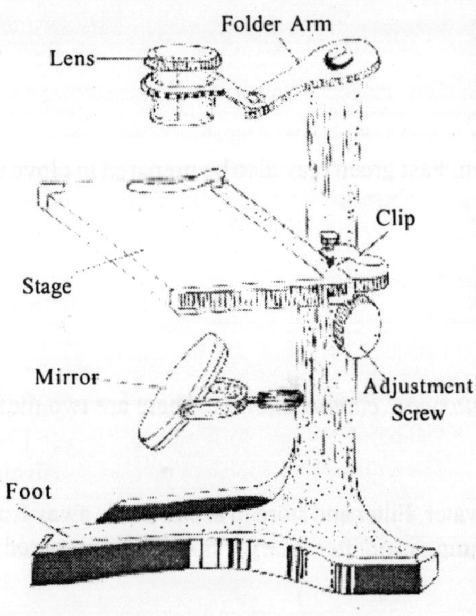

Fig.1.2 A dissecting microscope

Operation: The specimen to be observed / dissected is placed on the stage and located with the help of lens. Floral parts or others may be then dissected with the help of two needles, while still observing the specimen.

Stereoscopic Binocular Dissection Microscope

This is a combination of compound microscope and dissection microscope. While it has the higher magnification power of a compound microscope with two sets of lenses, it has retained the versatility and the manoeuvourability of a dissection microscope.

Phase contrast and Electron microscopes are not normally used by students.

Other Laboratory Equipments

Stains: These are colouring fluids. The specimens to be observed (algal or fungal material or sections of plant parts, would be generally decolourised due to preservation. These are to be coloured in order to observe them clearly. The following are the stains that are generally used in the laboratory.

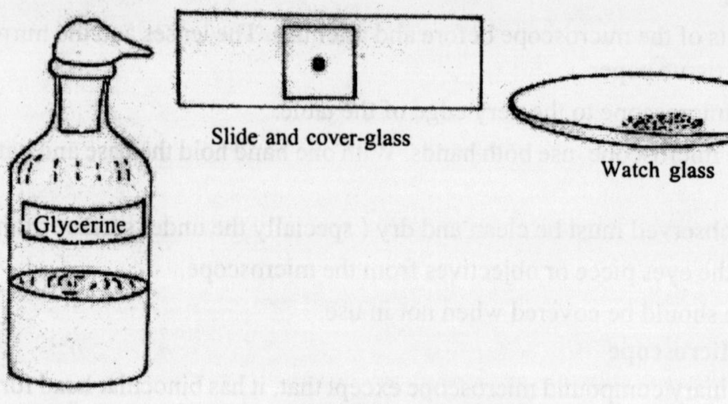

Fig.1.3 Dropping bottle

(a) Saffranin: This is the most commonly used stain. (1) It stains nucleus and lignified cell walls.
Preparation —

Saffranin —1gm.

95% (1) Ethyl alcohol –50ml.

Distilled water – 50ml.

An aqueous stain may also be prepared.

Saffranin – 18ml

Distilled water – 100ml.

(b) Fast Green

Fast green – 1gm.

90% Ethyl Alcohol – 100ml.

This stain is used to colour parenchyma cells and cytoplasm. Fast green may also be prepared in clove oil (75ml clove oil+25ml 100% alcohol).

(c) **Cotton blue.**

Cotton blue –1gm.

Lactophenol – 100ml.

(d) Haematoxylin

The dye is obtained from the heartwood of the tree *Hametoxylon campechianum,* There are two methods of preparation of the stain. These are:

1. *Heidenhain's Hametoxylin*

Dissolve 0.5 or 1.0 gm of the powder in 100 ml of distilled water. Filter and store in a dark bottle away from light. The stain is ready for use after 4-5 days of ripening. Ripening is not necessary if the stain is prepared in 70% ethyl alcohol.

2. *Delafelds Haematoxylin*

Sol.1 Aqueous solution of ferric ammonium sulphate 100ml).

Sol.2 One gm dye dissolved in 6ml of absolute alcohol.

mix solutions 1 and 2. To this mixture add 25 ml of glycerine and 25 ml absolute alcohol. The solution should be allowed to stand until it becomes dark red.

(e) Anilin blue

Also called cotton blue or China blue. It is prepared as follows:

Aniline blue – 1gm.

Ethyl alcohol (90%)

or — 100ml.

Distilled water

Hydrochloric acid – 2drops.

(f) Gentian Violet (Crystal Violet)

Gentian Violet – 1gm.

90% Ethyl Alcohol 100ml.

(g) Erythrosine

Erythrosine – 1gm.

90% Ethyl alcohol – 100ml.

Erythrosine may also be prepared in clove oil (Absolute alcohol 0.5 ml +95ml Clove oil).

(h) Light green

Light green – 1gm.

Clove oil – 95ml.

Absolute alcohol – 5ml.

Light green may also be prepared without clove oil (in 100ml Absolute alcohol).

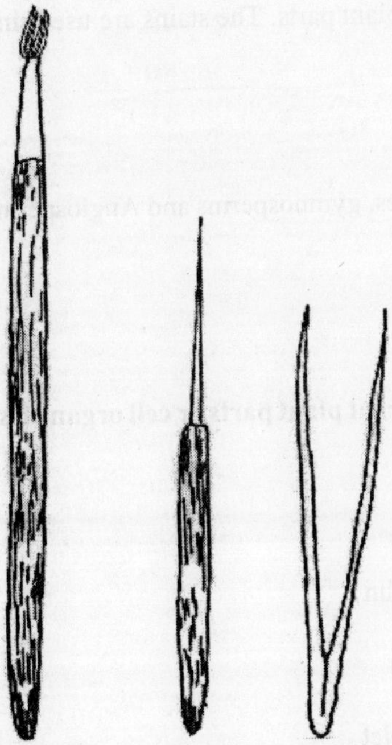

Fig.1.4 Brush Needle Forceps

(i) Acetocarmine: The stain is prepared as follows:

Add 1 gm of stain to 100 ml of boiling acetic acid (45%)or propionic acid. Cool and decant, A few drops of saturated ferric acetate solution is added and the mixture is cooled over ice. The solution is allowed to stand for about 12 hours. Filter and store in a dark coloured bottle under refrigeration.

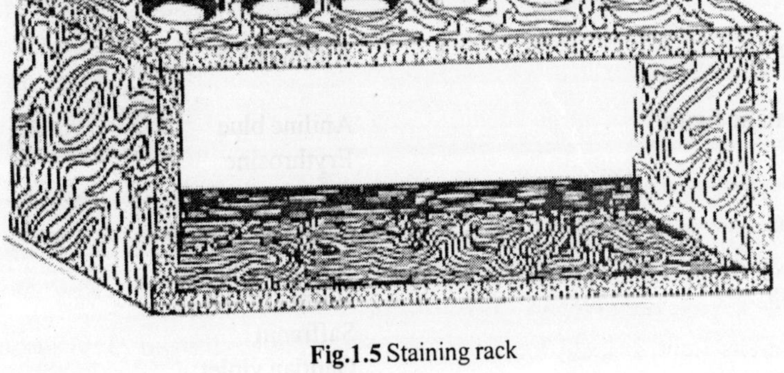

Fig.1.5 Staining rack

Specificity of Stains

Usually specific stains are employed in colouring different plant parts. The stains are used singly or in combination. The most popular stain combination are:

Saffranin - fast green.
Haematoxylin - Erythrosin.
Haematoxylin - light green.

While stain combinations are employed to study pteridophytes, gymnosperms and Angiosperms, single stains are used for Avascular cryptogams.

Green or Saffranin for Algae and Bryophytes.

Cotton blue for fungi.

The following table gives the specificity of stains to different plant parts or cell organelles.

Structure	Stain
Cellulose	Anilin blue Haematoxylin Fast green Light green
Cutinized cell wall–	Gentian violet Erythrosine Saffranin
Lignified cell wall	Gentian violet Saffranin
Suberised cell wall	Saffranin
Cytopalsm	Fast green Light green Aniline blue
Chitin	Saffranin
Callose	Aniline blue
Achromatic figure	Aniline blue Erythrosine
Proteins	Saffranin
Nucleus	Heamatoxylin Saffranin Gentian violet
Chromonemata	Heamatoxylin

Chromosomes	Saffranin
Plastids	Gentian violet Haematoxylin
Mitochondria	Gentian violet
Flagella	Cotton blue

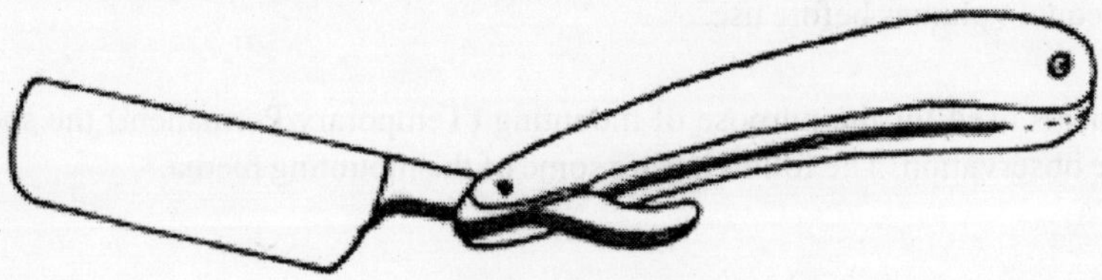

Fig.1.6 Razor

Fixing / Preserving Fluids

The plants or plant parts collected in nature are to be killed and preserved in the laboratory for the purpose of storage and study.

In killing the specimen, the various life processes are stopped at that instant, so that after preparation for microscopic study various stages may be observed.

When ever a material is collected, it is better to kill it in the field, so that there will be minimum or no distortion. Alternatively the material collected in the plastic bags or bottles should be brought to the laboratory and killed / preserved. In preservation, the material must be completely immersed in the fluid. The following are some of the killing/preservative fluids generally used:

1. *Formaline acetic alcohol (F.A.A)*

This is the most commonly used killing /preserving agent. It is popularly called standard preservative.

Ethyl alcohol (70%)	-90ml.
Glacial acetic acid	-5ml.
Formalin	-5ml

2. *Formalin - Propiono - Alcohol*

Formalin	-5ml
Propionic acid	-5ml.
Ethyl alcohol 70%	-90ml.

3. *Carnoy's Fluid - I*

Ethyl alcohol	-30ml
Glacial acetic acid	-10ml

4. *Carony's Fluid - II*

Ethyl alcohol	-30ml.
Glacial acetic acid	-30ml.
Chloroform	-15ml.

Carony's fluid is widely used in killing root tips, flower buds etc. for the purpose of cytological study.

5. *Randolph's Modified Navashin Fluid*

This consists of two solutions A and B.

Solution A

Chromic acid	- 5.0 gms.
Glacial acetic acid	- 50ml
Distilled water	- 320ml.

Solution B

Commercial formalin	- 200ml.
Distilled water	- 175 ml.
Saponin	- 3gms.

Mix A and B in equal volumes before use.

Mounting media

These are chemicals used for the purpose of mounting (Temporary/Permanent) the specimens on the slide for microscopic observation. The following are some of the mounting media.

Fig.1.7 Petri dish

1. *Acid water*

HCl – 10-15 drops

water – 100 ml.

2. *Lactophenol*

Phenol	-100gms
Lactic acid	-100ml.
Glycerine	-100ml.
Distilled water	-100ml.

3. *Glycerine*

Glycerine	- 15-20 drops
Formaline	- 1-2 drops
Distilled water	- 100ml.

4. *Canada Balsam:* This is a gum obtained from Gymnosperms. This fluid is readily available in the market.

5. *Euparol:* This is available in the ready to use form in the market.

Canada balsam and Euparol are used for permanent mounting, while the rest are used for temporary preparations.

Staining rack: This is a wooden rack used for holding the staining bottles.

Dropping bottles: This is a glass bottle having a narrow mouth fitted with a stopper. The stopper is grooved and has a beak. When the bottle is upturned the stain/fluid comes out in drops.

Slides: These are readily available in the market. The usual size 25mm x 75mm. The thickness is about 1mm. Special types of slides called cavity slides are available for specific study.

Cover slips (Cover glasses)

These are meant to cover the materials when the preparation is ready for observation. The coverslips may be round, square or rectangular.

Methods of Study of Plant Material

The following is the procedure for a scientific study in the laboratory.

Morphological study

The specimen given to study external detail should be observed either with or without a magnifying glass. Whatever details can be seen are to be recorded. If the specimen requires microscopic study, it should be suitably prepared for the purpose of microscopic observation. There are two types of microscopic preparation.

1. Whole Mount Technique

This is suitable for the purpose of external study of algal and fungal material. Take a little of the material, stain it with a suitable stain (staining procedure is explained later.) Wash the excess stain in water / alcohol. Mount in a suitable medium (temporary or permanent) and observe under the microscope.

2.Section cutting

For the purpose of anatomical study, sections are to be cut of plant parts. In algae, globule of *Chara*, stolons of *Caulerpa*., receptacles of *Sargassum* etc., can be cut. Similarly fruiting bodies of fungi, thalli of bryophytes also many be cut in a suitable plane. In higher plants cylindrical plant parts like root, stem, petiole etc. may be cut for the study of internal structure.

Planes of Cutting: In order to study the internal structure from various angles, sections are cut in various planes.

Transverse :The section is cut at right angles to the longitudinal axis.

Longitudinal : The section is cut parallel to the long axis (right angles to the transverse axis. The longitudinal section may be cut along a radius (Radial longitudinal section-RLS) or along a tangent (Tangential longitudinal section-TLS).

Cylindrical organs like roots and stems are cut transversely or longitudinally, while dorsiventral organs like thalli of bryophytes, leaves etc. are usually cut in vertical transverse plane.

Types of Section cutting

There are usually 2 types of section cutting methods employed for plant parts. These are (a)Free hand sectioning and (b) Microtome sectioning .

(a) Free hand Sectioning

For the purpose of laboratory study where slides are not made permanent, usually free hand sectioning method is employed. Free hand sections are cut mostly with the help of a good sharp razor (sometimes safety razor blades also may be used).

Honing the razor : The razor used for sectioning must have a sharp, even edge. There should be no nicks. Before cutting sections, the razor edge has to be honed to maximum sharpness either on a stone or on a leather strap. Oblique strokes, with the edge forward are to be given on the sone or strap. Usually oil drops are used on the honing medium to give maximum sharpness.

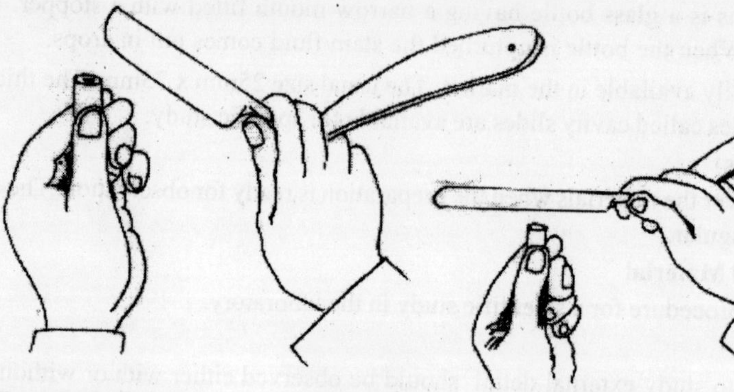

Fig.1.8 Method of holding pith

Methoding of holding razor

Method of section cutting

Method of cutting : Open out the razor and bend the handle backwards. Hold the razor between the index finger and thumb. The handle should be free. The remaining fingers of the hand will be touching the back edge of the razor blade.

In order to cut a transverse section of a cylindrical part like stem or root, follow the procedure given below.

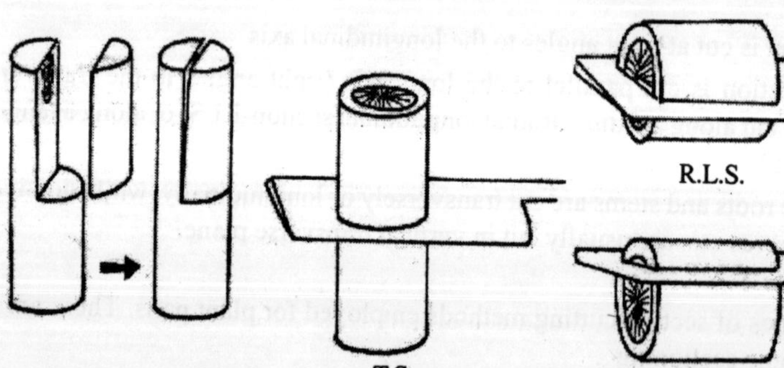

Fig.1.9 Method of using pith for insertion of material

The stem or root should be cut into a 2-3" long piece. It should be held in the hand in an erect position between the thumb and index finger. The material held in the hand should be exactly at right angles to the razor's edge. The thumb holding the specimen should be slightly at a lower level. Flood the edge of the razor with water. Now using the index finger as the platform move the razor quickly several strokes across the material. During this process sections are cut and they will be collected on the blade of the razor. The sections now may be carefully transferred with the help of a fine brush in to a petridish containing water.

For cutting sections of delicate and fragile materials like thalli or leaves, a suitable supporting medium has to be used. For this purpose a pith, carrot or potato may be used. These are to be cut into rectangular cubes and then split at one end. The material for sectioning should be inserted in the slit and sections are cut

After section cutting the razor should be wiped dry. It should be greased and encased.

Microtome Sectioning: This is generally employed very frequently in the study of bryophytes, pteridophytes and other higher plants. The microtome sectioning is a long and complicated process. It takes anywhere between 15-20 days to process a material and to obtain the sections.

Microtome sectioning involves the following steps:

1. Killing and preserving the material.
2. Dehydration and infilteration.
3. Paraffin embedding .
4. Section Cutting.

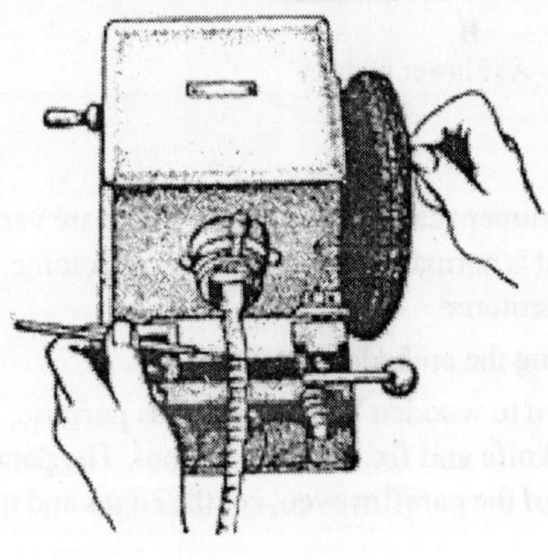

Fig.1.10 Rotary microtome

The material like stem, leaf, flower etc. is to be killed and preserved in a suitable fluid.

(a) Cut the material into bits of suitable size.

(b) Place the material in small vials containing a fixative or preservative.

(c) Label the vial giving the details of place of collection , name of the material, preservative used and date of collection.

The material should be in the killing fluid at least for six hours (preferably a day).

Dehydration: In order to dehydrate the material, remove it from the preservative and wash thoroughly in water and transfer it to a suitable tube or vial.

Dehydrate the material in alcohol series (Ethyl alcohol) as follows

Stage 1	50% Alcohol	First day
Stage 2	70% Alcohol	Second day
Stage 3	90% Alcohol	Third day
Stage 4	100% Alcohol	Fourth day
Stage 5	Absolute alcohol and xylol (50/50)	Fifth day
Stage 6	Absolute alcohol and xylol (25/75)	Sixth day
Stage 7	Xylol	Seventh day

When the material is transferred to pure xylol, there should be no fogging (i.e the solution should not turn milky). If there is fogging repeat the procedure starting from 4th stage.

The material in xylol will be completely dehydrated and looks transparent. It is now ready for infilteration by paraffin wax. The purpose of infilteration is to see that the wax enter into the material completely so that it can easily be cut.

Infilteration: Add a few pieces of paraffin wax everyday for about 4-5 days and then transfer the vial into a hot air oven, while temperature is adjusted to the melting point of wax. Everyday add a few pieces of wax. Adding of wax should be continued until all the xylol in the vial is replaced by paraffin wax. When the tube is smelt, there should be no smell of xylol. The material is now ready for embedding.

Paraffin embedding: In order to cut the material, it has to be embedded into a paraffin block. Prepare a paper boat or take a porcelain rectangular dish and smear glycerine to the inner surfaces. Pour molten paraffin into the porcelain dish and empty the contents of the vial into the dish. Before the paraffin solidifies, arrange the material in proper rows and in such a manner as desired. Leave it for cooling for 6-8 hours. For quick cooling the dish may be transferred to a bowl containing ice water. Carefully remove the solidified paraffin block from the dish and cut it into suitable pieces. The material is now ready for sectioning.

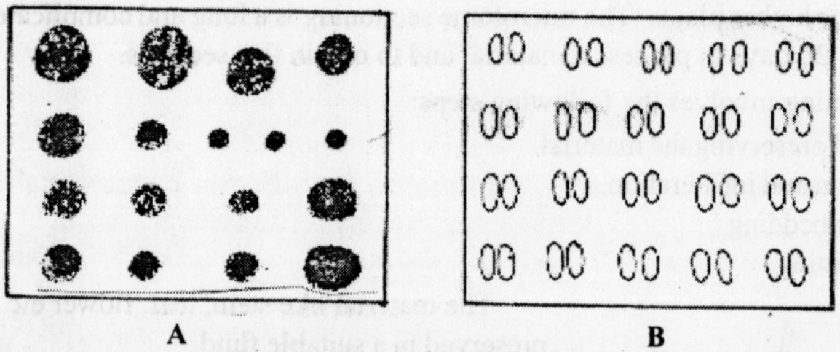

Fig.1.12 Arrangement of materials in paraffin blocks - A) Flower buds in groups B) flower buds in pairs

Sectioning: Sectioning is carried out with the help of an instrument called microtome. There are various types of microtomes like Rotary, sliding, Freezing etc. The one that is normally used is a rotary microtome. The following procedure can be adopted in the case of the rotary microtome.

1. The paraffin block is to be cut into small pieces each having the embedded material.

2. The pieces of paraffin containing the material is to be fixed to wooden 'riders'. For this purpose, heat the top of the rider and the base of the paraffin piece with a hot knife and fix. Allow it to cool. The paraffin piece is now firmly fixed to the rider. In order to make the edges of the paraffin even, cut the edges and make them parallel.

3. Fix the wooden rider into the microtome so that the paraffin piece with the material is exactly parallel to the edge of the microtome knife.

4. Adjust the knob of the microtome to give the required thickness to the sections.

5. Rotate the wheel of the microtome in a clockwise direction. The cylinder containing the rider moves up and down on the knife edge. Sections are cut in a sequence and get attached to each other forming a paraffin ribbon.

6. Keep a needle in your left hand to hold up the ribbon as it lengthens. When the ribbon is sufficiently long, cut it and transfer it on to a black cardboard. Cut the ribbons into suitable length and transfer to the slide.

7. Smear the slides with an adhesive like egg albumen or gelatin and rub it well.

8. Pour some water onto the slide with the help of dropping bottle.

9. Transfer the paraffin ribbons on to the slide and drain excess water.Retain a little water on the slide.

10. Place the slide on a slide warmer. The ribbon expands, and so also the sections of the material. Now draw out the water completely and allow the slides to cool.

11. The slides should be allowed to stand for atleast 24 hours whereupon the sections will stick to the slide permanently.

The slides are now ready for staining.

Staining procedure

The plant material (whole mount or sections) are to be stained and mounted properly before they are ready for microscopic observations.

Algal filaments, fungal mycelia, pollen grains and spores, prothalli of pteridophytes are usually stained with a single stain - saffranin or green. In some case Aniline blue is used to stain the filaments of Algae and mycelia of Fungi. The staining procedure is as follows.

Place the material either in a watch glass or cavity slide and add a few drops of stain. The period of staining varies with different materials. But usually a 5-10 minute period is sufficient. Drain out the stain and observe the material under the microscope. If the material has taken less stain, it may be kept in stain for some more period, or if it has taken excess stain it can be removed by washing either with water or 50% ethyl alcohol.

After the material is suitably stained pass it through a series of alcohol grades to dehydrate it. Usually different grades of alcohol (70%,90%,100%) taken in a filler are put on the material. After passing through absolute alcohol the material is kept in xylol and it is now ready for mounting.

Fungal mycelia can be stained with aniline blue. Usually the material taken on a clean slide is stained with a drop or two of the stain prepared in lactophenol. The material can bemounted straightway as lactophenol is also a mounting medium. If the stain is not prepared in lactophenol the material may be transferred to fresh lactophenol and mounted.

Double Staining Procedure

In the case of most of the algae and fungi there is little or no tissue differentiation and hence single staining is sufficient. In some cases of plants with relatively more tissue differentiation, a single stain is not sufficient to mark out the various tissues clearly. In such cases 2-3 stain combinations are used to stain the tissues. For example lignified and non-lignified tissues are stained with a combination of acidic and basic dyes.This type of staining is also called differential staining where different tissues take up different stains.

The following are some of the popular stain combinations.

Main stain		*Counter stain*
Haematoxylin		Saffranin
Haematoxylin		Erythrosine
Haematoxylin		Eosin
Haematoxylin	Orange G.	
Saffranin		Fast green
Saffranin		Light green
Saffranin		Aniline blue
Saffranin		Crystal violet
Crystal violet	Erythrosine	

Staining Procedure for free hand sections

(a) Procedure for temporary preparations

Free hand sections of root, stem etc., should be washed thoroughly in distilled water in a watch glass. After repeated washing the sections should be transferred to a watch glass containg the main stain. The sections should be kept in the stain for a period ranging from 30-40 minutes. After this period, the sections should be transfereed to a watch glass containing distilled water. The sections should be washed for 2-3 minutes whereupon excess of stain is removed. They should be observed under the microscope. If the main stain is saffranin only lignifed tissues should retain the stain, others should be destained. If the main stain is specific to non lignified tissues, only they should retain the stain. Then the sections should be washed thoroughly in water and repeatedly observed under the microscope. The sections can be mildly stroked with a fine hair brush, so that excess of stain comes out with distilled water. The sections are now ready for counter-staining. They should be transferred to a watch glass containing the counter stain (fast green etc). Counterstain acts quickly on the tissues. Hence one to two minutes of staining should be sufficient. The sections are now observed under the microscope. If there is excess of counter stain, it can be removed by washing sections with 15-20%glycerine. After the final stianing the tissues whould be demarcated clearly with no mixing of the stains. The sections are now ready for mounting.

(b) Procedure for permanent preparaions

If the sections of plant materials are to be mounted and made into permanent slides, the procedure should be slightly altered so as to include dehydration and clearing with suitable agents. The procedure is as fallows:

1.Stain with the principle stain (Saffranin, Haemotoxylin etc).

2.Allow the sections to remain in stain for about 15-20 minutes.

3.Transfer the sections to water and wash thoroughly. The sections should be stroked mildly until no more stain comes out, and the water should remain clear. If the main stain like saffranin is produced in 50% alcohol then destaining (washing) should be done in 50% alcohol.

4.The sections should be now dehydrated in alcohol grades of 50%, 70% and 100%.

5.In each grade of alcohol keep the section for about 5 minutes.

6.Transfer the sections from one grade of alcohol to the other with the help of a fine brush. Never use needles as they injure the sections.

7.Transfer the sections to the counter stain (fast green etc). The counter stains are usually prepared in 90% alcohol.

8.Keep the section in counter stain for about a minute. Wash the sections in 90% alcohol and transfer them to 100% alcohol.

8a. In some case counter stains are prepared in clove oil. In such cases counter staining should be done after the sections are completely dehydrated with absolute alcohol.

1. Sections are transferred completely dehydrated to a solution of 50:50 Absolute alcohol and xylol followed by 25:75 Absolute alcohol and xylol and ultimately the sections are transfered to pure xylol. Now xylol should not turn milky or turbid. If it turns milky, dehydrate the sections again through combination of absolute alcohol : xylol series.

The sections are now ready for mounting.

Staining procedure for microtome sections

The slides with the paraffin ribbons adhering to them have sections of plant material.The processing and staining is as follows:

1. Immerse the slides in a coupling Jar containing xylol. Leave them for 10 minutes . The paraffin dissolves in xylol and only the sections remain on the slides.

2. The slides are now kept in 50:50 absolute alcohol and xylol.

3. Pass the slides through a down grade series of alcohol. 100%, 90% and 70%

4. Transfer the slides to pure water and wash them thoroughly to remove traces of alcohol.

5. Keep the slides in a mordant like 4% iron alum. This is necessary only when Haematoxylin is used. After about 10 minutes wash the slides in water.

6. Transfer the slides to the main stain Haematoxylin. Saffranin etc. and keep them for 15-30 minutes.

7. Remove the slides from stain and wash in water.

8. Observe the slides under microscope. Usually the sections are overstained.

9. If the main stain used in Hametoxylin, destaining can be done with 2% Iron alum or with water containing few drops of concentrated HCI. Repetaedly observe the slides under the microscope. Stop destaining when only the nuclei have retained the stain.

Caution: When destaining is done with HCI, not more than 5-6 drops should be used for 75ml of water. Excess acid will eat away the adhesive and the sections come out of the slide.

10.Wash the slides in water overnight if the destaining agent is Iron alum, otherwise washing for 30 minutes is enough.

11.Now pass the slides through an upgrade series of alcohol (5 minutes in each grade) 70% 90% 100%

12. When the sections are completely dehydrated stain them with the counter stain (Fast green etc) for a minute or two. Wash the excess stain in clove oil and transfer the slides to pure xylol. The slides are now ready for mounting.

Mounting

(a) *Mounting Media*

Mounting media are of two categories: Temporary mounting media and permanent mounting media.

A temporary mounting medium may be just water or water mixed with glycerine. If the temporary sections are to be retained for 2-3 hours, dilute glycerine (15-20%) is recommended.

Permanent mounting media are used when the slides are to be retained indefinitely. The following are some of the permanent media.

(1) **Canada Balsam:** This is the most widely used mounting medium in botanical preparations. It is available in ready to use bottles. It is a gummy substance. If the solution is very thick add little xylol.

Canada balsam is a resinous product of a Gymnosperm, Abies balsamea. The gum has very good refractive index and as such it is highly reccommended.

(2) **Lactophenol:** Lactophenol can be prepared in the laboratorty. Mix equal quantites of phenol cystals, lactic acid glycerine and distilled water.

(3) **Glycerine gelly:B** This is prepared as follows:

Gelatine 1 part.
Water 6 parts.
Glycerine 7 parts.

Glycerine jelly is suitable even for mounting fresh specimens (even without dehyderation).

Add gelatine to water, stir thoroughly and warm. Add glycerine and cool the solution. After sometime the solution settles into a jelly.

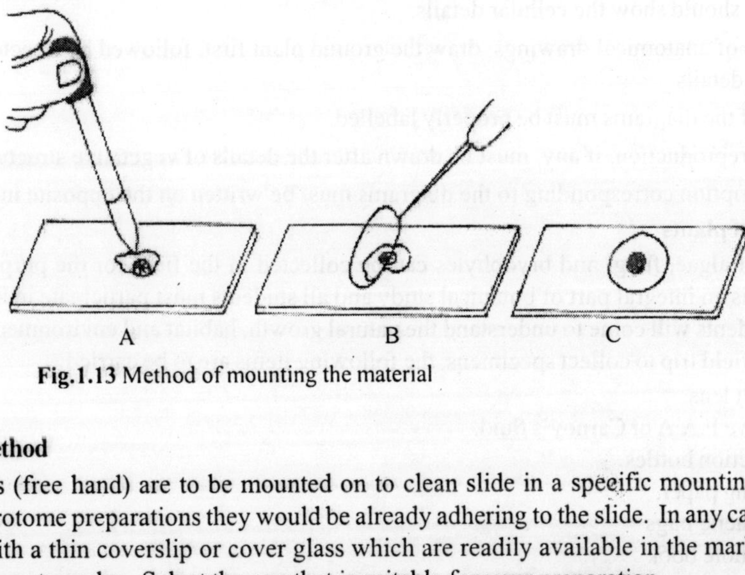

Fig.1.13 Method of mounting the material

(b) Mounting Method

The sections (free hand) are to be mounted on to clean slide in a speeific mounting medium. If the sections are microtome preparations they would be already adhering to the slide. In any case the sections are to be covered with a thin coverslip or cover glass which are readily available in the market. They may be round, square or rectangular. Select the one that is suitable for your preparation.

The sections or objects should be mounted on the centre of the slide. In order to ensure placing of the object on the centre of the slide, the following procedure may be adopted. On a white paper draw a rectangle of the same size as the slide. Join the opposite corners by drawing diagonal lines. The centre of the slide lies where the two lines cross. Place the slide on the rectangle and material or objects may be kept on the central point.

Place a drop or two of the mounting medium. Hold the coverslip at the edges (see the figure) and allow it to touch the edge of the mounting medium at an angle of approximately 45°. Slowly lower the cover glass on the mounting medium with the help of a needle. Take care to see that no air bubbles enter the mounting medium. If there are air bubbles, lift the cover glass and mount again.

Excess mounting medium may be carefully wiped with the help of a clean tissue. Allow the slides to dry. If necessary the slides may be transferred to a hot air oven for drying. The slides should be labelled at one end as shown in the figure. They are now ready for microscopic observation.

Ringing of Slides

Temporary preparations may be kept for few days if the borders are sealed. This procedure is known as ringing . Ringing is done by applying wax, canada balsam or even nail polish to the edges. Ringing may be done manually or by using a ringing table having a rotary metallic disc.

Record of Practical Work

Practical work carried out in the laboratory should be recorded. The botany practicval record book has two types of sheets viz., Interleaves and drawing sheets. Interleaves are meant to write the notes pertaining to the diagrams. The drawing sheets which are thicker than the interleaves are used for diagrams. The following points may be useful in maintaining a good record.

1. Use good pencils (HHB) with sharp points and a good quality eraser.
2. Leave a minimum of one inch margin on all the sides.
3. Always write diagrams of only one plant in a page. Do not mix up diagrams of other plants.
4. On the left hand corner of the drawing sheet write the date of the experiment.
5. Give the complete systematic position of the member on the top of the drawing sheet.
6. For microscopic organisms draw first the habit sketch followed by low power and high power view/s. High power view should show the cellular details.
7. In the case of anatomical drawings, draw the ground plant first, followed by a sector enlargement to show the cellular details.
8. All parts of the diagrams must be properly labelled.
9. Stages of reproduction, if any, must be drawn after the details of vegetative structure.
10. The description corresponding to the diagrams must be written on the opposite interleaf.

Field Collection of plants

Specimens of algae, fungi and bryophytes can be collected in the field for the purpose of laboratory study. Field trip is an integral part of botanical study and all students must participate in it. It is in the field study that the students will come to understand the natural growth, habitat and environmental association of plants. During a field trip to collect specimens, the following items are to be carried:

1. Pocket lens.
2. Fixative F.A.A or Carnoy's fluid.
3. Collection bottles.
4. Blotting paper.
5. Polythene bags
6. Field note book
7. Labels.

When the specimens are collected they can be fixed and preserved for the purpose of laboratory study. Use the required preservative and immerse the specimens in it. Label the bottle showing details of

Date of collection

Location

Preservative fluid (name) used.

Name of the plant

As it is necessary for the purpose of identification, details of habitat, associations with other plants if any, colour of the fresh specimen etc. should be recorded in the field note book.

PART ONE
PTERIDOPHYTES

Chapter 2

Classification of the plant Kingdom

The living world consists of myriads of plants and animals of various kinds. Even within the plant world many lakhs of different types of plants have been recognised and perhaps many more are yet to be discovered. If such huge number of individuals are not properly arranged their study will not only be difficult but also meningless because proper comprehension is impossible. Hence in order to bring an arrangement and discipline plants have been grouped into smaller or larger groups based on resemblences and differences (mostly morphological). This system of dividing plants into groups is known as "Plant classification or Systematics".

In general there are mainly two systems of classification viz.,

(1) Artificial system and

(2) Natural system.

In the artificial system of classification as proposed by earlier botanists importance is given to minor, superficial and frequently variable characters. For example division of plants based on their habit into herbs, shrubs and trees and using this as a criterion in the classification. The artificial system does not indicate the relationship existing between different plants.

The natural system of classification is based on the natural phylogenetic affinites listing between various types of plants. Natural system lays stress on the similarity in reproductive structures including anatomical characters, cell structure and physiology. Such characters are more stable than the ordinary vegetative characters and are not prone to frequent alterations by the environmental influence.

According to the natural system all the plants have been included in the plant kingdom. The plant kingdom is divided into *divisions*. Each division into several *sub-divisions*, each sub-division into *classes*, each class into *orders*, each order into *Families*, each family into *Genera* and each genera into *Species* and in some cases each species into *several varieties*.

Various natural systems of classifications have been proposed time and again. It is not the idea of the present attempt to give a detailed survey of all those systems. Only a few systems have been briefly discussed below so as to give an understanding to the students the relative positions of different groups of plants in the evolutionary sequence of the plant kingdom so that they will be in a better position to appreciate and understand the different groups of non-vascular and vascular plants.

In general two main categories can be identified in the *natural* systems. They are:-

(1) Old or Traditional system

(2) Revised or Modern system.

Traditional system.

Of all the old system perhaps the one that is followed even today is that of Eichles (1866).

He proposed that the plant kingdom be divided into -

A) Cryptogamae and B) Phanerogamae

(A) Cryptogamae

Cryptogams have their reproductive structures embeded in the plant body and do not bear flowers of seeds.

They include three divisions namely -

(1) Thallophyta

(2) Bryophyta and

(3) Pteridophyta

Thallophyta:- The division thallophyta is characterised by plants -

a) With an undifferentiated body called thallus (flat)

b) Absence of vascular tissue.

c) Unicellular reproductive structures.

d) Absence of embryonal stage in the life cycle.

e) Members generally aquatic (not a rule).

Thallophyta is further divided into two classes namely -

1) Algae and 2) Fungi.

Algae:- The class Algae includes majority of the members which are -

a) Aquatic in habitat.

b) Members are antotrophic because of chlorophyll green photosynthetic pigment, and other pigment.

c) Plant body may be unicellular or multicellular.

Fungi:- This includes all achlorophyllous thallophytes. As the members are pigmentless they are not antotrophic their mode of nurtition is heterotrophic. Fungi follow two methods of heterotrophic nutrition viz

1) Parasitic and 2) Saprophytic.

In the parasitic method, fungi obtain their food nutrition from living organisms (plats and animals) while in saprophytic method fungi devise their nutrition from dead and decaying organisms. (For details see introductory chapter on Fungi).

Bryophyta:- Plants are amhibious in their habitat. Plant body is mostly an undifferentiated thallus except in some. They differ from thallophytes in the following respects -

1) Possession of multicellular sex organs.

2) Presence of an Embryonal stage in the life cycle.

3) Occurence of heteromorphic alternation of generations in the life cycle regularly.

4) Sporophyte being dependent on the gemetophyte.

Pteridophyta:- Pteridophytes are called vascular cryptogams. Because of all the cryptogams only they have the vascular tissue or conducting tissue which helps in the transportation of food materials and also offers mechanical support to the plant body. They are seedless. Unlike in Bryophytes, saprophyte is the predominant plant body but never the less the gametophyte is independent (with very few exceptions).

(B) Phanerogams:-

These are the plants in which the reproductive structures are not embeded. They produce seeds. Phanerogams are divided into two divisions namely -

Division (1) *Gymnospermae* - seeds are naked, not covered by any envelope, no fruit formation.

Divison (2) *Angiospermae* - Seeds enclosed by fruit wall, fruit being formed by the mature every which is a special envelope for the ovule (seed) Angiosperms are divided into two classes namely -

1) Dicotyledonae - seeds with two cotyledoms.
2) Monocotyledonae - seeds with one cotyledon.

Some Botanists divide the entire plant kingdom into four major divisions viz -

1) Thallophyta
2) Bryophyta
3) Pteridophyta
4) Spermatophyta (includes Gymnosperms and Angiospersm).

Modern system

According to According to one of the modern system there are only three major divisions in the plant kingdom. They are -

(1) Thallophyta
(2) Bryophyta
(3) Tracheophyta

The last division includes all vascular plants, which are classified into the following sub divisions -

1) Psilopsida 3) Sphenopsida
2) Lycopsida 4) Pteropsida

Pteropsida includes three classes viz -

a) Filicineae

b) Gymnospermae

c) Angiospermae

Angiospermae has sub class 1) Dicotyledonae and sub class 2) Monocotyledonae.

In another system only Thallophyta and Embryophyta have been recognised as two major divisions. In Embryophyta there is the formation of an embryonal stage in the life cycle while it is absent in thallophytes. Embryophyta includes bryophytes pteridophytes, Gymnosperms and angiosperms. (Details not discussed).

Many modern botanists consider that the plant kingdom consists of 18 major division. They are -

1. Chlorophyta
2. Charophyta
3. Englenophyta
4. Pyrrophyta
5. Chrysophyta
6. Phaeophyta
7. Cyanophyta
8. Rhodophyta
9. Schizomycophyta (Schizomycota)
10. Myomycophyta (Myxomycota)
11. Eurnycophyta (Eumoycota)

12. Bryophyta
13. Psilophyta
14. Lycophyta (Lepidophyta)
15. Arthrophyta (Calamophyta)
16. Filicophyta (Pterophyta)
17. Gymnophyta
18. Angiophyta (Anthophyta)

Chapter 3

General Characters And Classification

Pteridophyta constitute a significant and important group in the plant kingdom. As the first true land plants they offer a very favourable material for the study of various adaptations that have made the colonization of the land possible for the plants.

Pteridophytes have a long geological history on our planet. They are known from as far back as 380 million years. Fossils of pteridophytes have been obtained from rock strata belonging to Silurian and Devonian periods of the Palaeozoic era.

The life cycle of a typical member comprises a regular heteromorphic alternation of generations in which both the gametophyte and the sporophyte exist as independent individuals. Nevertheless the sporophyte is the predominant generation. In terms of duration in the life cycle gametophyte is insignificant. In this respect pteridophytes are totally different from bryophytes. The reasons for the predominance of the sporophyte in the life cycle are not far to seek. Being a more robust body a sporophyte is suitable to effectively meet the challenges of a terrestrial environment.

As a group the position of the pteridophytes is between bryophytes and spermatophytes. They share the characters of both these groups in addition to their own unique features.

Sporophytic Generation

As has already been pointed out it is the sporophytic generation which constitutes the main plant body in pteridophytes. In any description whenever we refer to the plant body of a pteridophyte we mean the sporophyte.

There is a great variety in the nature and organiszation of the sporophytic plant body in pteridophytes. The simplest sporophyte in an extinct member is to be found in *Psilotum*. The plant body here is a nake branched axis with no evidence of roots. The axis is distinguishable into an underground prostrate system and an erect aerial system. Rhizoids help in the anchorage of the plant to the substratum. There are no leaves. The stem itself takes up the function of photosynthesis.

From such a type of simple plantbody further evolution (mainly necessiated by the environment) resulted in the complex and diverse types of sporophytes that we see today.

From a naked leafless plantbody the first step is the differentiation of photosynthetic laterals. (leaves) and roots.

The leaves of pteridophytes are basically of two types viz; microphylls and megaphylls (macrophylls).

Microphylls are those which are provided with a single unbranched mid vein. Further when a leaf trace departs from the main vascular cylinder to provide the leaf, no gap is left.

In megaphylls there is a branched venation. The branching may be dichotomous or reticulate. Further there invariably will be a leaf gap in the main vascular cylinder above the leaf trace.

Microphylls are seen in members like *Lycopodium, selaginella* etc. Megaphylls are found in members like *Adiantum*, *Pteris* etc.

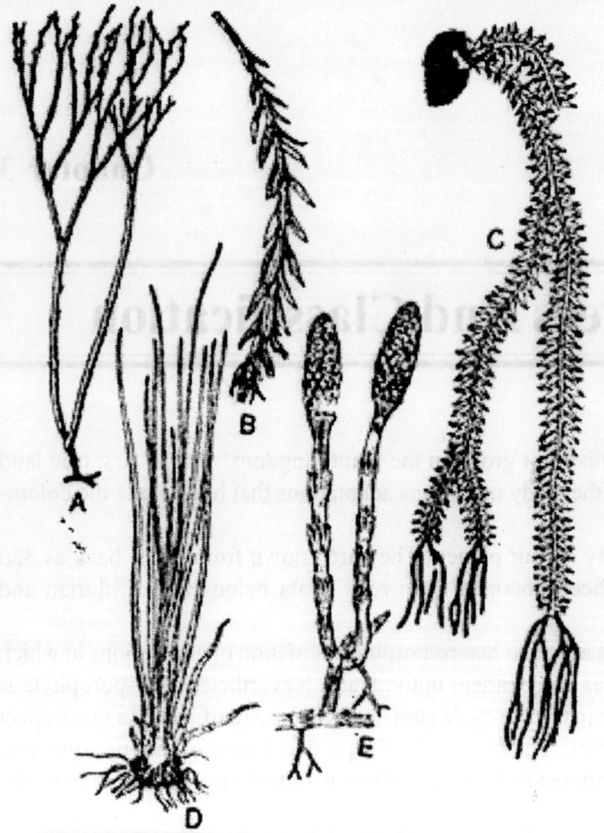

Fig. 3.1. Examples of some living Pteridophytes
A. Psilotum, B. Tmesepteris, C. Lycopodium, D. Isoetes, E. Equisetum

Vasculature in Pteridophytes.

All the Pteridophytes possess a vasculature in the center of the axis traversing from one end to the other end branching with all the branches.

The two main vascular elements are xylem and phloem. The xylem consists of mostly tracheids and rarely vessels (*Selaginella, Pteris* etc). The tracheids have different types of thickenings like scalariform, pitted, annular etc. The tracheids or vessels are of two types viz, protoxylem and metaxylem. The former matures early and has a narrow lumen while the latter matures later and has a wide lumen.

The relative distribution of protoxylem and metaxylem elements forms the basis of the classification of xylem group into three types viz exarch (centripetal protoxylem pointing towards the periphery), endarch (centrifugal- protoxylem pointing towards the center), and mesarch (protoxylem having metaxylem on either side).

The pholem is composed of sieve tubes and phloem parenchyma. A sieve tube consists of a series of long, living cells with sieve plates. Sometimes a distinction could be made between proto and metaphloem. But there is no pronounced morphological difference between the two.

Types of vasculature (Stele): A transverse section of the stem or root of the plantbody shows an outer epidermis middle cortex and central stele. The cortex is either parenchymatous or may have sclerenchyma also.

The central region consists of the concentration of the vasculature known as the stele. A stele may be defined as a vasculature sorrounded by pericycle and endodermis.

Originally plant anatomists thought that only the vascular bundles (xylem and phloem) formed the fundamental unit of the vasculature. Stele as the fundamental unit of the vasulature was first proposed by Van Tieghem and Douliot (1886). According to them the cortex and the stele are seperated by the endodermis. (The endodermis consists of barrel shaped cells and possesses thickenings known as Casparian thickenings on the radial and tangential walls).

There are mainly three types of steles in pteridophytes. These are 1) Protostele 2) Siphonostele and 3) Dictyostele. A protostele has the vasculature (xylem) occupying the central region of the stele. There is no pith in a siphonostele the vasculature is like a tube or siphon. Here the central region consists of non vasculated cells (parenchyma) forming the pith. In dictyostele the siphon breaks up into a number of individual bits called meristeles.

Protostele:- Basically all protostele are alike in not having a pith. They may be classified into the following types based mainly on the configuration of the xylem.

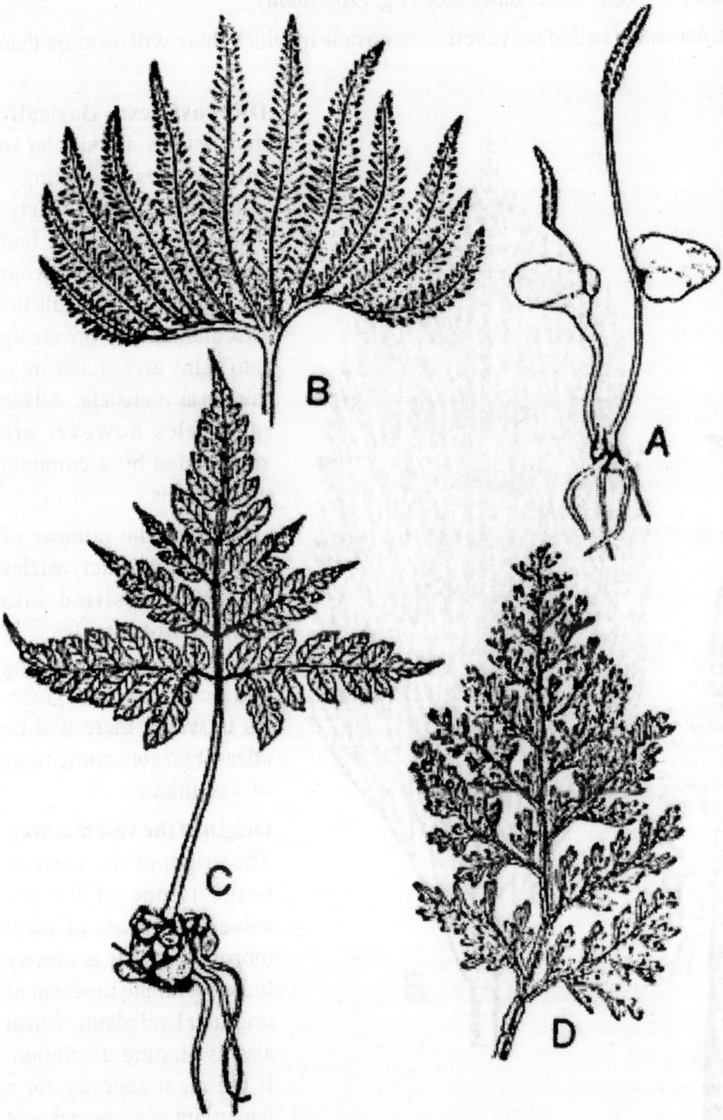

Fig. 3.2 Examples of some living Pteridophytes.
A. Ophioglossum, B. Matonia, C. Marattia, D. Hymenophyllum

a) *Haplostele:-* This is the simplest type of stele that could be visualized for a vascular plant. The central region of the stele consists of a smooth core of xylem sorrounded by phloem. This type of stele is seen in *Rhynia*.

b) *Actinostele:-* The xylem is star shaped and the phloem completely surrounds the xylem. The contour of the stele is smooth. Eg. *Psilotum, Lycopodium phlegmaria* etc.

c) *Plectostele:-* The xylem breaks up into a number of bands or plates arranged parallely. The phloem not only surrounds the xylem but is distributed between the sylem plates. Eg. *Lycopodium wightianum*.

d) *Mixed Protostele:-* In this type the xylem breaks up into a number of small masses. The phloem is intermixed with the xylem. Eg. *Lycopodium cernuum*.

e) *Polystele:-* Generally the stem is traversed by a single stele. But in some cases as in some species of *Selaginella* several independent steles run parallely in the stem. Each stele has its own endodermis. The individual steles are haplostelic.

Siphonosteles:- The steles having a pith in the centre are reffered to as siphonosteles. These are further classified into two types viz, simple siphonostele and solenostele.

a) *Simple siphonostele:-* Here the central region of the stele consists of a parenchymatous or sclerotic pith. Leaf traces that depart from the main vascular cylinder do not leave any leaf gaps (a parencymatous cavity found immediately above the leaf trace in the vascular cylinder). This type of siphonostele is also called a medullated protostele or a cladosiphonic stele. Eg. *Psilotum* sp.

b) *Solenostele:-* In Solenonosteles leaf gaps are successive so that there is only one break in the vascular cylinder at any one given point. Solenosteles are also known as phyllosiphonic steles. Depending on the pattern of distribution of phloem solenosteles may be classified into two viz, *Amphipholic solenostele* in which the phloem lines the xylem both on its inner and outer face (Eg. *Adiantum, Marsilea* etc) and *Ectopholic solenostele* in which phloem lines the xylem only on its outer face (Eg. *Osmumda*).

There is also a third category of solenostele called polycyclic solenostele in which there will be more than one concentric ring of vasculature.

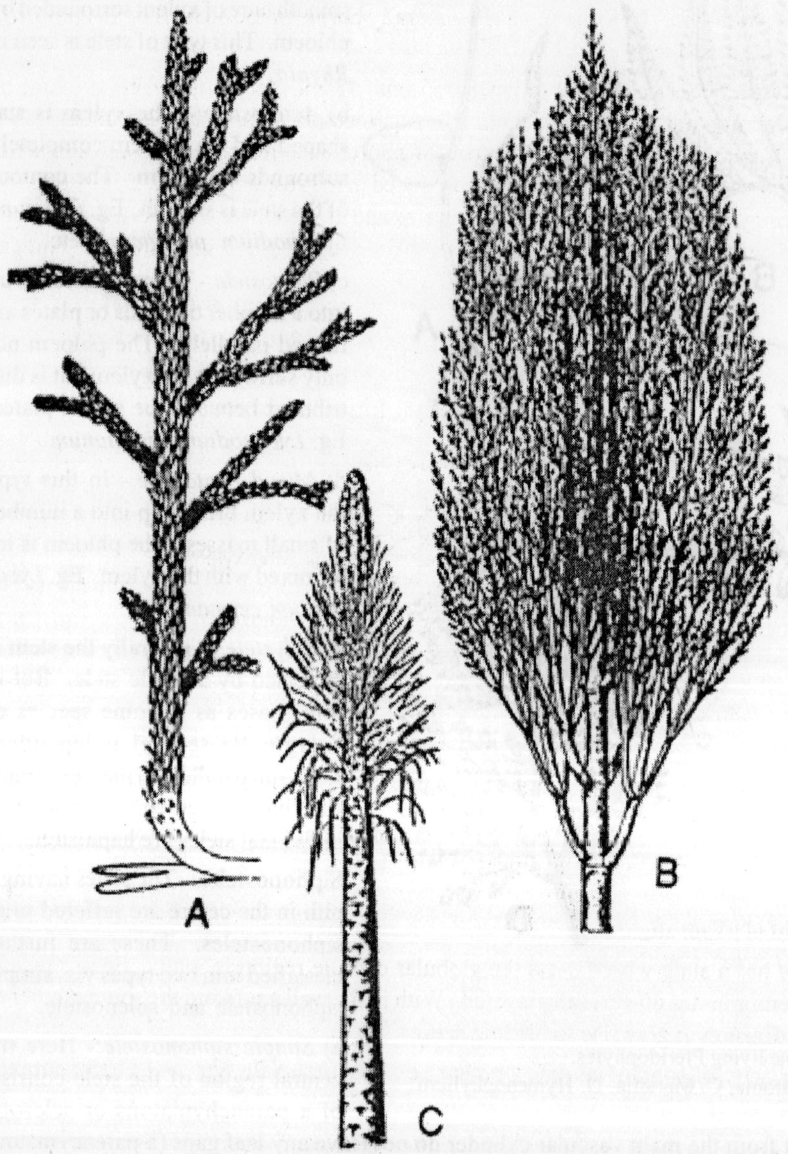

Fig. 3.3 Examples of some fossil Pteridophytes. (Contd...)
A. Asteroxylon, B. Calamites (reconstruction), C. Pleuromia

Dictyosteles:- Basically dictyosteles are similar to solenosteles in having a pith and leaf gap. In dictyosteles however the leaf gaps overlap (many occur at a point), as a result the vasculature ring breaks up into many arcs. Each arc is known as meristele. All the meristeles however are sorrounded by a common endodermis.

Based on the number of vascular rings dictyosteles may be classified into simple diotyosteles. in which there is only one ring and polycyclic dictyosteles in which there will be atleast two concentric rings of vasculature.

Origin of the vasculature:- The origin of the vascular tissue is one of the unsolved problems of plant morphology. It io oloooly linked with the problem of origin of land plants, which also is eluding a solution. If the algal ancestry for a land plant is proposed then there are no steps indicating the origin of vasculature. If one belives in the bryophytic ancestry of land plants, the columella in the sporophyte of *Anthoceros* seems to be a good starting point for the origin of vasculature. Indeed in its ideal location and to some extent in its function (though not in structure) the columella of the sporophyte of *Anthoceros* tempts any one to consider it as a precursor of the vasculature.

REPRODUCTION:- The sporophyte of pteridophytes generally reproduces by two methods viz, Vegetative propagation and Spore production.

Vegetative propagation takes place by a variety of methods such as fragementation, bulbil formation, resting bud formation, persistant apices etc. (The deatails of thes are given with the life histories of individual forms).

Spore production:- The characteristic method of reproduction is by the formation of spores. Spores are haploid and are produced in sporangia. The sporangia are generally borne on the adaxial or abaxial surface of a leafy appendage called the sporophyll.

Pteridophytes are homosporous (Isosporous) or heterosporous. Homosporous members produce only one type of a spore in the sporangium which develops into a gametophyte bearing both antheridia and archegonia. Heterosporous members produce two types of spores i.e., microspores and megaspores (macrospores). The microspores are smaller in size and are produced in microsporangia develop into the male gametophyte (microgametophyte). Megaspores are bigger in size and are produced in megasporangia borne on megasporophylls. On germination megaspores develop into the female gametophyte (mega or macrogametophyte).

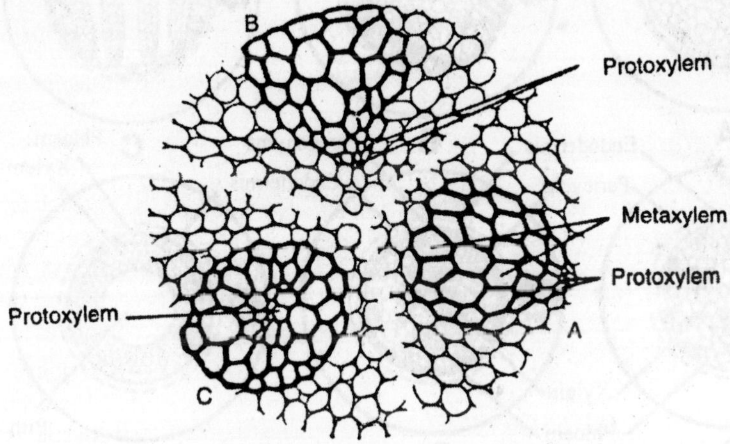

Fig. 3.4 . Direction of maturation of Xylem Elements
A. Exarch. B. Endarch. C. Mesarch.

Structure and development of a sporangium.

A sporangium usually has a stalk which bears the globular capsule region in which are produced the spores. The wall of the sporangium is one to many layered. With in the wall are found the spore mother cells which undergo reduction division to give rise to haploid spores.

There are mainly two types of sporangial development i.e., Ensporangiate type and Leprosporangiate type.

In Eusporangiate type the sporangium is derived from a group of cells. Further after the initial periclinal division the sporogenous tissue is derived from the inner daughter cell. In Leptosporangiate type, the sporangium is derived from a single initial cell. Further the sporangenous tissue is derived from the outer daughter cell of the sporangial initail.

A mature sporangium dehisces either vertically or transversely. Special types of cells may or may not be present in the wall to help in the dehisence. Usually in ferns the sporangia have a thick walled annulus and a thin walled stomium. The annulus may be vertical, shield shaped or cap shaped. By the differential hygroscopic response of these cells the sporangium breaks open liberating the spores. The spores are usually wind disseminated.

Gametophytic generation.

The haploid spore is the starting point in the development of the gametophyte. In heterosporous forms two types of spores are formed (mega and microspores) hence there will be two types of gametophytes (male and female). In homosporous forms there will be only one type of gametophyte.

The spores rarely exceed a few mm in size. Their shape varies from triradiate to spherical. They have two layered wall. The outer sculptured layer is known as the exine (exospore) and the thin inner layer is known as

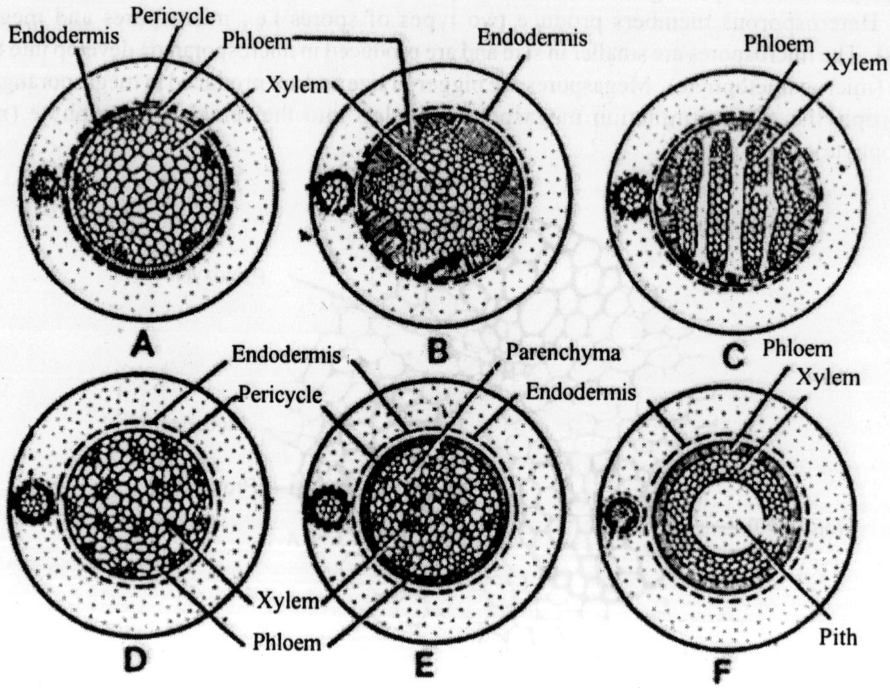

Fig. 3.5 Types of Vasculature in Pteridophytes.
A. Haplostele, B. Actinostele, C. Plectostele, D. Mixed Protostele,
E. Mixed Protostele (with pith), F. Ectopholic Siphonostele.

the intine (endospore). Sometimes there will be an outer most epispore and a middle mesospore in addition to exine and intine.

The germination and development the gametophyte is different between homosporous and heterosporous forms.

In heterosporous forms the gametophytes are extremely reduced and their size is limited to the confines of the spore. The development of the gametophyte here is called *endosporic*.

In homosporous forms the gametophytes are comparitively bigger and they break open the spore to grow independently on the soil. This type of development is said to be *exosporic*. Homosporous gametophytes are filamentous, cordate or tuberous in shape. Some of them exhibit dorsiventral symmetry. The gametophytes are composed mainly of parenchyma. The nutrition is either autotrophic (When chlorophyll is present) or saprophytic (when mycorrhiza is present).

Reproduction:- Gametophytes also reproduce by two methods viz, vegetative propagation and sexual reproduction. Vegetative propagation is uncommon and it takes place by the formation of gemmae or brood bodies (see *Lycopodium*).

Sexual reproduction takes place by the formation of antheridia (Male) and archegonia (female). Both the sex organs are extremely reduced in comparison with their counter part in Bryophytes.

An antheridium is a globose structure partially or completely embeded in the gametophytic tissue. It has a single layered Jacket enclosing the androgonial cells. The antherozoids are spirally coiled and may be biciliate or multiciliate.

An archegonium is a flask shaped structure having a basal venter and a short neck. Within the archegonium are found one or two neck canal cells one venter canal cell and an egg cell.

Fertilization:- Even in pteridophytes fertilization is dependent upon external moisture. The antherozoids that come out of the antheridium are attracted chemotactically by the nucilagenous mass produced by the disintegrating cells (neck canal and venter canal cells) of the archegonium. Many antherozoids enter the archegonium. Ultimately one succeeds in fusing with the egg resulting in the formation of a zygote.

Embryogeny:- The zygote represents the first cell of the sporophytic generation. The first division of the

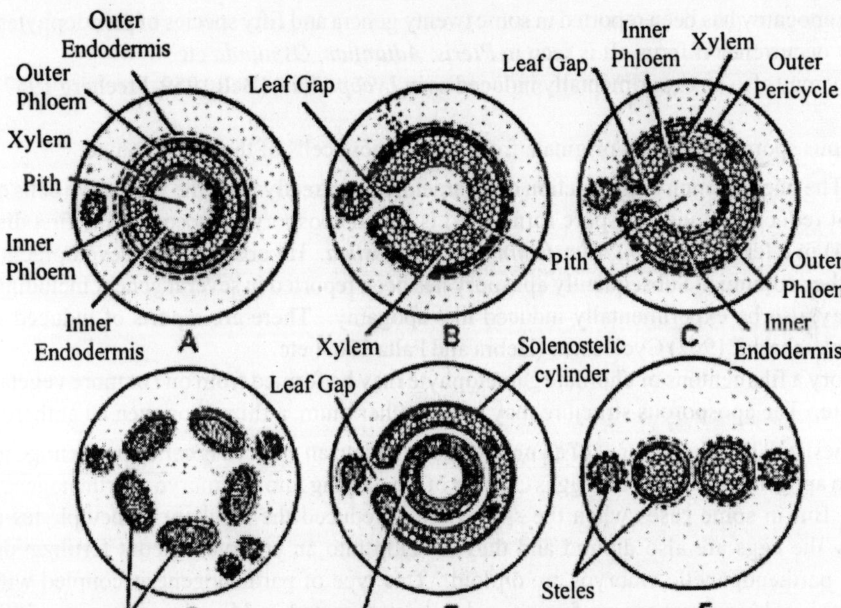

Fig. 3.6 Types of Vasculature in Pteridophytes (Contd....)
A. Amphipholic Siphonostele, B. Ectophloic Solenostele, C. Amphipholic Solenostele (with leaf gap), D. Dictyostele, E. Polycyclic Solenostele F. Polystele

zygote is either transverse or vertical. When the division is transverse two superposed cells are formed, of which the one nearer to the archegonial neck is the epibasal cell and the one away from the archegonial neck is the hypobasal cell. The embryo proper may be derived from either the epibasal cell (exoscopic) or the hypobasal cell (endoscopic). In the latter case the epibasal cell forms a tubular structure known as the suspensor. The function of the suspensor is to push the developing embryo deep into the gametophytic tissue so that it can easily absorb nutrition.

Further divisions in the embryonal cell are variable. After the first few divisions the characteristic parts of the embryo like foot, root, cotyledons and stem apex are differentiated. Untill the root establishes itself on the soil the foot behaving like a hanstorium absorbs food from the gametophyte. After the root grows into the soil the sporeling seperates itself from the gametophyte and grows independently.

Aberrvations in the life cycle.

In the normal type of life cycle showing alternation of generations the sporophyte gives rise to a gametophyte and vice-versa. Syngamy and reduction division occur at specific points in the life cycle so that the gametophyte is always haploid and sporophyte is always diploid.

In some cases there are deviations or aberrations in the life cycle disturbing the chromosome constitution of the sporophyte and the gametophyte. The aberrations are of three types viz, apogamy, apospory and parthenogenesis.

Apogamy:- This was first discovered by Farlow (1874) in *Pteris cretica*. Winkler (1908) defines apogamy as follows-it is the formation of sporophyte directly from the vegetative cells of the gametophyte without the act of syngamy or gametic fusion.

In nature apogamy has been reported in some twenty genera and fifty species of pteridophytes. Apogamy is of frequent occurrence if ferns. It is seen in *Pteris, Adiantum, Osmunda* etc.

Apogamy may also be experimentally induced as in *Lycopodium* (Bell 1959, Freeberg 1957) *Equisetum* (1959) etc.

Apogamous sporophytes may originate from one or more cells of the gametophyte.

Apospory:- The phenomenon of production of a gametophyte directly from the vegetative cells of the sporophyte without reduction division (spore formation) is called apospory. Apospory was first discovered by Drucry (1894) in a fern *Atheyrium fillex femina* var. *clarissima*. He observed the development of gametophytes from the sporangia. Subsequently apospory has been reported in several genera including *Osmunda*. Apospory may also be experimentally induced like apogamy. There are reports of induced apospory in *Pteridium* (Tak- ahashi, 1962) Cyclosorus (Hebra and Palta 1969(etc.

In apospory a filamentous or chordate gametophyte may be formed from one or more vegetative cells of the sporophyte. The aposporous structure may be an antheridium, a rhizoid or even an antherozoid.

Parthenogenesis:- The development of an unfertilized egg into an embryo is called parthenogenesis. This is different from apogamy here only the egg is capable of developing into an embryo. Parthenogenetic embryos are haploid. But in some cases when the spores are unreduced the resultant gametophytes are diploid. Consequently the eggs are also diploid and they develop into an embryo without fertilization. In these instances the parthenogenetic embryos are diploid. This type of parthenogenesis coupled with failure of reduction division during megaspore formation has been reported in *Marsilea drummondii* (Strassburger 1907).

Classification of Pteridophytes

Since a long time vascular plants are customarily divided into pteridophyta and spermtophyta. This classification was based on the assumpion that the former lack the seeds while the latter produced them. But the discovery of pteridosperms (seed bearing ferns) broke down this artificial classification. Simmott (1935) has introduced the term Tracheophyta to include all vascular plants. Tracheophyta are further divided into four main groups viz, Psilopsida, Lycopsida, Sphenopsida and Pteropsida. But it is not certain whether these are divisions or classes. Hampt (1953) considers them as classes in the division Tracheophyta. While Limmermann (1930) Arnold (1974) Wardlaw (1952) etc., regard them as divisions.

Reimers (1954) considers pteridophyta as a division and divides it into five classe viz, Psilophytopsida, Lycopsida, Psilopsida, Artienlatae and Filices.

Smith (1955) Bold (1957) and others divide vascular cryptogams into five divisions viz Psilophyta, Lycophyta, Sphenophyta, Noeggerathiophyta and Pterophyta. Some people have changed the name Lycophyta to Lepidophyta (Smith 1955), Sphenophyta to Arthrophyta or Calamophyta (Smith 1955) and Pterophyta to Filicophyta (Parihar 1965).

In this book the classification is mainly based on the one proposed be Reimers in the 1954 edition of Engler's Syllubus der Pflanzen familien. This classification simple and easy to follow. Below is given the outline classification and the members discussed in this book.

Pteridophytes

1. Psilophytopsida
   Psilophytales. — Eg: *Rhynia**, *Asteroxylon**.

2. Psilotopsida
   Psilotales — Eg: *Psilotum*

3. Lycopsida
a) Protolepidodendrales
b) Lycopodiales — Eg: *Lyocpodium, Phylloglossum.*
c) Lepidodendrales — Eg: *Lepidodendron**, *Lepidocarpon**.
d) Selaginellales — Eg: *Selaginella.*
e) Isoetales — Eg: *Isoetes.*

4. Sphenopsida
a) Hygeniales
b) Sphenophyllales — Eg: *Sphenophyllum*
c) Calamitales — Eg: *Calmites**, *Calamostachys*
d) Equisetales — Eg: *Equisetum*

5. Pteropsida
A) Primofilices — Eg: *Zygopteris, Stauropteris, Botryopteris*
a) Cladoxylales
b) Coenopteridales

B) Eusporangiatae
a) Ophioglossales — Eg: *Ophioglossum*
b) Marathiales — Eg: *Angiopteris*

C) Osmundidae
   Osmundales — Eg: *Osmumda*

D) Leptosporangiatae
a) Filicales — Eg: *Hymenophyllum, Adiantum, Pteris, Pteridium*
b) Marsilcalcs — Eg: *Marsilea, Azolla, Salvinia*
c) Salviniales

(Asterisk mark indicates the fossil members).

Chapter 4

FOSSILS - A BRIEF ACCOUNT

A study of the Geological history of the Earth reveals that a sequenced succession of plants occurred on the face of this planet. During such a process diverse types of plant groups appeared and vanished in course of time. Some of them attained predominant position in the past flora, thereafter met an extinction while new plant groups emerged.

The study of succession of plants in the geological history forms an exciting study of botany called **Paleobotany**

Paleobotany in its preview includes the study of the Flora of the past. Such a study is made by the observations of plant remains called **fossils.**

What are fossils? It is difficult to define the word *fossil* precisely. The word fossil is derived from the Latin verb *foedre* meaning "to dig". Hence basically a fossil is meant to include anything that is excavated from the Earth. In the present usage however the term fossil is restricted to organic remains obtained from earth and excludes inorganic objects or objects fashioned by human beings.

A fossil includes the remains of a whole organism or some part of the organism or the direct evidence of the prior existence of some organism in the sediments of earth. In addition to this, any thing resulting from or indicating the prior existence of organisms such as the imprints tracks, trails, borings and in some cases even chemical precipitations may be regarded as fossils. Chemical fossils include things such as sulphur and iron deposits indicating the former bacterial acitivity and lime stones representing the deposition of calcium carbonate by algae.

TYPES OF FOSSILS:

Several types of fossils have been recognised based on the structure, process of fossilization etc. In general the following types of fossils have been recognised.

Compressions: Of all the types of fossils, these are the most common ones found in the rocks. As the very name suggests, the plantbody gets flattened due to the pressure of the sediments. Due to this a thin carbonaceous film is retained outlining the surface features of the plant parts. Generally little or no internal structure is retained. Rarely however the cell pattern of cutinised epidermis is retained. Sediments of various types are involved as matrices for compressing the plant parts. These are shale, sand stone, volcanic ash, diatomaceous earth etc. Compressions help us to understand the external features of the plant. They provide reliable clues in establishing the affinities.

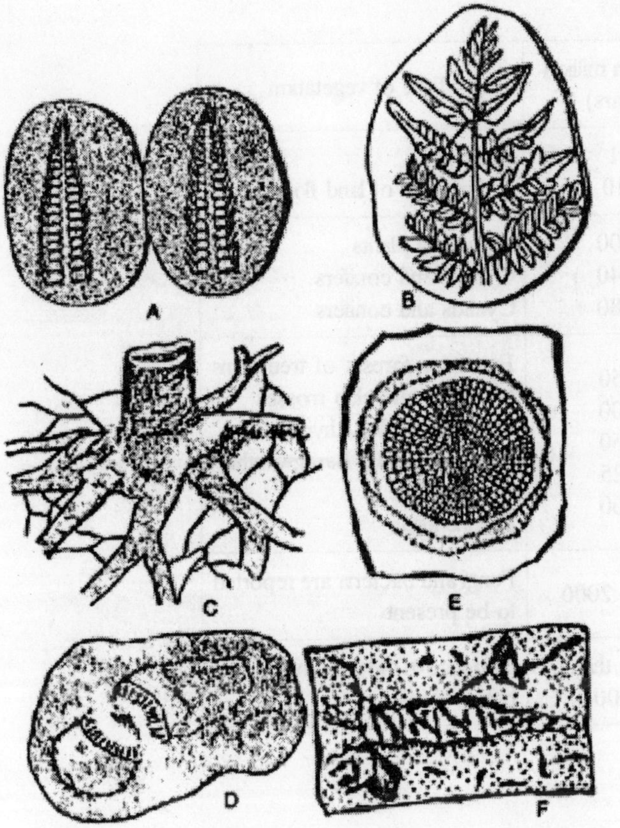

Fig. 4.1 Types of Fossils
A. Compression (cone of Lepidodendron), B. Impression (Leaf of Neuropteris), C. Cast (Stump of Stigmaria), D. Section of a coal ball showing transverse section of a petrified stem (Sphenophyllum). E. Sections of stems within coal balls. F. Compression (spike of Zosterophyllum)

Impressions: An impression may be defined as a negative of a compression. The rock strata that enclose the plant parts become hardened and when split open, they show the negative imprint of the plant part. There is no organic matter in an impression; therefore cellular details are not available. But the outlines of plantsparts are very well preserved. Of all the plant parts leaves yield good impressions.

Moulds, incrustations and casts: After the plant part is entombed in the sediment, if there is no compression, the surrounding sedimentary material hardens to form a three dimensional structure called *mould*.

Incrustations may be defined as external moulds of a plant, usually in some incompressible material like sand stone, iron stone or terfa.

After the formation of mould, sometimes the plant part may decay. This space is later filled by crystalline substances such as iron pyrites, sphalarites, chalerite, opal, agate etc. Such hardened structures are called casts. Casts exhibit the same external configurations of the original plant fragment. No organic material is seen in a cast.

Petrifactions: Of all the types of fossils, petrifactions are the best. Among the fossils they are also rare to obtain. In these fossils there is preservation of external form and internal structure and sometimes the substances of original plants are also preserved. Cellular details are preserved due to infilteration of some minerals into the plant tissues.

During petrifaction, the entombed plant is impregnated with about twenty minerals such as carbonates, sulphates, silicates, phosphates, iron pyrites etc. Solidification of plant tissues takes place immediately.

Petrifactions are of great importance, because they can be cut into sections for anatomical studies.

Geological time table: This is a table which gives in a chronological order, the different ages in the history of the Earth together with the fossils found in that period.

Era	Period	Age (in million years)	Type of vegetation
Cenozoic	Quaternary	1	Modern
	Tertiary	10	Culmination of land flora
Mesozoic	Cretaceous	100	Flowering plants
	Jurassic	140	Cycads and conifers
	Triassic	180	Cycads and conifers
Paleozoic	Carboniferous	250	Extensive forests of tree ferns horsetails and club mosses.
	Devonian	300	Land plants like Rhynia.
	Silurian	350	Land plants like Baragwanthia.
	Ordovician	425	Marine algae
	Cambrian	500	Marine algae
Proterozoic		550 - 2000	Fungi and bacteria are reported to be present
Archeozoid		more than 2000	Evidence of the existence of life; but no fossils.

Chapter 5

PSILOPHYTOPSIDA

The division psilophytopsida constitutes a very primitive group among pteridophytes. Members of the division are all fossils and no living members of the group are available.

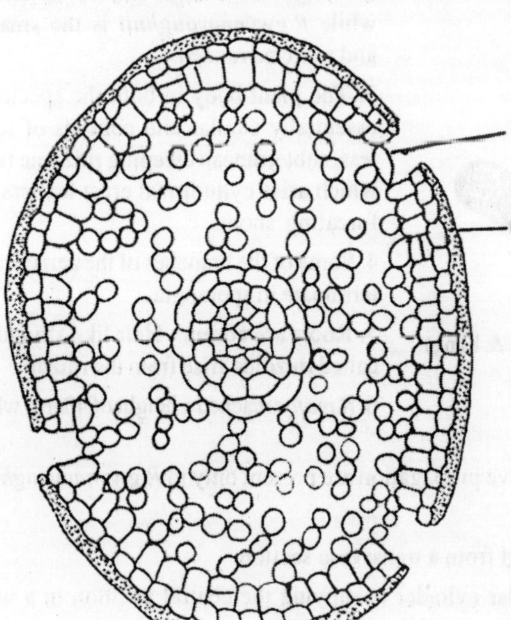

Fig. 5.1. T.S. of Aerial Shoot of *rhynia major*

Fossils of the group were first discovered in 1859 by Dawson a Canadian geologist. But the discoveries of Kidston and Lang in the year 1917 of well preserved plant specimens firmly established the affinities of Psilophytopsida with other pteridophytes. Kidston and Lang described certain petrified fossils from old red sand stone beds near Rhynie, in Aberdeenshire, Scotland.

The general characters of psilophytopsida may be listed as follows -

1. Simple plantbody with no differntiation into root, stem and leaf.
2. The naked green stems are spiny having cuticles and stomata. Occasionally leaf like organs may be present.
3. Vasculature is of the simplest type.
4. Sporangia are borne at the tips of ultimate branches.

Classification: The class consists of a single order Psilophytales and five families viz, Rhyniaceae, Psilophytaceae, Zosterophyllaceae, Asteroxylaceae and

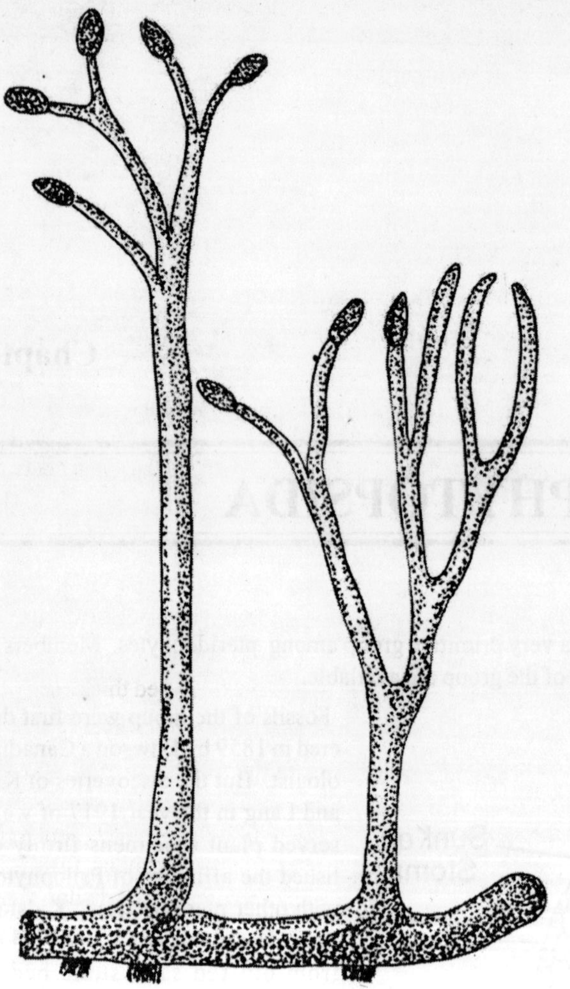

Fig. 5.2. Reconstruction of *Rhynia major* (After Kidston & Lang)

Pseudosporochnaceae. Of these families Rhyniaceae and Asteroxylaceae are discussed here.

RHYNIA

Occurrence and Distribution

Rhynia is a fossil member. The petrified fossils were first discovered in Rhynie chert in Aberdeen shire, Scotland from old Red sand stone beds.

Practical work

1. Study the external morphology from the model of the plant body.
2. Study the internal sturcture of the stem.
3. Study the reproductive structures (Sporangia) from a long section of the sporangium.

Morphology of the plantbody

1. The genus *Rhynia* consists of two species viz, *R.gwynnevaughnii* and *R.major*.
2. *R.major* is the larger and less advanced, while *R.gwynnevaughnii* is the smaller and more advanced.
3. The plant body in both the species is essentially similar and consists of rootless, subterranean creeping rhizome from which arise cylindirical erect leafless bifurcating shoots.
4. Some of the branches of the aerial shoot terminate in sporangia.
5. Roots are absent. Hair like structures called rhizoids arise from the rhizome.
6. *R.major* reached a height of 40cms while *R.gwynnevaughnii* measured 17-26cms in height.
7. Hemispherical bulges, probably helping in vegetative propagation are present only in *R.gwynnevaughnii*.

Anatomy

1. Internal structure of the aerial shoot may be studied from a transverse section.
2. A transection of the stem shows a smooth vascular cylinder occupying the central position in a wide cortex.
3. The stele is a protostelic haplostele with xylem forming a smooth central core.
4. There is no pith.
5. Phloem surrounds the xylem; pericycle and endodermis are absent.
6. The cortex is divided into a broad inner cortex and a narrow outer cortex.

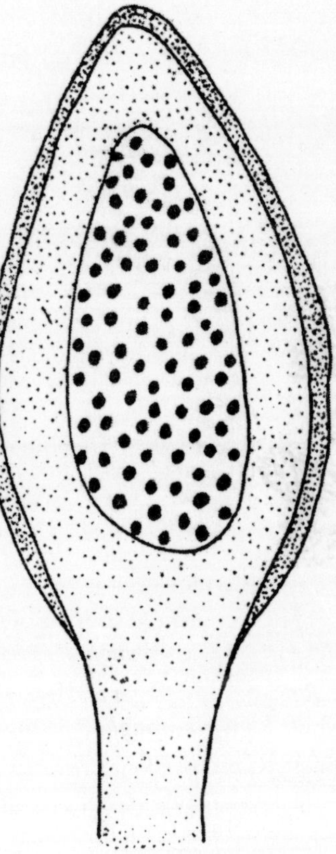

Fig. 5.3. A Sporangium of *Rhynia major*

7. The inner cortex consists of radially arranged rounded cells with intercellular spaces.

8. The inner cortical cells are chlorophyllous constituting the chief photosynthetic tissue of the plant.

9. The outer cortex is one to four layered thick and composed of large clear cells.

10. Outermost layer is the epidermis.

11. Stomata are present in the epidermal layer.

12. There is sub stomatal cavity which brings the assimilatory zone in direct communication with the stoma.

Reproduction

1. All the fossil specimens of *Rhynia* obtained so far are only sporophytes.

2. Gametophytic generation, has not been preserved as a fossil.

3. The sporophyte reproduces by means of sporangia. The sporangia are borne distally on certain branches of the stem.

4. The sporangium is cylindrical but tapers towards the distal end. It has a stout stalk at the base.

5. The sporangial wall is several layered thick.

6. Cuticle is deposited on the outermost layer (epidermis).

7. The innermost layer of the sporangial wall may be called tapetum.

8. A number of spores (often in tetrads) are found lying in the sporangial cavity.

9. The spores are homosporous.

10. Spores have cutinised cell walls.

Diagrams to be drawn

1. Habit sketch (Reconstruction)

2. Transverse section of Aerial shoot.

3. Long section of sporangium.

Systematic position

PSILOPHYTOPSIDA

PSILOPHYTALES

RHYNIACEAE

RHYNIA

ASTEROXYLON

(Astero = star)

Occurrence and distribution

Asteroxylon is a fossil member. The petrified fossils (*A.mackie*) have been obtained from *Rhynie* chert as well as from middle Devonian rock strata from Germany (*A.elberfeldens*)

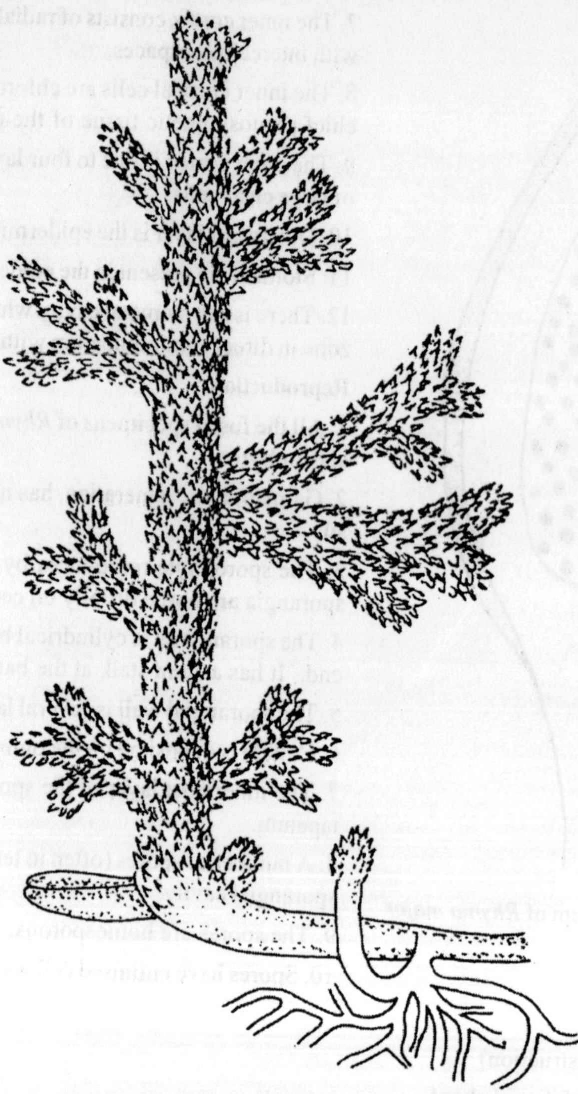

Fig. 5.4. Reconstruction of *Asteroxylon mackie*

Practical work

1. Study the external features of the plantbody from the reconstruction.

2. Roots as well as rhizoids are absent.

3. Some of the rhizomatous branches dichotomised freely and produced slender leaf like organs which penetrated the substatum.

4. The aerial branches have spirally arranged non casculated leaves.

5. At the base of the aerial shoot a transition zone is present studded with scales.

6. The branching of the aerial shoot is lateral and exogenous.

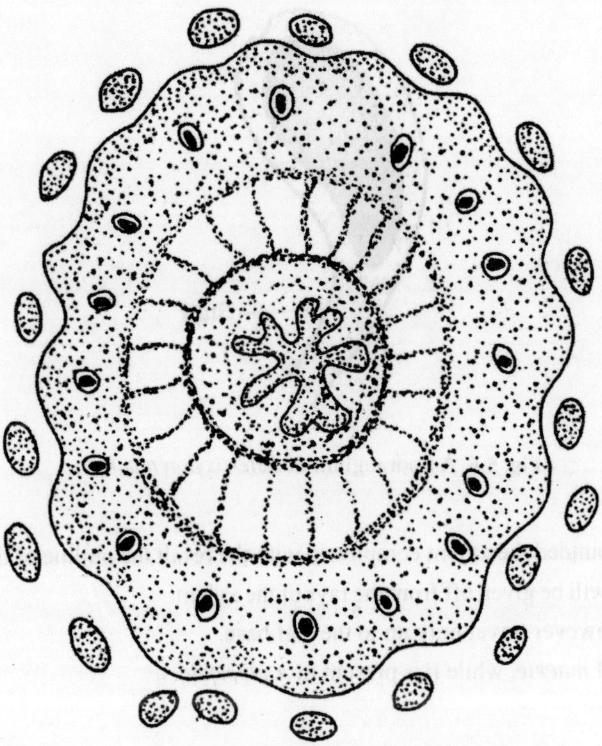

Fig. 5.5. T.S. of Aerial Shoot of *Asteroxylon mackie*

Anatomy

1. Internal structure of the aerial shoot can be studied in a cross section.
2. A cross section of the stem shows three regions viz., Epidermis, Cortex and Stele.
3. The outline of the section would be wavy due to the presence of leaf bases.
4. Epidermis is single layered. Stomata are present disturbing the continuity of the epidermis. Stomata are sunken unlike in *Rhynia*.
5. The cortex has three zones.
6. The inner cortex has compactly arranged parenchyma cells.
7. Middle cortex has elongated trabecular cells consisting of radiating plates of cells separating the lacunae.
8. Outer cortex has tangentially elongated cells.
9. Cortical cells contain aseptate fungal hyphae and resting spores belonging to the type genus *Paleomyces*.
10. The central region of the stem is occupied by the vascular cylinder.
11. Pericycle and endodermis are absent.
12. The stele is an actinostelic protostele.
13. The xylem strand is star like with four or more sometimes forked arms.
14. Protoxylem elements terminated the tips of the xylem arms.

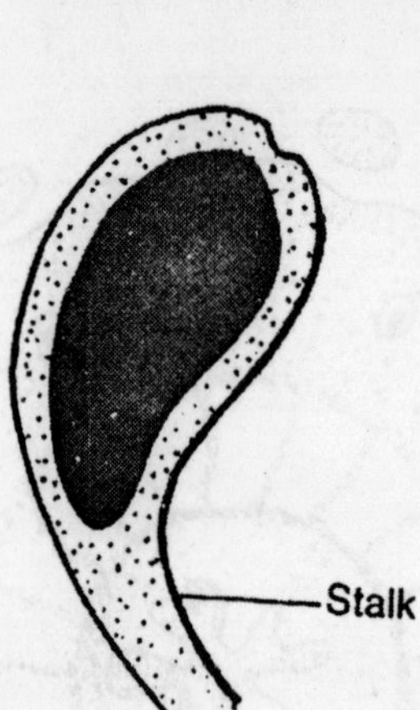

Fig. 5.6. A Sporangium of *Asteroxylon mackie*

15. The phloem surrounded the xylem completely with the result the outline of the stele is smooth.
16. Small leaf traces will be given off from the rays of the xylem.
17. The leaf traces however travel only up to the leaf base.
18. Pith is absent in *A.mackie*, while it is present in *A.elberfeldens*.

Reproduction

1. The fertile shoots were not in organic contact in *A.mackie*, while they were connected in *A.elberfeldens*.
2. The Sporangia are borne on slender naked leafless branches.
3. The sporangia are ovoid or ellipsoid in shape.
4. An annulus is present at the wider free end of the sporangium marking the line of dehiscence.
5. Spores are homosporous.

Diagrams to be drawn

1. Habit sketch (Reconstruction).
2. Transverse section of the stem.
3. Reproductive branches and sporangia.

Systematic position

PSILOPHYTOPSIDA

PSILOPHYTALES

ASTEROXYLACEAE

ASTEROXYLON

Chapter 6

PSILOTOPSIDA

This class consists of most primitive living members among pteridophytes. The members show close resemblance to psilophytales.

In general, the class is characterised by the following features.

1. The sporophyte is a rootless, leafless structure. Rhizoids help in the absorption of water.

2. Branching of the sporophyte is always dichotomous.

3. The plantbody is usually differentiated into an underground prostrate system and an aerial erect shoot system.

4. The spore bearing organs are bi or trilobed and are borne in the axils of small leaf like appendages.

5. Gametophytes are cylindrical, branched subterranean and achlorophyllous. Nutrition is saprophytic.

The class Psilotopsida has a single order psilotales and a single family psilotaceae. The family includes two genera viz., *Psilotum* and *Tmesepteris*. Of the two genera *Psilotum* is discussed here.

PSILOTUM

Occurence and distribution:

The genus *Psilotum* has two species viz., *P.nudum* and *P.flaccidum* and is widely distributed in the tropics and subtropics. The genus is included in the list of endangered species in India.

Practical work

1. Observe the morphological structure of the plant.

2. Study the anatomy of rhizome and aerial shoot.

3. Study the structure of the reproductive organs.

Morphology of the plant

1. The plantbody is differentiated into two parts viz., a horizontal undergroundrhizome and an erect aerial shoot.

2. The rhizome is brownish in color and dichotomously branched.

3. Roots are absent. Rhizoids develop from the rhizome to help in absorption.

4. Some of the branches of the rhizome grow up and form erect or pendulous aerial shoots.

5. The aerial shoots are 20-70cms long usually ribbed and multiangular.

6. Here and there on the aerial shoot are found a number of scaly leaves.

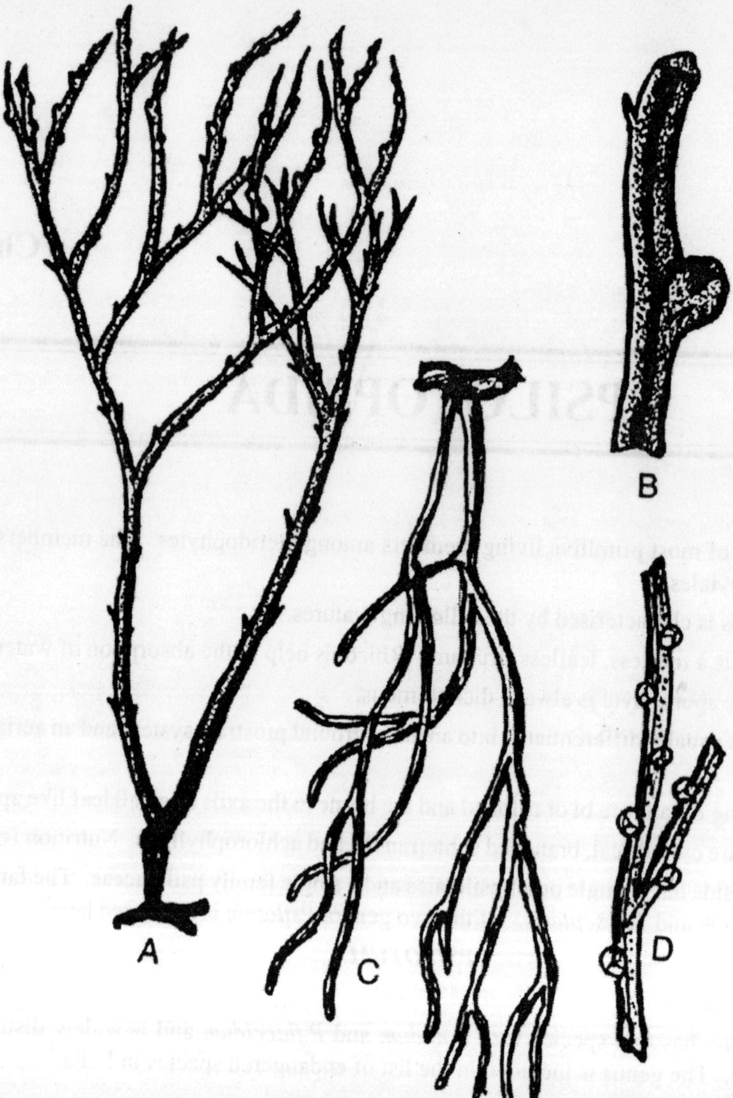

Fig. 6.1. ***Psilotum*** **: Habit of** ***Psilotum nudum*** **(A) and a Part of the Shoot Bearing Synangia (B); Habit of** ***Psilotum flaccidum*** **(C) and a part of the Shoot bearing Synangia (D)**

7. The scaly leaves are of two types fertile and sterile.

8. The sterile ones are found all along the length of the aerial shoot while the fertile ones are restricted only to the upper portions.

9. The fertile ones bear in their axils, trilobed spore bearing structures called Synangia.

Anatomy of rhizome

1. A transverse section of the rhizome shows an epidermis, a cortex and stele.

2. The epidermis is single layered.

3. The cortex is divided into three zones. outer cortex is parenchymatous and the cells have the mycorrhizal association.

4. The middle cortex is also parenchymatous with cells rich in starch grains.

5. The inner cortical cells are brown in color due to the deposition of a compound called phlobaphene.

6. The stele is protostelic surrounded by pericycle and endodermis.

7. The xylem is exarch and star shaped.

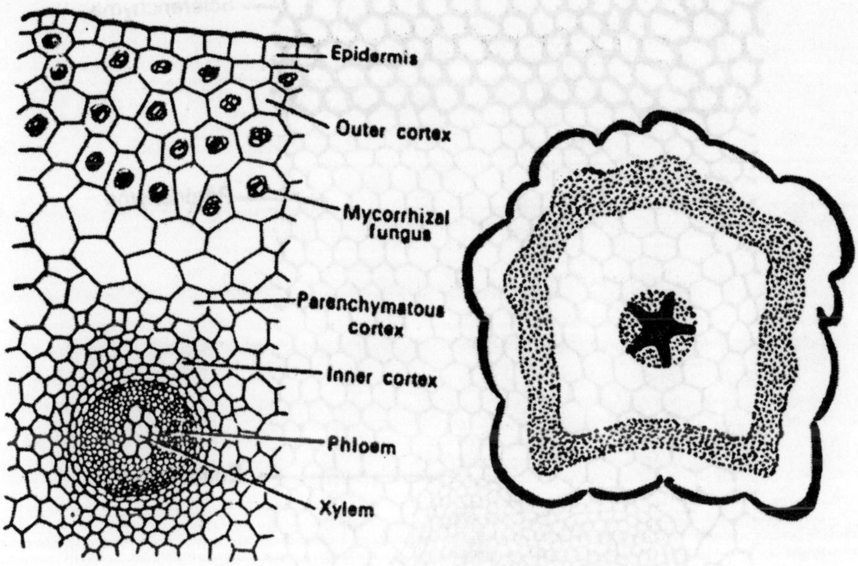

Fig. 6.2. ***Psilotum***: T.S. of Rhizome.
A. Sector Enlarged, B. Ground Plan.

Anatomy of aerial shoot

1. A transverse section of the aerial shoot shows the usual three regions viz., epidermis, cortex and stele.

2. Epidermis is single layered and has closely packed cells.

3. Stomata are present in the epidermis and are restricted to the grooves. They are sunken in nature.

4. A thick cuticle is deposited above the epidermis.

5. Next to the epidermis is the cortex. It is divided into three zones.

6. The outer cortex is chlorenchymatous. It is 2-5 layered thick with loosely arranged cells. These cells constitue the chief photosynthetic tissue of the plant (leaves are reduced).

7. The middle cortex consistes of 4-5 layeres of sclerenchyma. This offers mechanical support to the plant.

8. The inner cortex has a few layers of parenchyma cells.

9. Occupying the central region of the stem is the stele. Outermost layer of the stele is endodermis.

10. Next to the endodermis is an ill defined pericycle.

11. The stele is usually an actinostelic protostele. But in the basal portions the stele may be a siphonostele with a sclerenchymatous pith.

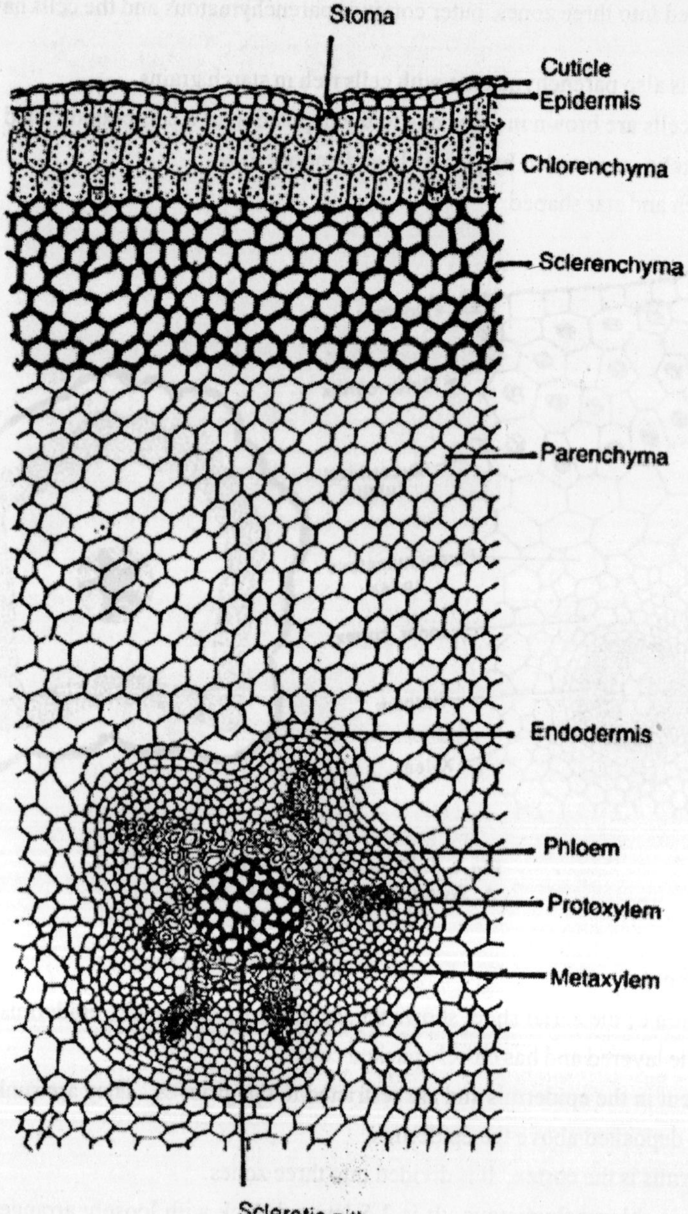

Fig. 6.3. ***Psilotum*** **: T.S. of Aerial Shoot (a sector enlarged)**

Reproduction

1. The sporophyte reproduces by means of spores. The spores are produced in trilobed structures called synangia.

2. The synangia are usually present in the upper part of the aerial shoots, in the axils of fertile scale leaves.

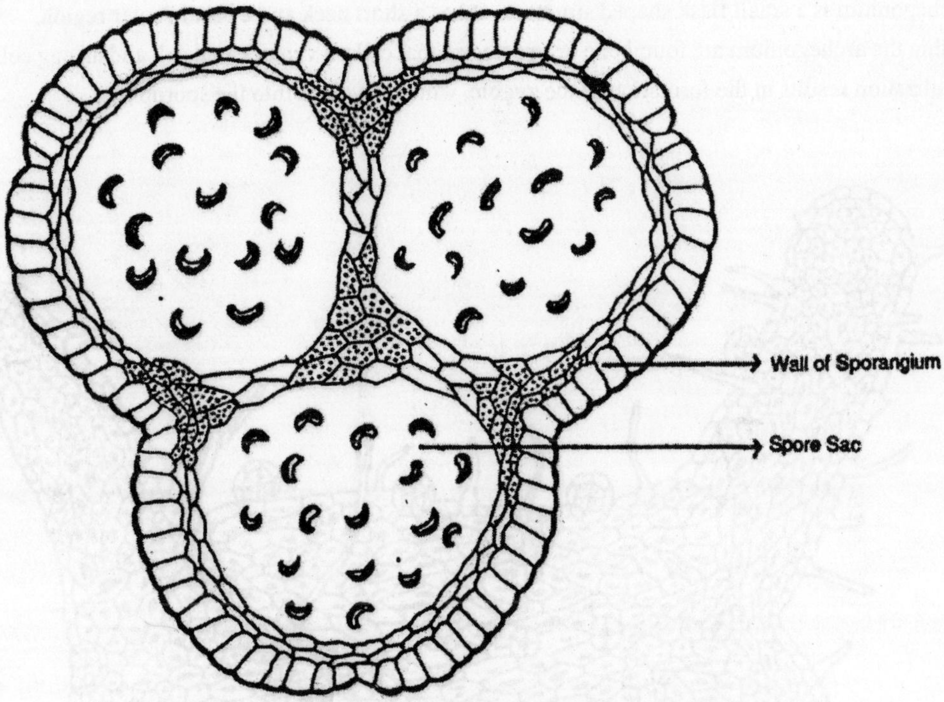

Fig. 6.4. *Psilotum*, T.S. of Mature Synangium with Spores

3. The wall of the trilobed synangium is made up of 4-5 layers of cells. The outermost layer of the wall is prismatic.

4. Within the synangium, ther are three chambers or spore cavities containing spores. The spores are homosporus.

5. The spores on germination develop into the gametophyte.

Structure of the gametophyte.

1. The gametophyte of Psilotum is a cylindrical structure with dichotomous branches.

2. The gametophyte is partly or totally subterranean.

3. The size of the gametophyte ranges from 0.5 to 2mm.

4. The gametophyte is usually dark brown in color. This is due to the mycorrhizal association.

5. Parenchyma is the only tissue found in the gametophyte. The cell walls of the outer layer are strongly cutinised.

6. A number of rhizoids develop from the outer layer of cells.

7. The cells are achlorophyllous and the nutrition of the gametophyte is saprophytic.

8. The gametophyte reproduces asexually as well as sexually.

9. Asexual reproduction of the gametophyte is brought about by means of gemmae.

10. Sexual reproduction is brought about by means of antheridia (male) and archegonia (female).

11. A mature antehridium is somewhat a spherical structure. It has a jacket (wall) surrounding many antherozoids.

12. Archegonium is a small flask shaped structure. it has a short neck and a basal venter region.
13. Within the archegonium are found one or two neck canal cells, a venter canal cell and an egg cell.
14. Fertilization results in the formation of the zygote, which develops into the sporophyte.

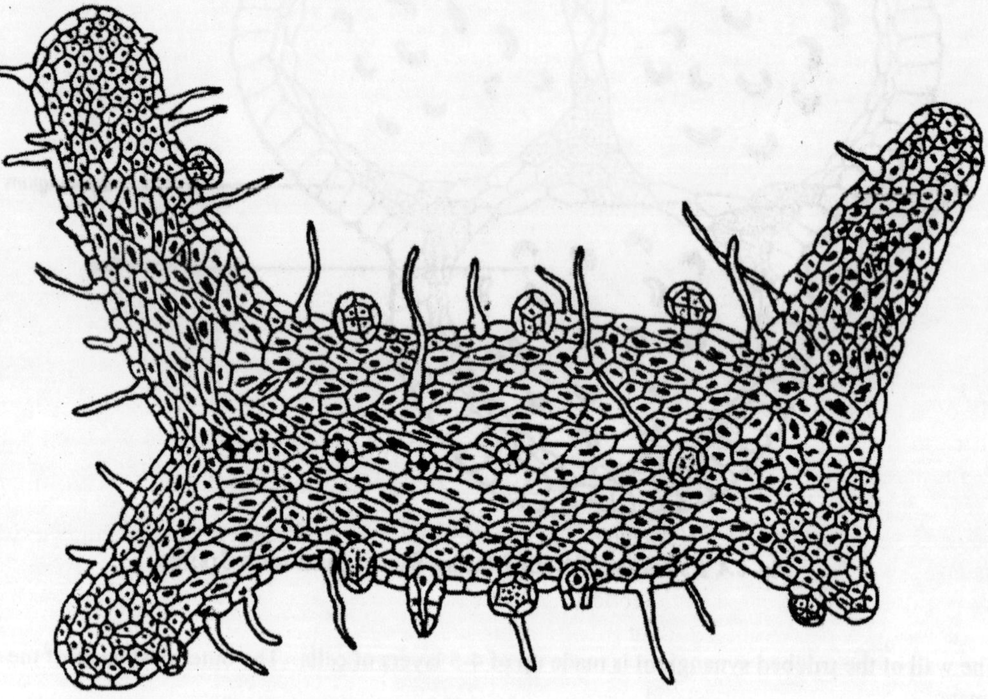

Fig. 6.5. ***Psilotum*, Gametophyte of *P.nudum***

Diagrams to be drawn

1. Habit of the plant.
2. A part of the aerial shoot with synangia.
3. Transverse section of the rhizome.
4. Transverse section of the aerial shoot.
5. Cross section of the syangium.
6. Habit of the gametophyte.

Systematic position

PSILOTOPSIDA

PSILOTALES

PSILOTACEAE

PSILOTUM

Chapter 7

LYCOPSIDA

The division Lycopsida includes both living as well as extinct plants. The Geological history of the group dates back to the Devonian period. In early periods, members of Lycopsids were tree like in habits. There were herbaceous members also.

The living members of the group are commonly called club mosses. There are only five living genera represented by over nine hundered species. They occupy a wide variety of habitat and enjoy universal distribution.

Lycopsids are characterised by the following features

1. Sporophytes are distinguished into roots, stems and leaves.
2. The branching of the stem as well as of the root is always dichotomous.
3. Leaves are spirally arranged and microphyllous (They have an unbranched midvein).
4. Leaves may have a small appendage (ligulate) occasionally. The leaves may be eligulate (no appendages) also.
5. Stele is protostelic with a number of variations in the configuration of the xylem.
6. Sporangia may or may not aggregate into strobili. They are always borne on the adaxial surface of the sporophyll.
7. Members are both homosporous as well as heterosporous.

Classification:

Lycopsida is classified into two sub classes Eligulopsida and Ligulopsida. In the former, the leaves are eligulate (a ligule is a small leafy outgrowth present on the adaxial surface of the leaves) and in the latter the leaves are ligulate.

Eligulopsida comprises two orders viz., Protolepidodendrales and Lycopodiales.

Ligulopsida has three orders viz., Lepidodendrales, Selaginellales and Isoetales.

In this book the following genera are discussed -

Lycopodium, Phylloglossum, Lepidodendron, Lepidocarpon, Selaginella and Isoetes.

LYCOPODIUM

Occurence and distribution

The genus *Lycopodium*, popularly called club moss, is world wide in distribution. The genus has 180 species. The species are so diverse in their nature, that many botanists divide the genus into two sub Genera viz., *Urostachya* and *Rhopalostachya*. The major differences between the two are as follows.

Fig. 7.1. *Lycopodium* : Habit of *L. phlegmaria* (*L. macrostachyum*)

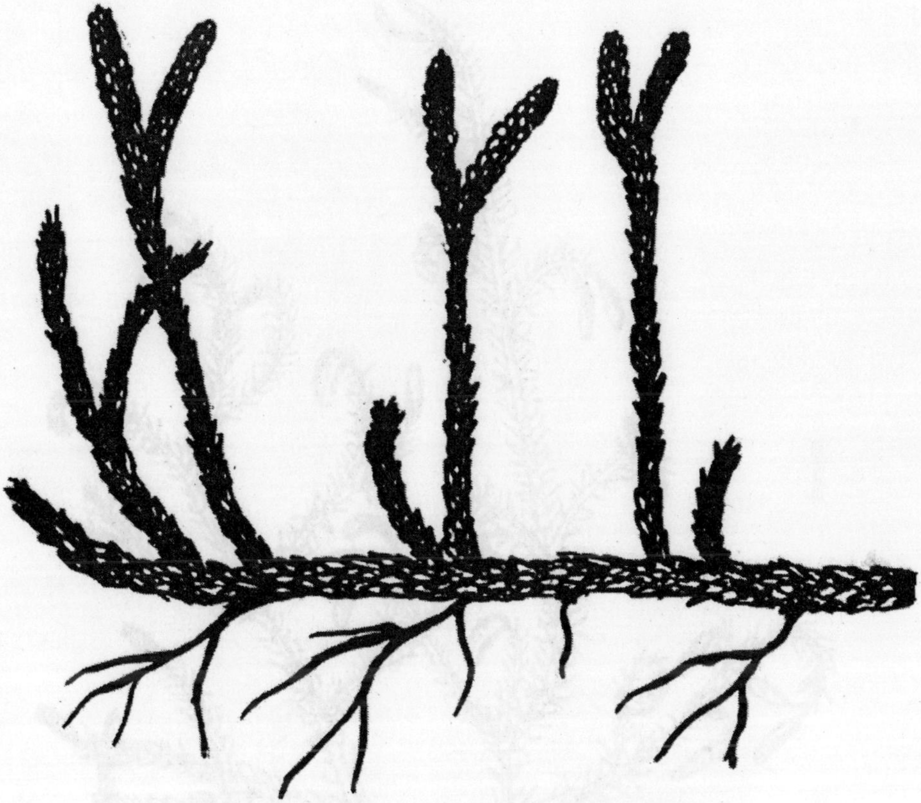

Fig. 7.2. *Lycopodium* : Habit of *L. wightianum*

UROSTASHYA	**RHOPALOSTACHYA**
Stem: Erect or pendent	Prostrate or semi erect
Branching: Dichotomous	Dichotomous are apparently monopodial.
Roots: Arise at the base of the axis	Arise all along the length of the stem.
Strobilus: May or may not be organised	Well organised.
Examples: *L.selago, L.phlegmaria,* L. lucidulum etc.	L.wightianum, L.cernum, L.complanatum, L.inundatum etc.

Practical work

1. Study the morphological characters of various species.
2. Cut transverse sections of root, stem and leaf.
3. Mount a single sporangium and observe.
4. Study the details of sporangium from a long section of the strobilus.
5. Study the structure of the gametophyte.

Fig. 7.3 *Lycopodium* : Habit of *L. cernuum*

Morphological features of Lycopodium phlegmaria

1. The plantbody is differentiated into root stem and leaves.
2. The plant is an epiphyte and generally grows in humus packets in the crevices of the branches.
3. The stem is pendant hanging down from the branches of tress.
4. The stem is branched and the branching is dichotomous.
5. Roots arise only at the base of the stem. The primary root being short lived, the adult sporophyte has only adventitious roots.

Fig. 7.4 . *Lycopodium* : Habit of *L. lucidulum*

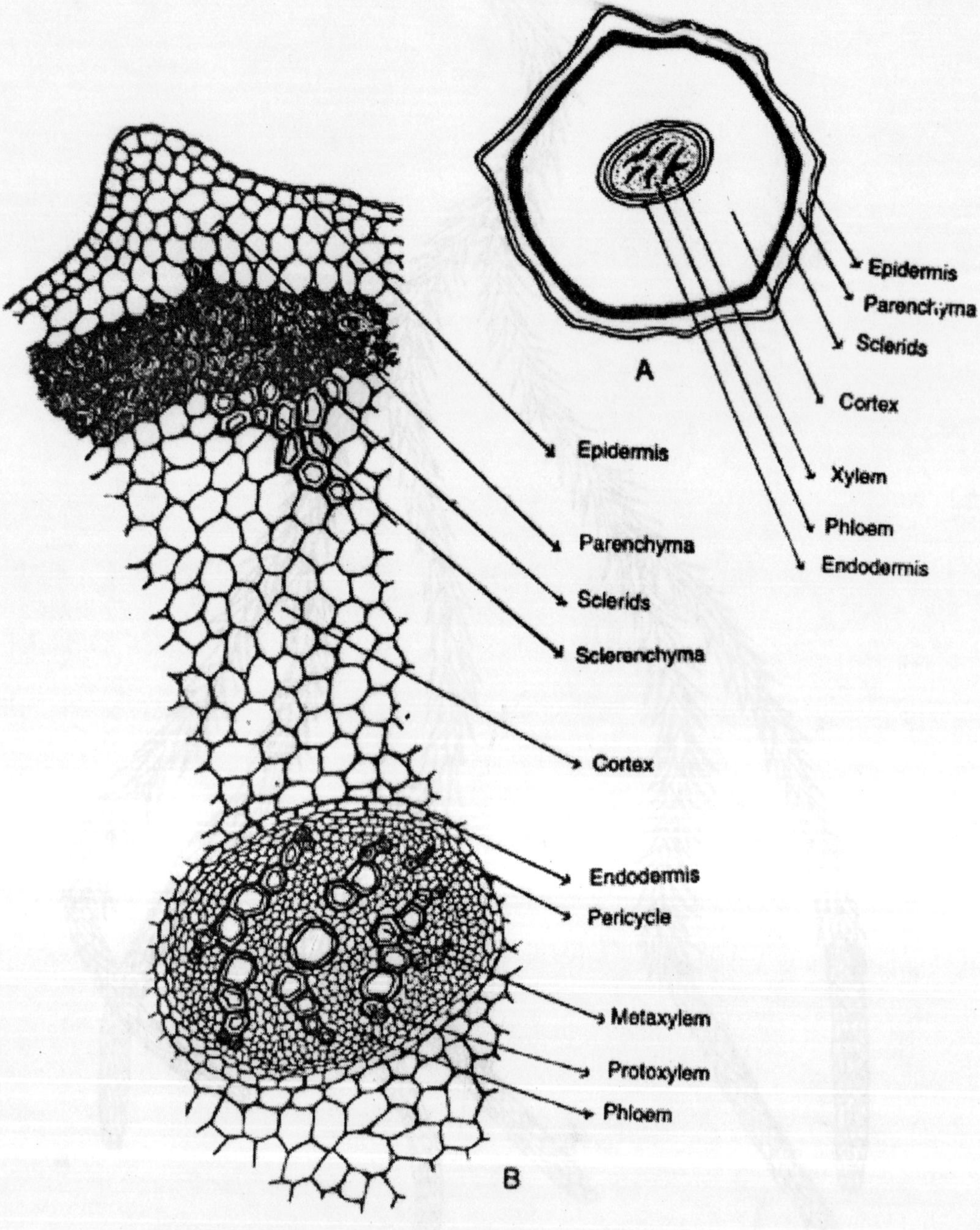

Fig. 7.5. *Lycopodium* : Internal Structure of Stem of *L. cernuum*
A. Ground Plan, B. A Sector Enlarged

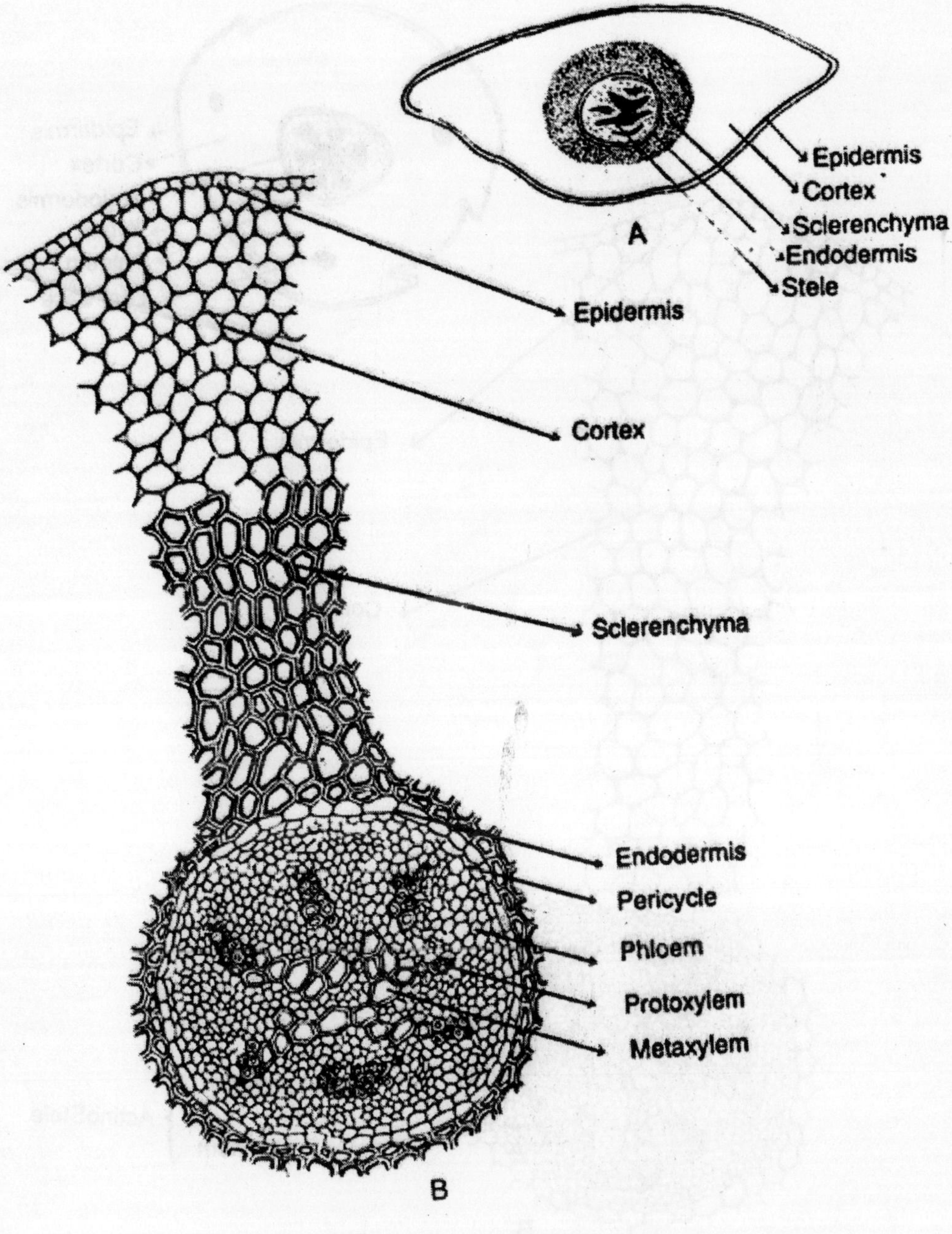

Fig. 7.6 *Lycopodium* : Internal Structure of Stem of *L. wightianum*
A. Ground Plan, B. A Sector Enlarged

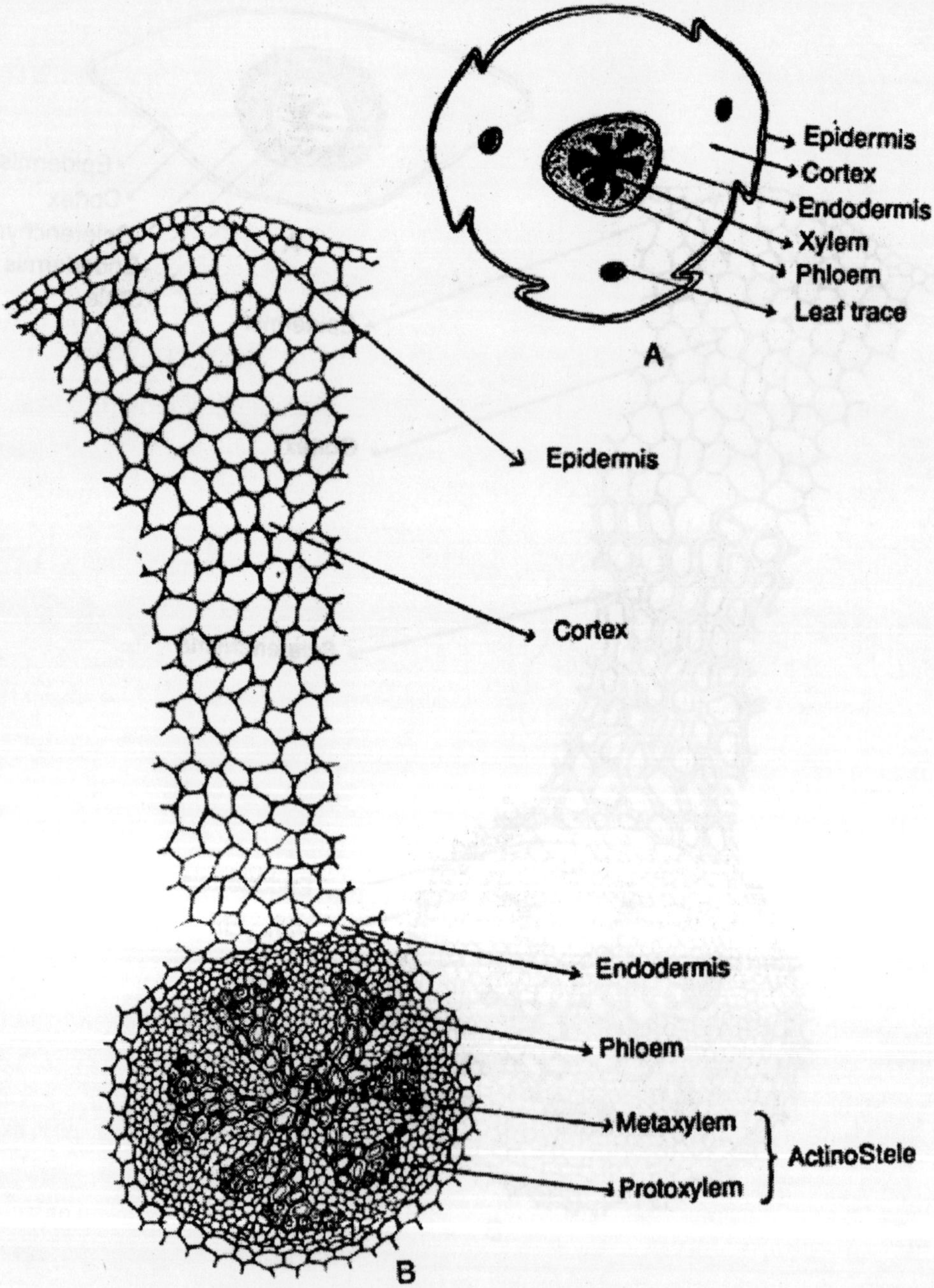

Fig. 7.7. *Lycopodium* : Internal Structure of Stem of *L. serratum*
A. Ground Plan, B. A Sector Enlarged

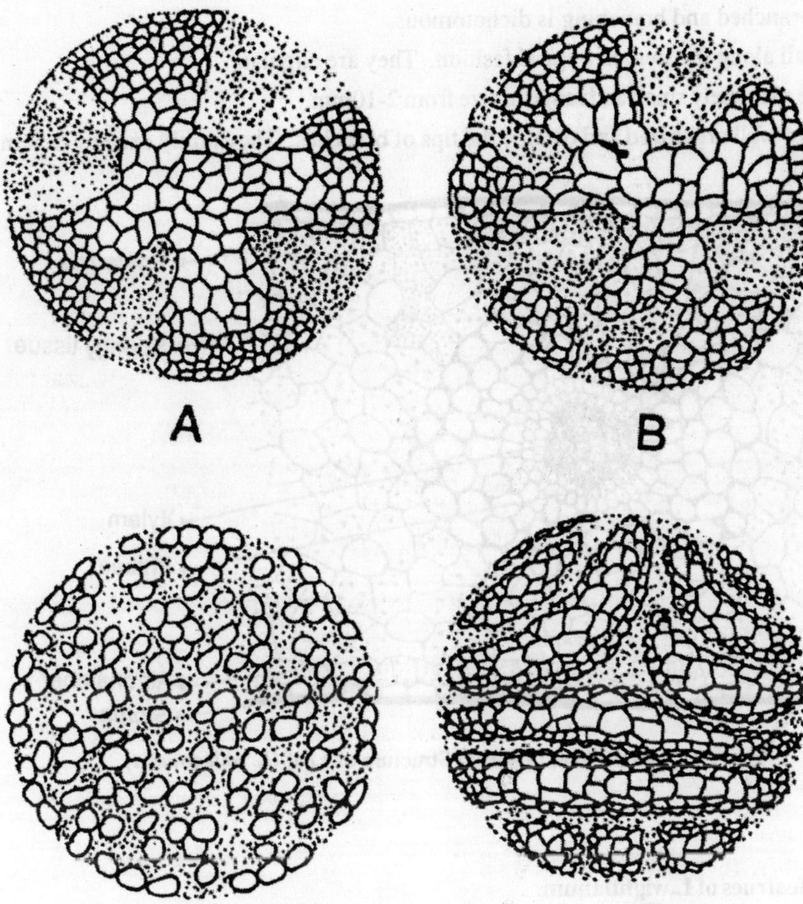

Fig. 7.8. *Lycopodium* : Various Types of Stelar Organisation in Stem
A. *L. serratum*, B. *L. lucidulum*, C. *L. cernuum* and D. *L. volubile*

6. Leaves arise all along the stem and are arranged in a spiral manner. They are microphyllous.

7. There is a clear distinction between foliage leaves and sporophylls.

8. The foliage leaves have a central unbranched vein. They are small, eligulate and sessile with a broad base.

9. The leaves in *L.phlegmaria* may sometimes be 2-3cms long.

10. The reproductive structures (sporangia) accumulate at the tips of branches in the form of strobili. The strobili are also branched.

Morphological features of Lycopodium cernuum

1. The plant belongs to the sub genus Rhopalostachya.

2. The plant body (sporophyte) is differentiated into root, stem and leaves.

3. The plant body is semi erect to erect in habit.

4. The plant grows in the soil and roots arise all along the length of the stem. They are adventitions.

5. The stem is branched and branching is dichotomous.
6. Leaves arise all along the stem in a spiral fashion. They are eligulate.
7. The leaves are extremely small and range in size from 2-10mm.
8. The strobili are well organized and occupy the tips of branches. They rarely exeed 1-1.5cms in length.

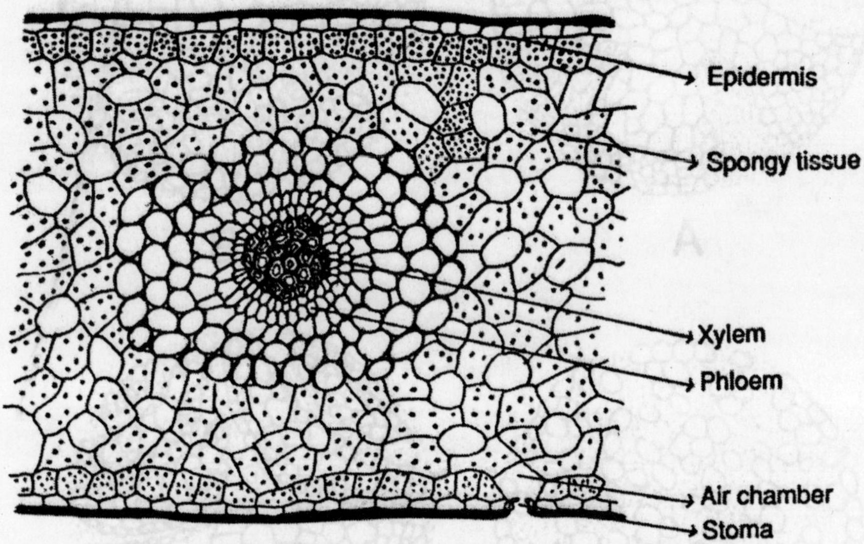

Fig. 7.9. *Lycopodium* **: Internal Structure of Leaf (*L. phlegmaria*)**

Morphological featrues of L.wightianum

1. The sporophytic plant body is distinguished into root, stem and leaf.
2. The stem (rhizome) is horizontal and it produces upright branches.
3. The adventitious roots arise all along the stem.
4. The plant belongs to the sub genus Rhopalostachya.
5. Leaves are small, eligulate, microphyllous (larger than those of *L.cernuum*) and spirally arranged.
6. Strobilus is clearly demarcated from the rest of the plant body.
7. Sporophylls are much smaller than those of foliage leaves and are compactly arranged.
8. The strobili are also branched.

Morphological features of L.hamiltonii

1. The plant belongs to the sub genus Urostachya.
2. The stem is erect and roots are produced at the basal region only.
3. The branching of the stem is dichotomous.
4. The leaves are small (but larger than those of L.cernuum), eligulate and microphyllous.
5. There is no distinction between the sporophylls and foliage leaves. No strobilus is found.
6. Some of the leaves bear the sporangia.

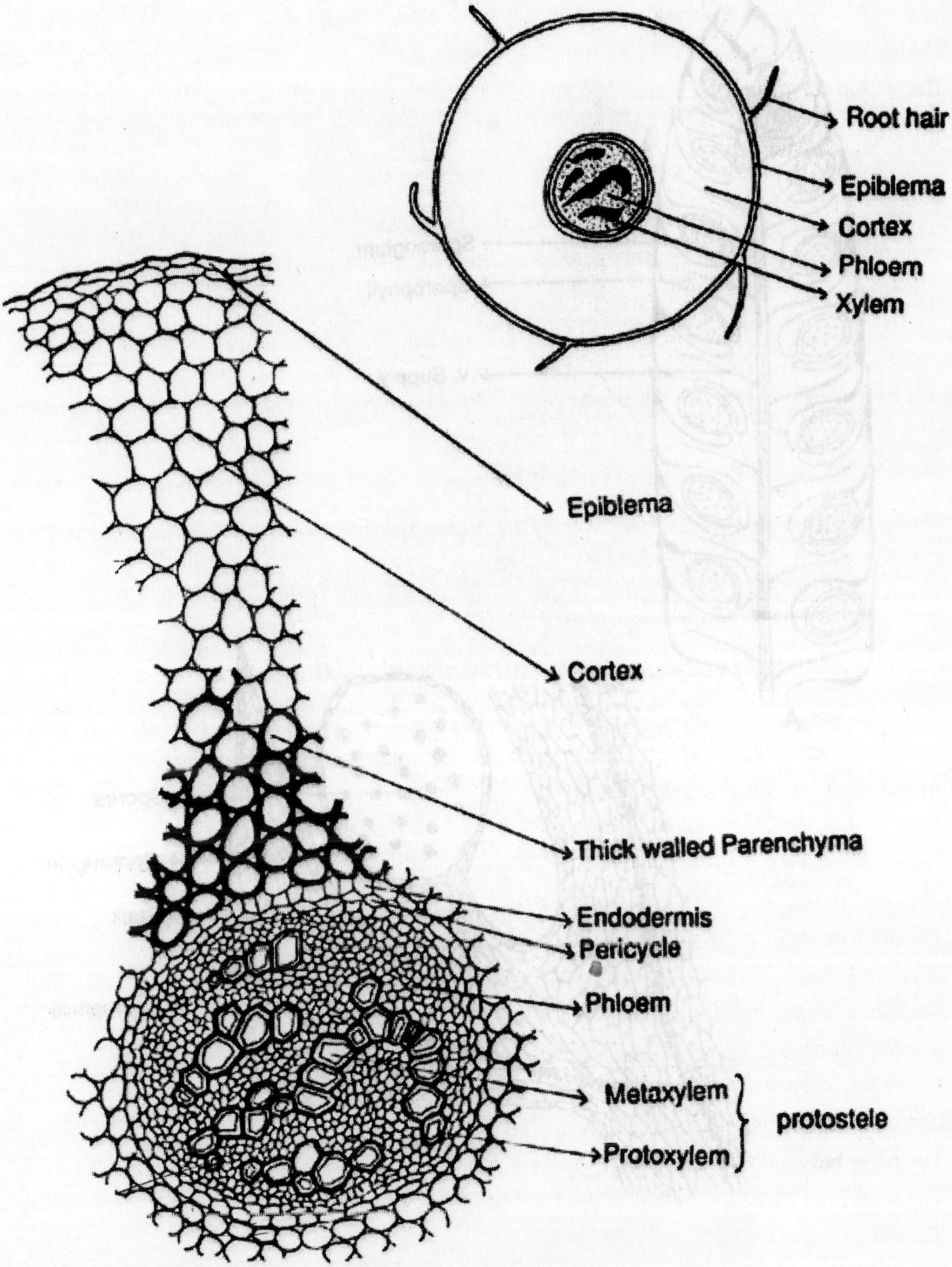

Fig. 7.10. *Lycopodium* : Internal Structure of Root (T.S.)

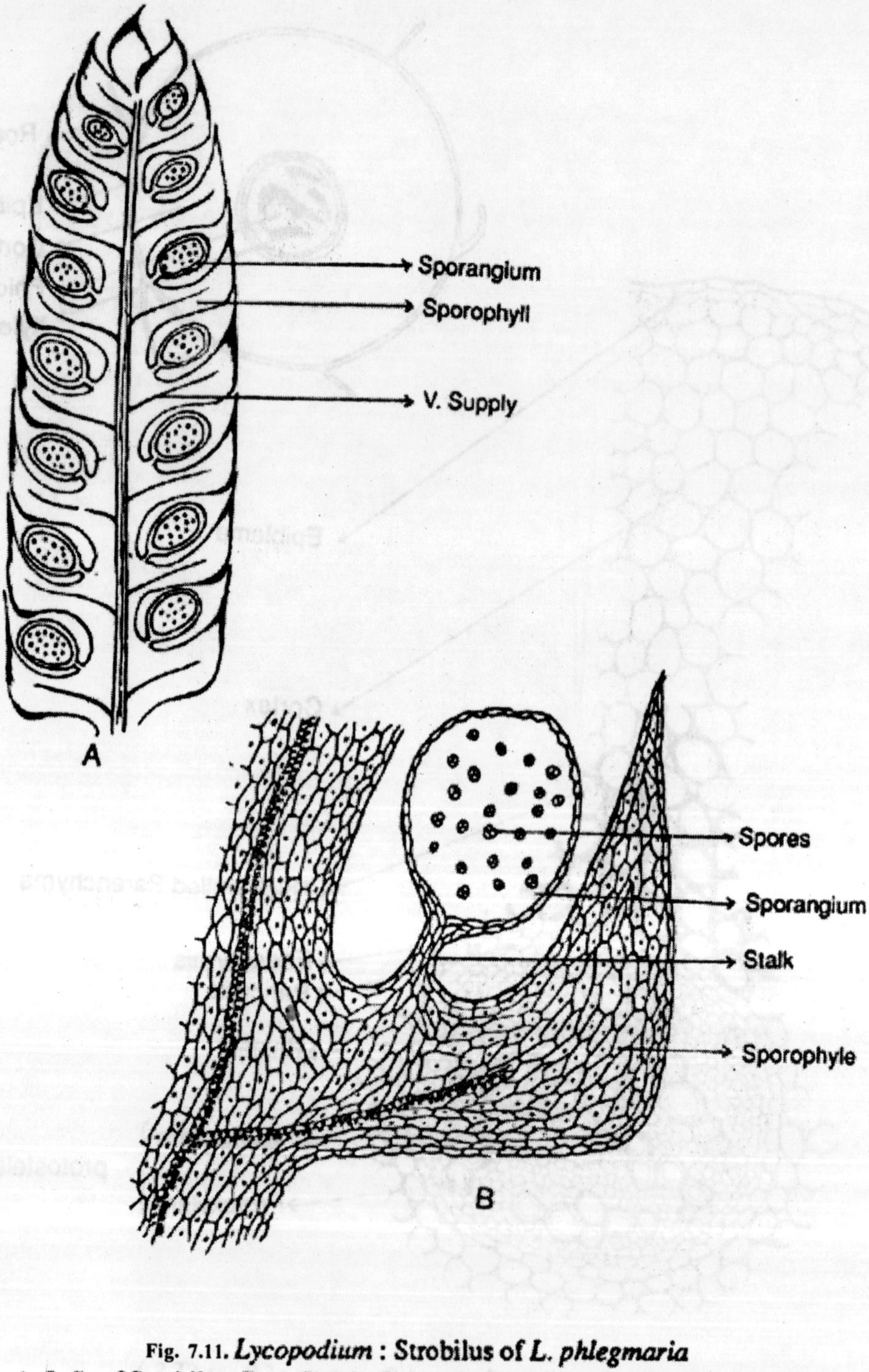

Fig. 7.11. *Lycopodium* : Strobilus of *L. phlegmaria*
A. L.S. of Strobilus, B. A Single Sporophyll (enlarged) with Sporangium

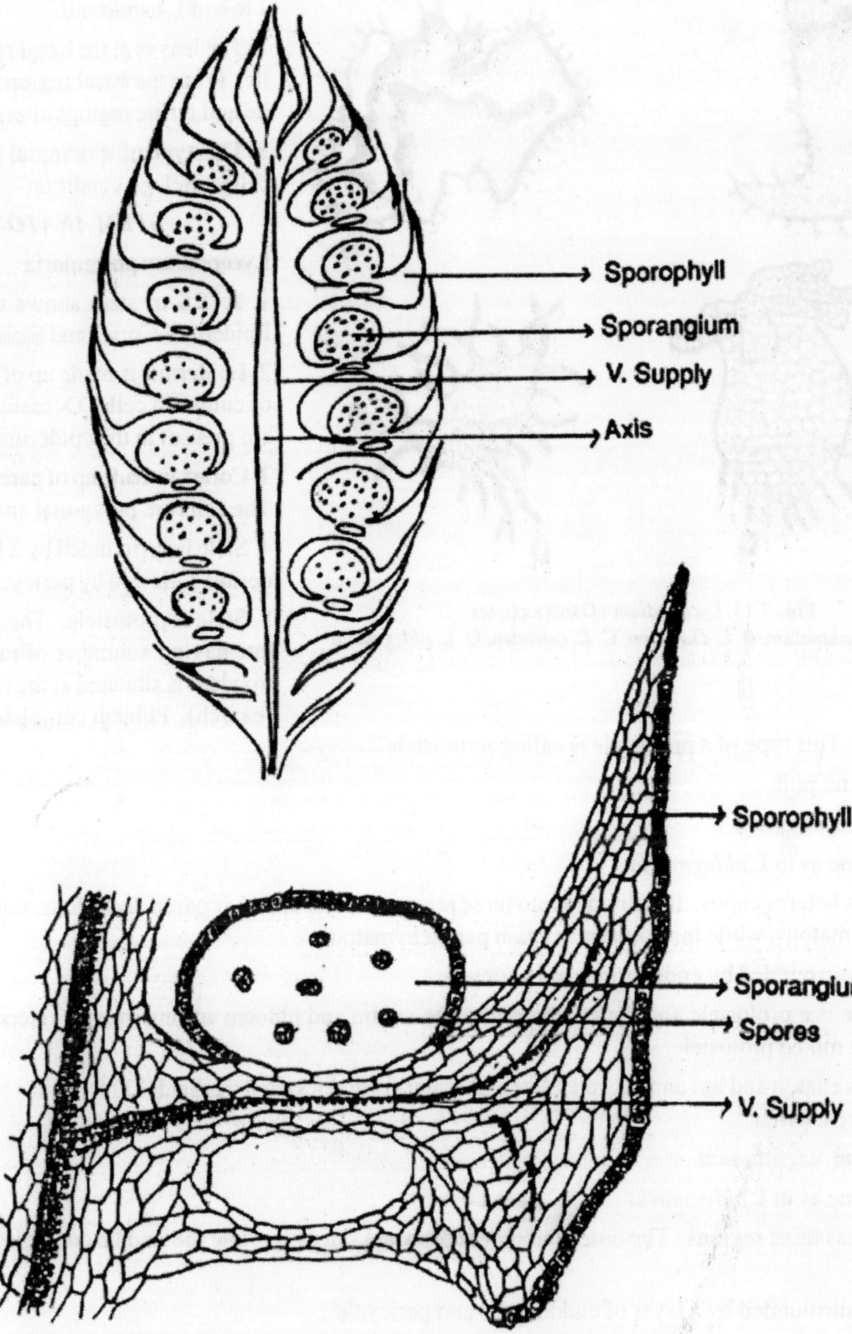

Fig. 7.12. *Lycopodium* : Strobilus of *L. cernuum*
A. L.S of strobilus B. A single sporophyll (enlarged) with Sporangium

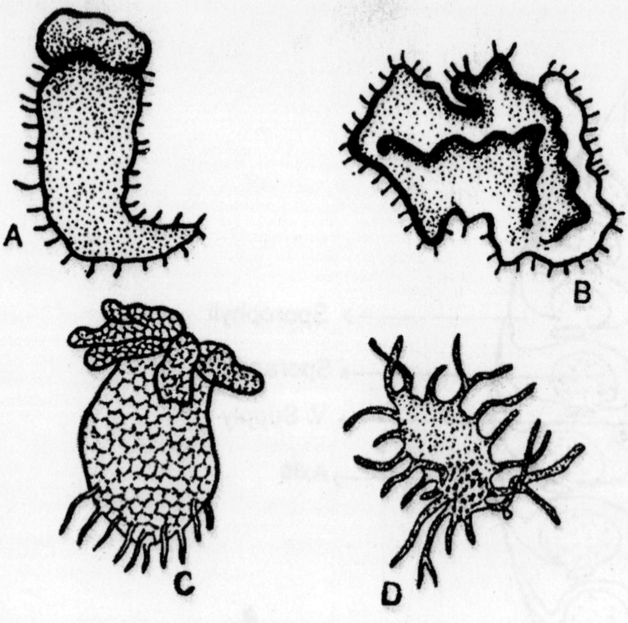

Fig. 7.13. *Lycopodium* : Gametophytes
A. *L. complanatum*, B. *L. clavatum*, C. *L. cernuum*, D. *L. phlegmaria*

Mophological features of L.selago

1 to 6 of L.hamiltonii.

7. The leaves at the basal region are sterile. From the basal region upwards sterile and fertile regions alternate.

8. This type of sporangial distribution is called :Selago condition".

STEM ANATOMY

Lycopodium phlegmaria

1. A T.S of stem shows three regions. Epidermis, Cortex and Stele.

2. Epidermis is made up of a single layer of cutinised cells. Occasionally stomata are present in the epidcrmis.

3. Cortex is made up of parenchyma cells. The cells are polygonal and thin walled.

4. Stele is surrounded by a layer of endodermis followed by pericycle.

5. Stele is protostele. The xylem is stellate having a number of rays. The protoxylem is situtated at the tips of the rays (exarch). Phloem completely surrounds the xylem. This type of a protostele is called actinostele.

6. There is no pith.

Lycopodium cernuum

1 and 2 same as in *L.phlegmaria*.

3. Cortex is heterogenous. It is divided into three regions. Outer cortex is parenchymatous, middle cortex is sclerenchymatous, while inner cortex is again parenchymatous.

4. Stele is surrounded by endodermis and pericycle.

5. The stele is a protostele and there is no pith. The xylem and phloem are intermixed. Hence the stele is known as a mixed protostele.

6. Xylem is exarch and has annual, reticulate and scalariform thickened tracheids. Phloem has seive cells and phloem parenchyma.

Lycopodium wightianum

1 and 2 same as in *L.phlegmaria*.

3. Cortex has three regions. The outer and inner cortex are sclerotic while the middle cortex is parenchymatous.

4. Stele is surrounded by a layer of endodermis and pericycle.

5. Stele is a protostelic plectostele. Xylem is divided into a number of parallely arranged plate like structures. It is exarch.

6. Phloem not only surrounds xylem but is also distributed between the plates.

7. There is no pith.

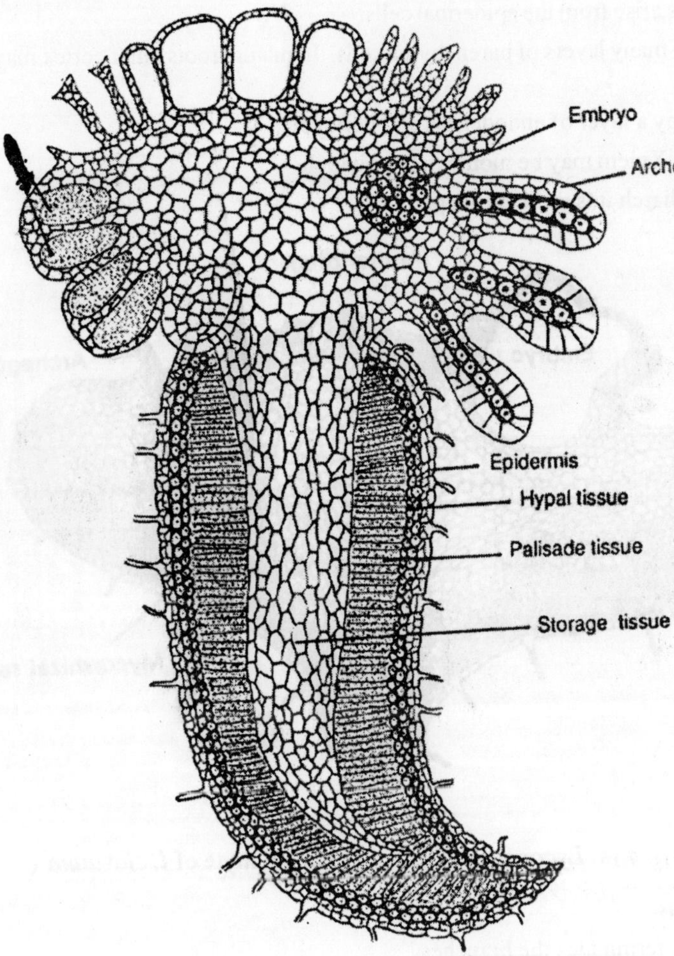

Fig. 7.14. ***Lycopodium*** **: Long Section of the Gametophyte of** ***L. complanatum***

Anatomy of leaf

1. T.S. of leaf shows epidermis, mesophyll and a concentric vein.
2. Upper and lower epidermis made up of a single layer of cells each.
3. Leaves may have stomata only on the lower side(hypostomatic) or have stomata on both surfaces (amphistomatic).
4. Mesophyll is undifferentiated. Cells may be angular or rounded.
5. The vein is concentric. Xylem is surrounded by phloem.
6. Vascular bundle may or may not be surrounded by endodermal sheath.

Anatomy of root.

1. A T.S of root shows epidermis cortex and stele.
2. Epidermis is single layered and is composed of thin walled cells.

3. Unicellular root hairs arise from the epidermal cells.

4. Cortex is made up of many layers of parenchyma cells. In mature roots outer cortex may have thick walled cells.

5. Stele is surrounded by a layer of endodermis and pericycle.

6. Stele is a protostele. Xylem may be monarch to polyarch.

7. When the xylem is diarch it is crescent shaped.

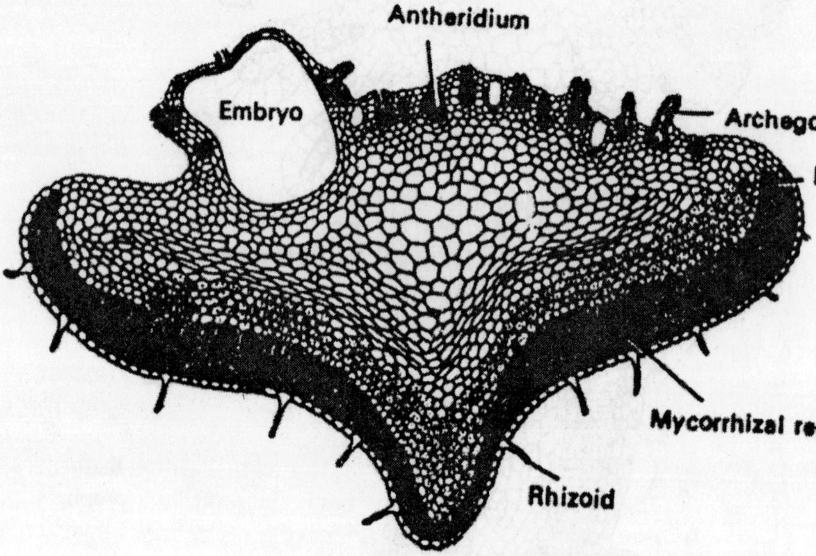

Fig. 7.15. *Lycopodium* : L.S. of Gametophyte of *l. clavatum*

Stucture of the strobilus.

1. The strobilus or cone terminates the branches.

2. Each strobilus has a strobilar axis on which are arranged sporophylls in a spiral fashion.

3. Each sporophyll bears a single sporangium on its adaxial surface. The sporophyll are eligulate.

4. The sporangium has a short multicelluar stalk.

Long section of the strobilus.

1. A long section of the strobilus at the median region shows a central axis and the sporophylls on either side. Vasculature can be seen in the cone axis.

2. Each sporophyll is cut length wise and the top is folded up.

3. The sporophylls are eligulate.

4. Each sporophyll bears a single sporangium on its adaxial surface.

5. Each sporangium has a multicellular stalk.

6. The wall of the sporangium also is multicellular.

7. A number of spores (spore mother cells, or spore tetrads are found when the sporangium is young).

8. All the spores are of the same type (homosporous).

9. In a mature sporangium certain thin walled cells called stomial cells are present marking the future line of dehiscence.

Structure of the gametophyte.

Usually three types of gametophytes are found in *Lycopodium* based on life span, growth pattern and nutrition.

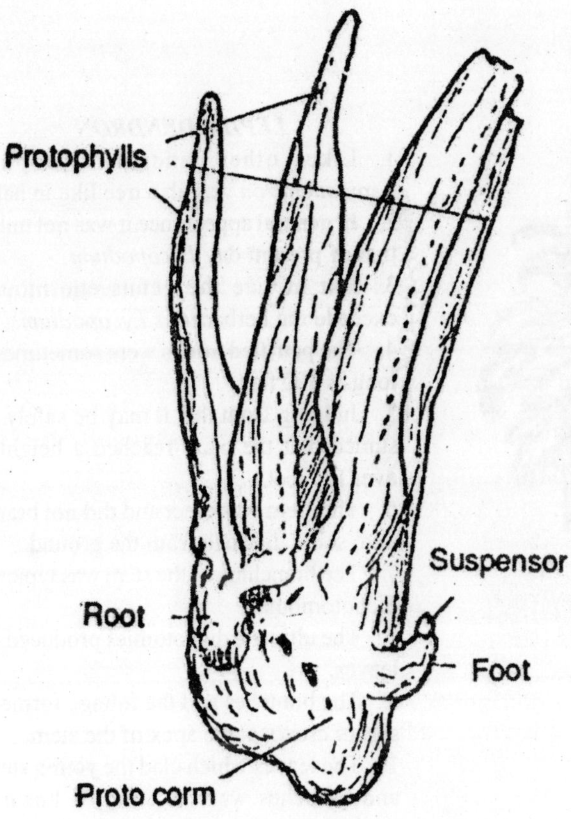

Fig. 7.15. Protocorm of Lycopodium

First type (Eg: *Lycopodium cernuum, L.innundatum* etc)

1. The gametophyte grows up to only one season.
2. The size of the gametophyte is limited. It ranges from 2-3 mm.
3. The gametophyte usually grows above ground but the basal portion is buried in soil.
4. The upper part of the gametophyte has a number of small finger like projections forming the crown.
5. The finger like projections help in photosynthesis.
6. Basal portion of the gametophyte has a mycorrhizal association. Rhizoids also arise from this region.
7. Sex organs (archegonium and antheridium) are found in between the photosynthetic lobes.
8. Nutrition of the prothallus is both photosynthetic as well as saprophytic.

Second type (Eg: *L.clavatum, L.complanatum* etc).

1. Gametophytes are long lived. They grow for many seasons.
2. In size the gametophytes range from 1-2cms. They may be top shaped, disc shaped or carrot shaped.
3. The gametophytes are brownish to yellowish in colour and completely subterranean.
4. A number of rhizoids are produced from the lower surface.
5. Anatomically the gametophyte has only parenchyma cells but they vary in size, shape and structure.
6. Tissue diversificaion is seen in the lower cylindrical portion only, while the upper portion has got only polygonal parenchyma cells.
7. Central region has vertically elongated cells constituting the storage tissue.
8. Surrounding the storage tissue is the palisade tissue. The cells at this region are closely packed.
9. External to the palisade tissue is the hyphal tissue. The cells have fungal hyphae within them.
10. Outside the hyphal tissue is the epidermis. Rhizoids are produced from some of the epidermal cells.
11. Sex organs are borne on the upper surface of the gametophyte.

Third type (Eg: *L.phlegmaria*)

1. Gametophytes are small. They grow for only one growing season.
2. Nutrition is saprophytic. Parenchyma cells have the association of fungal hyphae in them.
3. Gametophytes are usually found on rotting barks close to the soil.
4. The gametophyte is an irregularly shaped tuber.
5. From this central tuber arise several cylindrical branches. The length of the branches varies from 1-6mm.
6. Sex organs (antheridia and archegonia) are borne on the upper surface of these branches.

Systematic position
LYCOPODIALES
LYCOPSIDA
ELIGULOPSIDA
LYCOPODIALES
LYCOPODIACEAE
LYCOPODIUM

Fig. 7.16. *Lepidodendron* : Reconstruction of *L. obovatum*

LEPIDODENDRON

1. Like other ancient lycopods, *Lepidodendron* was also tree like in habit.
2. In general appearance it was not unlike that of present day *Lycopodium*.
3. But in size the genus enormously exceede the herbaceous *Lycopodium*.
4. The petrified trunks were sometimes as long as 100 feet.
5. Judging from this it may be safely assumed that the plant reached a height of over 120 feet.
6. The stem was erect and did not branch upto some distance from the ground.
7. The branching of the stem was typically dichotomous.
8. The ultimate dichotomies produced the leaves.
9. The branches and the foliage formed a sort of crown at the apex of the stem.
10. The leaves which clad the young stems and branches were acicular or linear in shape having a length of 5-9 inches.
11. The arrangement of the leaves was spiral or very rarely they showed a whorled arrangement.
12. The leaves were ligulate.
13. Each leaf had a single vein with the stomata situated in two bands on the ventral surface.
14. The leaves were deciduous.
15. Upon abscission a flat rhomboidal scar persisted on the stem resembling a small cushion.
16. The base of the stem had a stigmarian type of root system.

Internal Structure

Stem:

1. In majority of thespecies, secondary growth is characteristic.

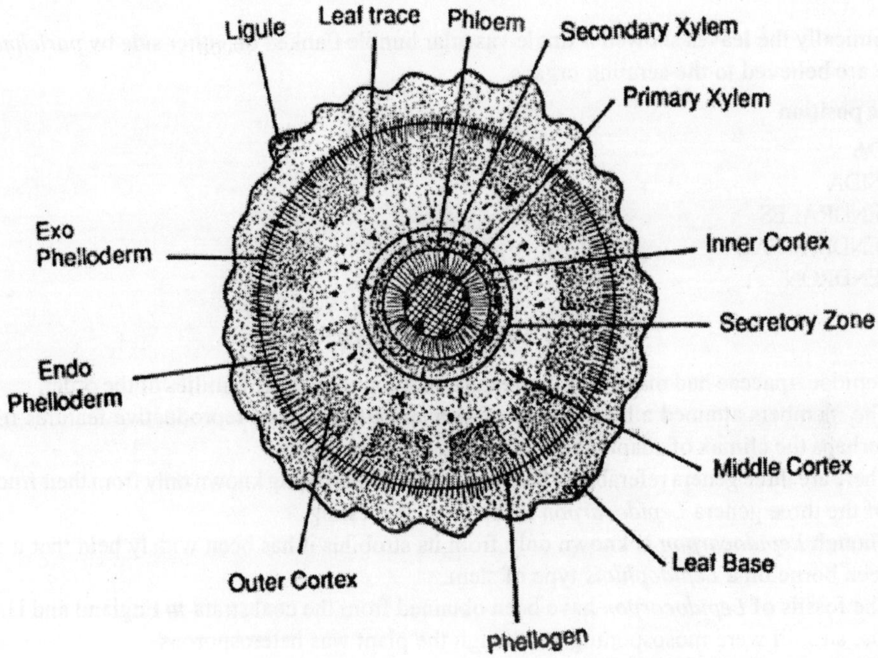

Fig. F7.17. *Lepidodendron* : **T.S. of Stem of *L. vasculare***

2. But some species seem to lack a cambium.
3. A transverse section of the trunk of L.vasculare shows three regions, stele, cortex and a periderm.
4. In the primary structure there was an epidermis but soon i.e., even before the initiation of secondary growth in the vasculature, it was replaced by the periderm.
5. The periderm was produced by a phellogen which produced phelloderm towards the interior and phellem towards the exterior.
6. The outline of the bark was wavy due to the presence of leaf bases.
7. Internal to this was the cortex.
8. It consisted of four regions viz. (1) outer cortex consisting of alternating bands of sclerotic and parenchymatous cells, (2) a middle cortex having a homogenous mass of parenchyma cells. Interspersed with the parenchyma cells were the leaf traces, (3) secretory zone consisting of glandular cells which were filled with a dark coloured substance. They pprobably secreted the waxy material which covered the surface of the stem, (4) An inner cortex having parenchyma cells.
9. The central region of the stem was occupied by the stele which was either protostelic or siphonostelic.
10. The protoxylem was exarch and polyarck.
11. In many species (*L. vasculature*) there was a secondary growth initiated by the cambium.
12. This produced secondary xylem to the interior and secondary phloem to the exterior.
13. The cambial activity was not uniform, as a result there was a tendency for the formation of an eccentric vascular ring.

14. The secondary xylem had radial rows of tracheids separated by *xylem rays* which were uniseriate.

Leaf:

1. Anatomically the leaves showed a single vascular bundle flanked on either side by *parichnos cavities*.
2. These are believed to the aerating organs.

Systematic position

LYCOPSIDA
LIGULOPSIDA
LEPIDODENDRALES
LEPIDODENDRACEAE
LEPIDODENDRON

LEPIDOCARPON

1. Lepidocarpaceae had many striking features not found in other families of the order.
2. The members attained a high degree of specialization in their reproductive features representing perhaps the climax of adaptation of the palaeozoic lycopods.
3. There are three genera referable to this family, all of them being known only from their fructifications.
4. Of the three genera *Lepidocarpon* is the most interesting.
5. Though *Lepidocarpon* is known only from its strobilus it has been widely held that it could have been borne on a *Lepidophlois* type of stem.
6. The fossils of *Lepidocarpon* have been obtained from the coal strata in England and U.S.A.
7. The strobili were mososporangiate though the plant was heterosporous.
8. The microsporophylls were like those of *Lepidostrobus*.
9. Each microsporophyll was peltate and had an elongate microsporangium.
10. The megasporophylls bore elongate microsporangia on their adaxial surface.
11. The jacket of the megasporangium extended at the apex to form a projection.
12. The sporangium was enveloped by the flaps of a tissue which is termed as integument.
13. But it is doubtful how far this terminology is appropriate.
14. Because in reality this inegument is nothing but an upturning of the megasporophyll (Amold 1947; Hosins and Cross 1941) though many people (Scot 1901, Schopf 1941) believe it to be a special outgrowth artising from the megasporophyll.
15. There were four spores in a sporangium of which only one developed into the female gametophyte.
16. The size of the gametophyte and the fact as to whether it completely or incompletely filled the sporangium are specific characters.
17. Andrews and Pannel (1941) have reported several fertile megagametophytes with archegonia at the apex.
18. The megasporangial cavity also showed many spores of which at least some of them were the microspres of *Lepidocarpon*.
19. This suggests that *Lepidocarpon* had something like an incipient pollination which is also seen in the extant genus *Selaginella*.
20. The one unique feature of *Lepidocarpon* which is not seen in any lycopods extinct or extant was the permanent retention of the megagametophyte within the sporangium thus marking a significant step towards the evolution of seed habit.
21. The megasporophylls were, ultimately liberated and on falling upon a suitable substratum the embryo developed into the sporophyte.
22. It will be very interesting to analyse whether *Lepidocarpon* could be regarded as having attained a seed habit.
23. Most of the attributes of *Lepidocarpon* - an integument, a permanent retention of the female game-

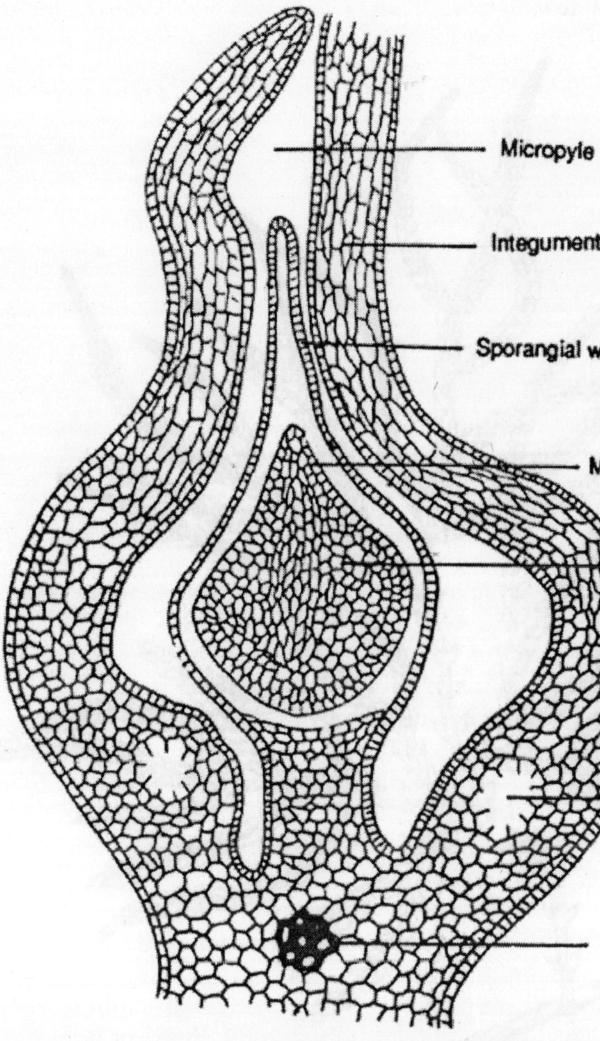

Fig. 7.18. ***Lepidocarpon*** **: L.S. of Seed (diagrammatic)**

tophyte, development of only one spore in the megasporangium, incipient pollination, all point out towards the seed like habit.

24. But a closer scrutiny reveals that all the seed like features are only superficial, for, the so called integument is a part of the megasporophyll and not of the megasporangium.
25. Further there is no formation of pollen tube, no seed coat and in all probability the embryo did not have dormancy characteristic of true spermatophytes.
26. All these clearly show that though Lepidocarpaceae apparently possessed the attributes of a seed-habit (particularly in the permanent retention of the female gametophyte but the additional feature to support such a retention were not sufficient) it did not reach the level of a seed habit and hence it could at best be regarded as a pseudospermatophyte.

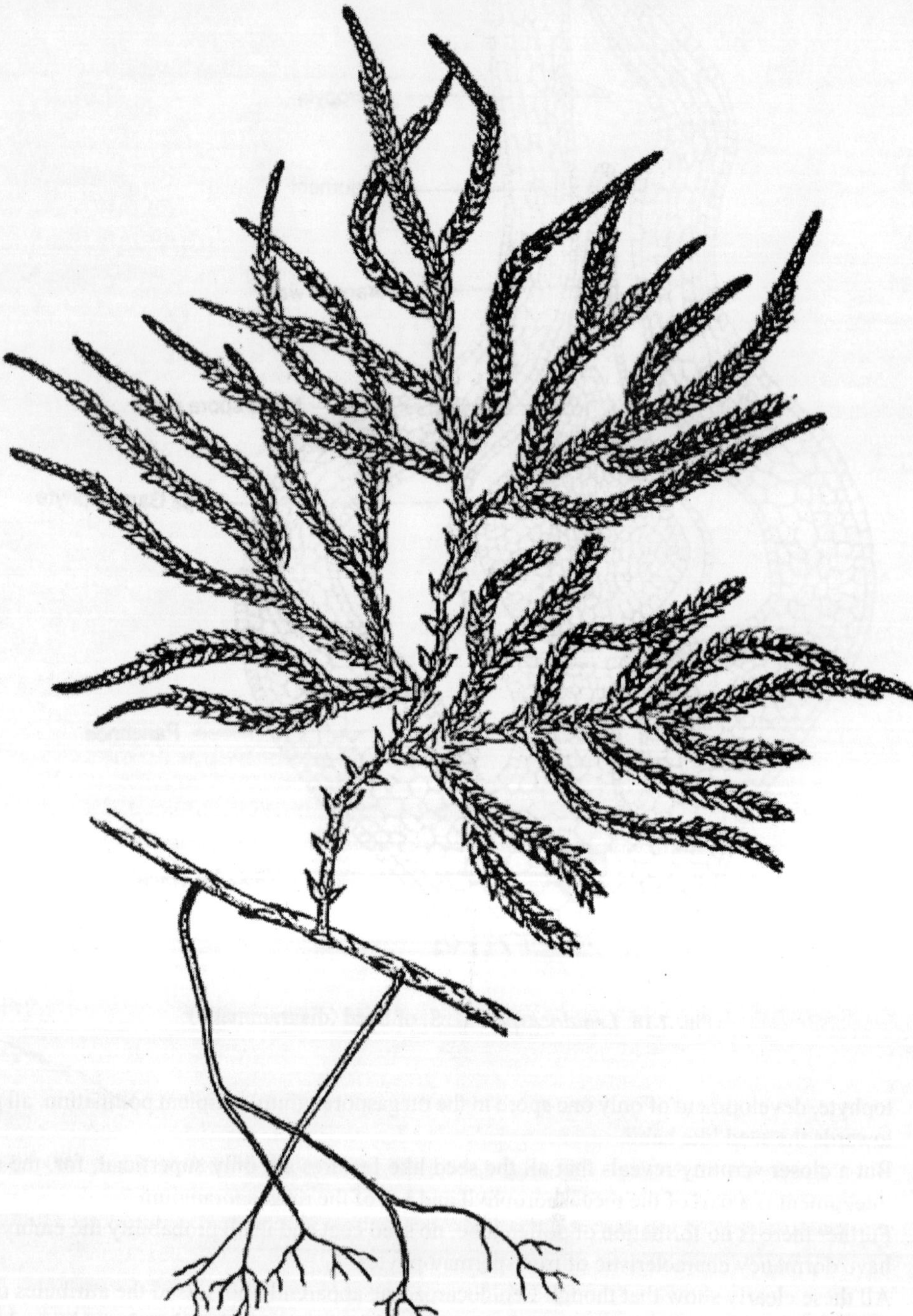

Fig. 7.19 Habit of Selaginella martant

Systematic position

LYCOPSIDA

LIGULOPSIDA

LEPIDODENDRALES

LEPIDOCARPACEAE

LEPIDOCARPON

SELAGINELLA

Selaginella is commonly called the spike moss. There are about 700 species in this genus with world wide distribution.

Practical work to be done

1. Study the habit of the plant.
2. Cut transverse sections of stem, root and leaf and observe under the microscope.
3. Observe the structure of the strobilus from a long section of the slide.
4. Observe the structure of the gametophyte from the slide.

Structure of the plantbody

1. The plantbody is herbaceous with a soft branched stem.
2. The branching of the stem may be dichotomous or lateral.
3. Leaves are found on the stem. These are small measuring a few mm in size.

Fig. 7.20. *Selaginella* : Habit of *S. kraussiana*

4. Leaves are spirally arranged.

5. In some species leaves are *dimorphic* i.e of two sizes. They are arranged in four rows; of these two rows are smaller and two rows are larger.

6. Leaves are simple and lanceolate to oval in shape.

7. In all the species leaves are ligulate. Ligule is a membranous outgrowth arising from the adaxial surface of the leaf.

8. Ligule has a basal region called glossopodium placed in a depression called ligular pit.

9. Ligule varies in shape. It may be tongue shaped, fan shaped or lobed.

10. The function of the ligule is not known. It may be secretory secreting water or mucilage.

11. Certain curious branches arise from the ramifications of the stem. These are called rhizophores. One to

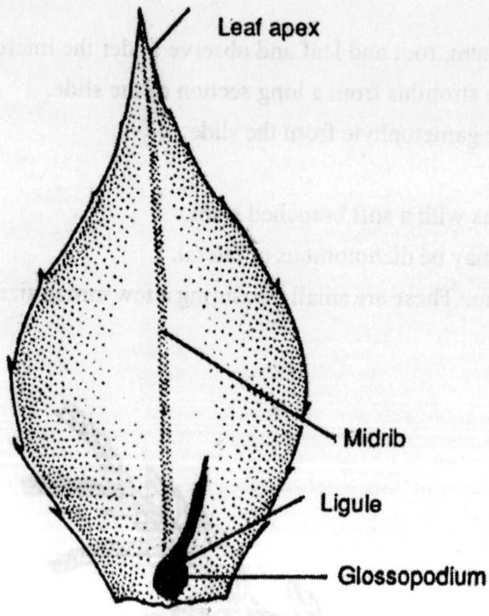

Fig. 7.21. ***Selaginella*** **: Surface View of Leaf showing Ligule**

many rhizophores arise from a single point.

12. The morphological nature of the rhizophore is controversial. It is regarded as a root, stem, organ of a special kind (organ sui generis) etc.

13. Rhizophores arise from special meristems called angle meristems.

14. Root system is adventitous. Roots may arise directly on the stem or on rhizophores. Branching of the root is dichotomous.

15. Reproductive structures generally aggregate at the apices of the branches to form the strobilus.

Classification of the genus

The genus is classified into two - *Homeophyllum* and *Heterophyllum*. The following table lists the differences between the two sections.

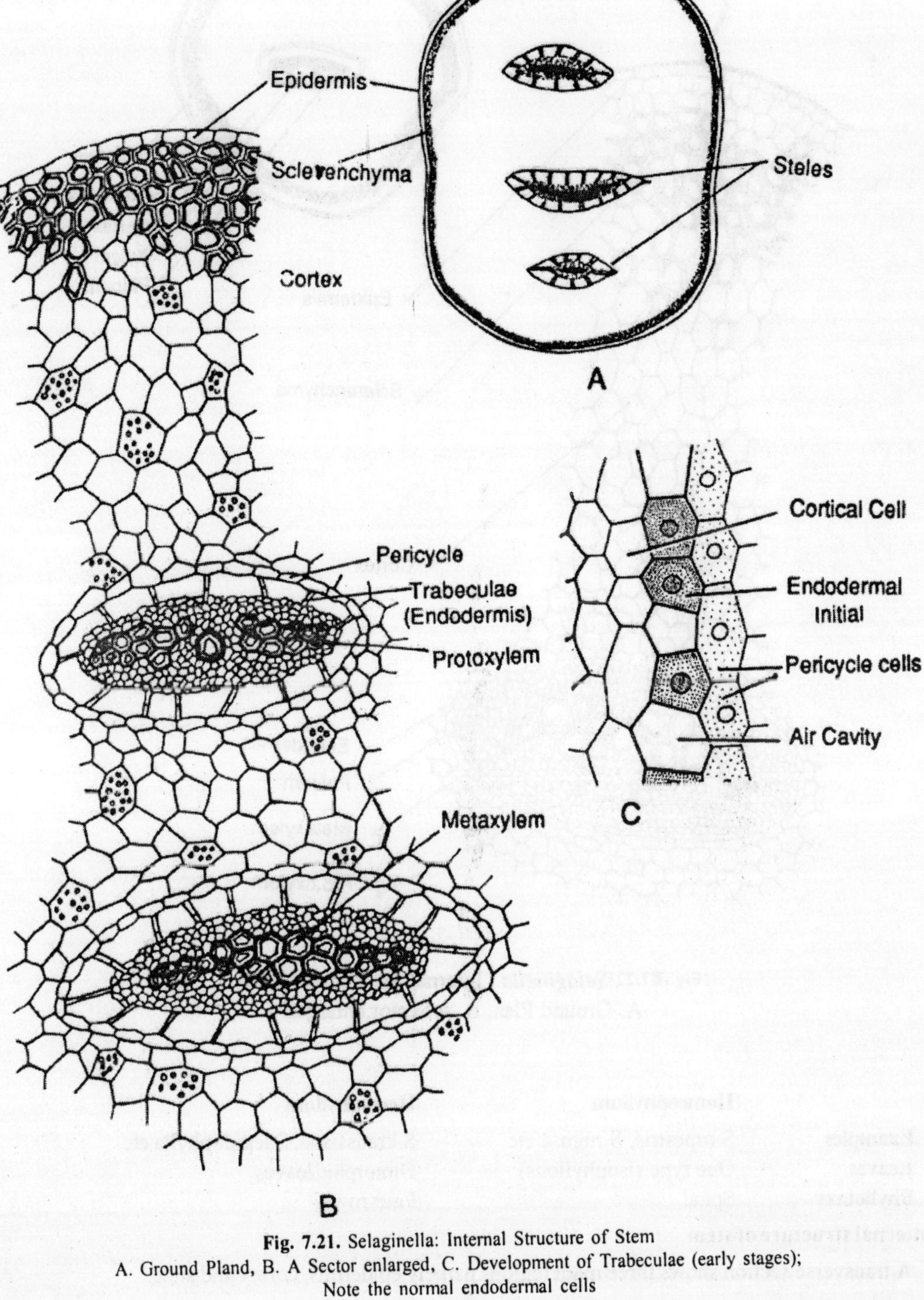

Fig. 7.21. Selaginella: Internal Structure of Stem
A. Ground Pland, B. A Sector enlarged, C. Development of Trabeculae (early stages);
Note the normal endodermal cells

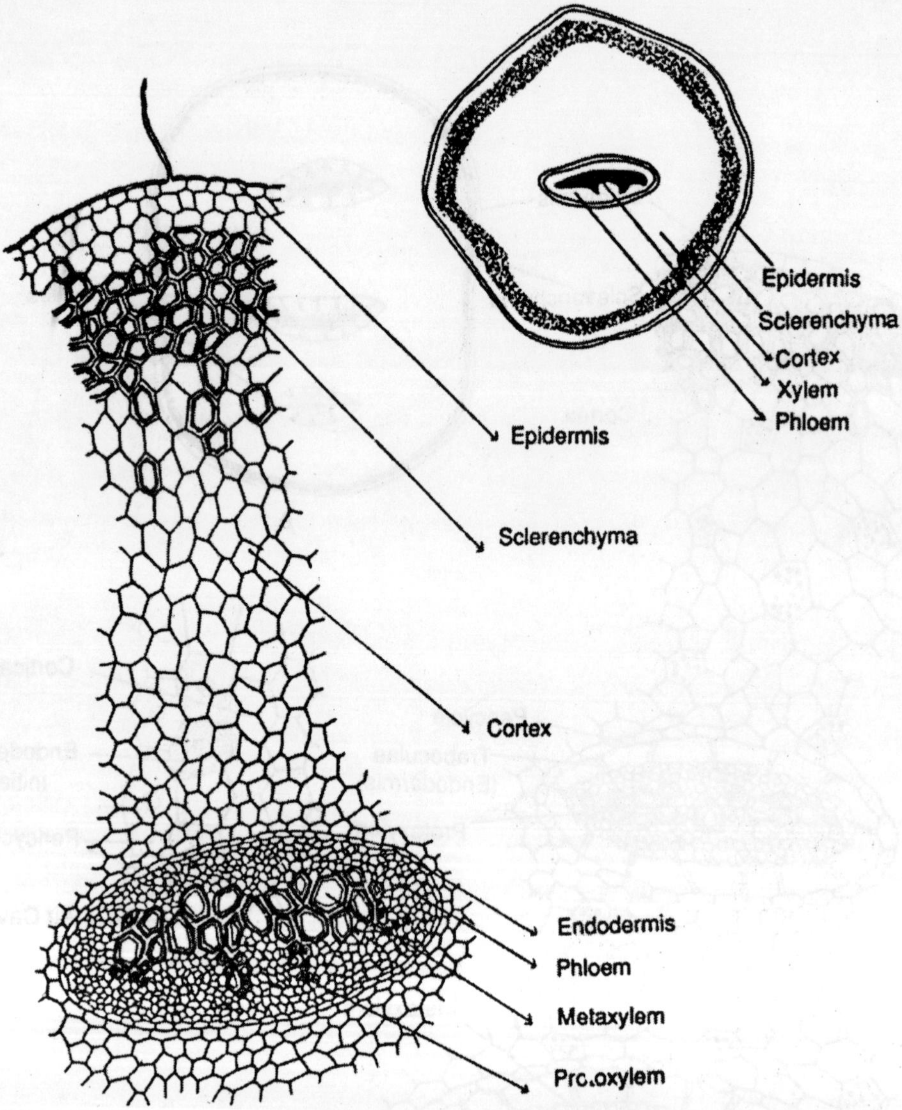

Fig. 7.22 *Selaginella* : Internal Structure of Root
A. Ground Plan, B. A Sector enlarged

	Homeophyllum	**Hetrophyllum**
1. **Examples**	S.rupestris, S.rugosa etc	S.kraussiana, S.lepidophylla etc.
2. **Leaves**	One type (isophyllous)	Dimorphic leaves.
3. **Phyllotaxy**	Spiral	Four rows.

Internal structure of stem

1. A transverse section shows three main regions namely epidermis, cortex and stele.

2. Epidermis is single layered. The cells are closely packed. Stomata are not present.
3. Cortex is parenchymatous completely (*S.flagellata*) or partially (*S.kraussiana*), when there will be a hypodermal sclerenchyma region.
4. Stele is protostelic. It has an outermost endodermis followed by pericycle and vasculature.

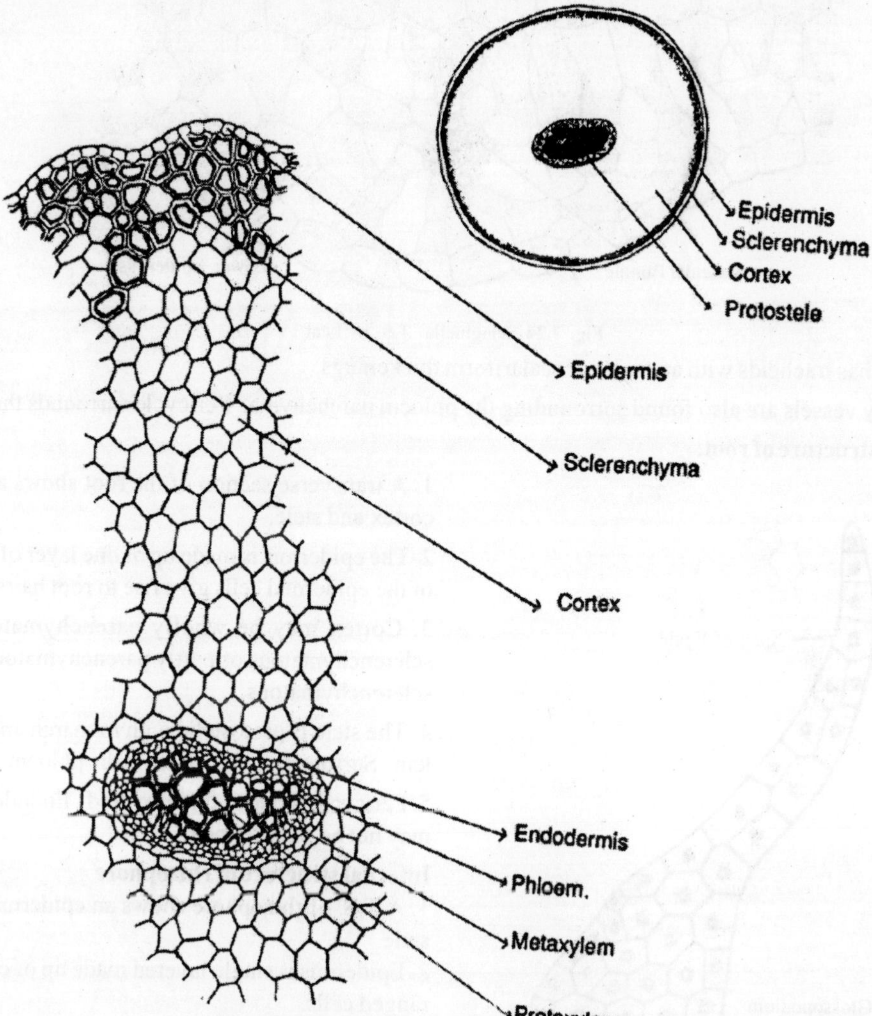

Fig. ▮7.23. *Selaginella* : Internal Structure of Rhizophore
A. Ground Plan, B. A Sector enlarged

5. Endodermis is called trabeculae. There is a lacuna between the cortex and the stele.
6. The trabecular cells (endodermis) are radially elongated and are transversely divided.
7. The stele is normally polystelic. Usually 2-3 steles are present.
8. The xylem is exarch and diarch or rarely polyarch.

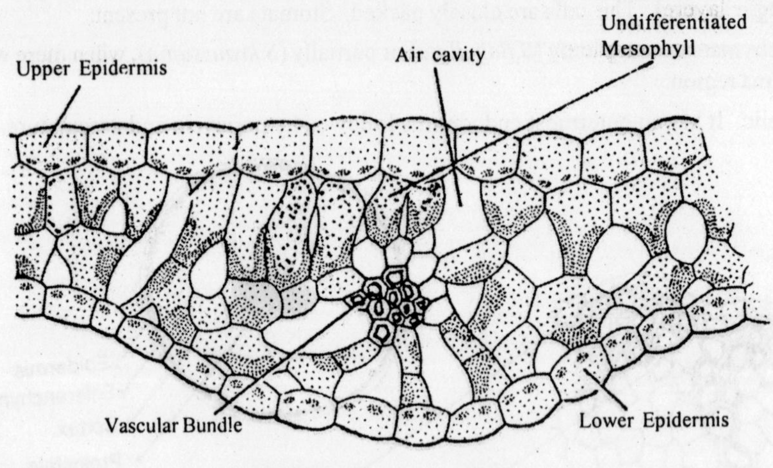

Fig. 7.24 Selaginella: T.S. of Leaf

9. Xylem has tracheids with annualar or scalariform thickenings.

10. Rarely vessels are also found surrounding the phloem parenchyma. Pericycle surrounds the phloem.

Internal structure of root.

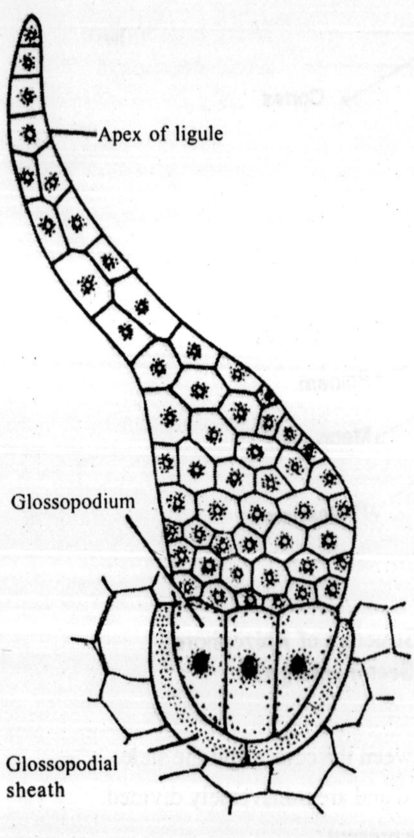

Fig. 7.25. Selaginella: A Ligule enlarged

1. A transverse section of the root shows an epidermis cortex and stele.

2. The epidermis is made up of one layer of cells. Some of the epidermal cells give rise to root hairs.

3. Cortex may be wholly parenchymatous, wholly sclerenchymatous or partly parenchymatous and partly sclerenchymatous.

4. The stele is protostelic with monarch and exarch xylem. Surrounding the xylem is the phloem.

5. Pericycle is one to three layered. Endodermis may or may not be well defined.

Internal structure of rhizophore

1. A T.S. of rhizophore shows an epidermis cortex and stele.

2. Epidermis is single layered made up of compactly arranged cells.

3. Cortex is parenchymatous made up of several layers of cells.

4. The stele is surrounded by non trabecular endodermis followed by a layer of pericycle.

5. Stele is protostelic. There will be a single stele in the rhizophore even though the stem may be polystelic.

6. Xylem is exarch and monarch.

Internal structure of the leaf.

1. Anatomically the leaf shows epidermis, mesophyll and a vascular bundle.

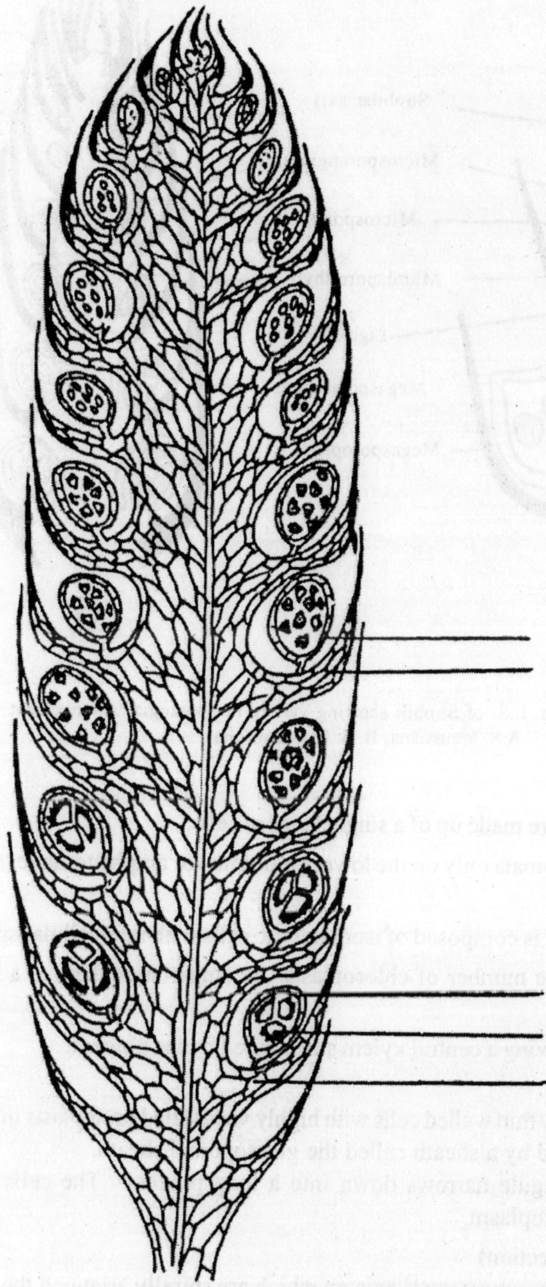

Fig. 7.26. *Selaginella* : L.S. of Strobilus (megasporangia at the base and microsporangia at the top)

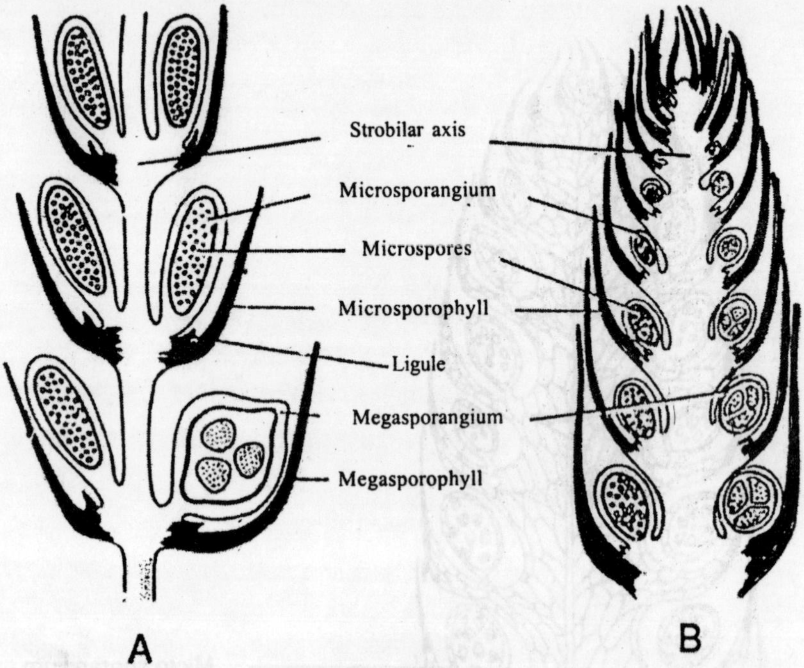

Fig. 7.27. Selaginella: L.S. of Strobili showing variety in Sporangial Arrangement A.S. kraussiana, B. S. inaequalifolia

2. The upper and lower epidermis are made up of a single layer of cells.

3. Leaves may be hypostomatic (stomata only on the lower epidermis) or amphistomatic (stomata present on both the epidermal layers).

4. Mesophyll is undifferentiated. It is composed of isodiametric cells with intercellular spaces.

5. Mesophyll cells contain variable number of chloroplasts. In some cases there is a single cup shaped chloroplast in each cell.

6. Vascular bundle is concentric having a central xylem surrounded by the phloem.

Anatomy of ligule.

1. A transverse section reveals many thin walled cells with highly vacuolated protoplasts in the glossopodium.

2. The glossopodium is surrounded by a sheath called the glossopodial sheath.

3. Above the glossopodium the ligule narrows down into a long process. The cells at this region are polygonal and filled with dense cytoplasm.

Structure of the strobilus (Long section)

1. A long section of the strobilus shows a central axis on which are spirally arranged the sporophylls.

2. Each sporophyll bears a single stalked sporangium arising in its axil or a little above.

3. The sporangia are of two types megasporangia and microsporangia. Megasporangia have less number of larger spores, while microsporangia have large number of smaller spores.

Structure of a mature micro sporangium

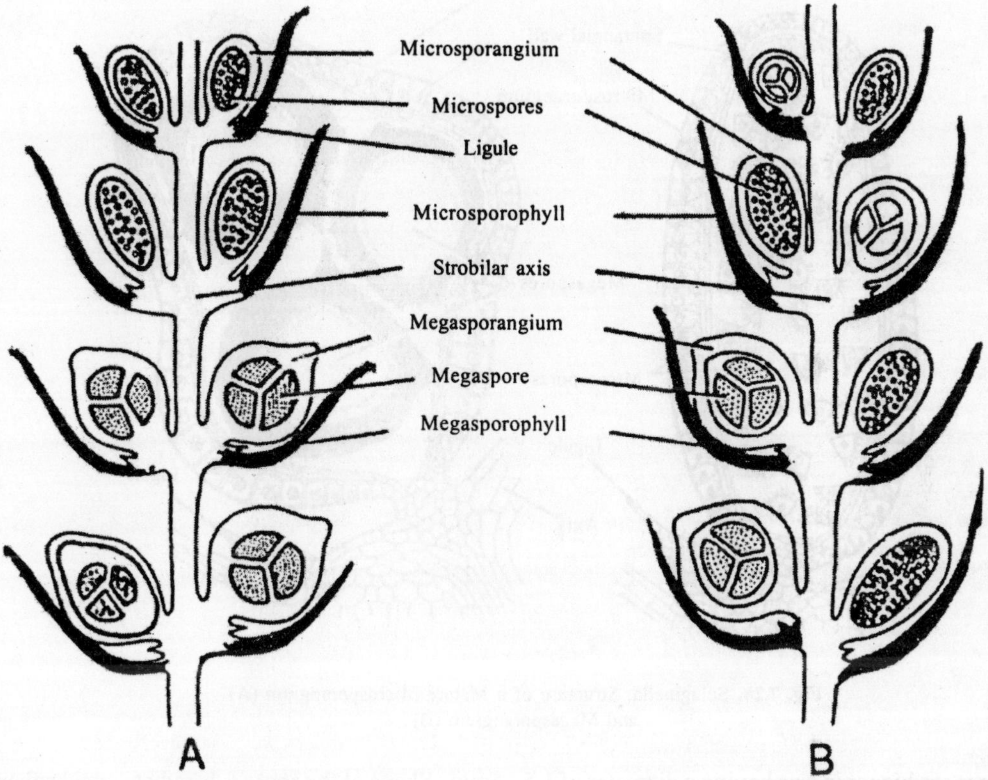

Fig. 7.28. Selaginella: L.S. of Strobili showing variety in Sporangial Arrangement (contd...)
A.S. rupestris, B. S. martensii

1. The sporangium is red, yellow or brown in colour.
2. The stalk of the sporangium is short.
3. The wall of the sporangia is two layered. The outer wall has columnar chlorophyllous cells. The inner layer has thin walled cells.
4. Sometimes a thin tapetal layer is persisting in the mature sporangium.
5. There are a large number of microspores in the sporangium. Their size ranges from 0.015 to 0.05mm.
6. The spore wall is two layered. The outer wall (exine) is very thick and sculptured. The inner wall is thin and delicate.
7. Reserve food material in the form of oil globules are found in the spore.

Structure of the mega sporangium

1. The mega sporangium is yellow orange or chalky white at maturity.
2. The stalk of the sporangium is short and multicellular.
3. The sporangial wall is two layered. The outer layer has columnar chlorophyllous cells. The inner layer has thin walled cells.
4. Sometimes a tapetal layer will be persisting in the mature sporangium.
5. The megaspores are usually 3-4 in number per mega sporangium.

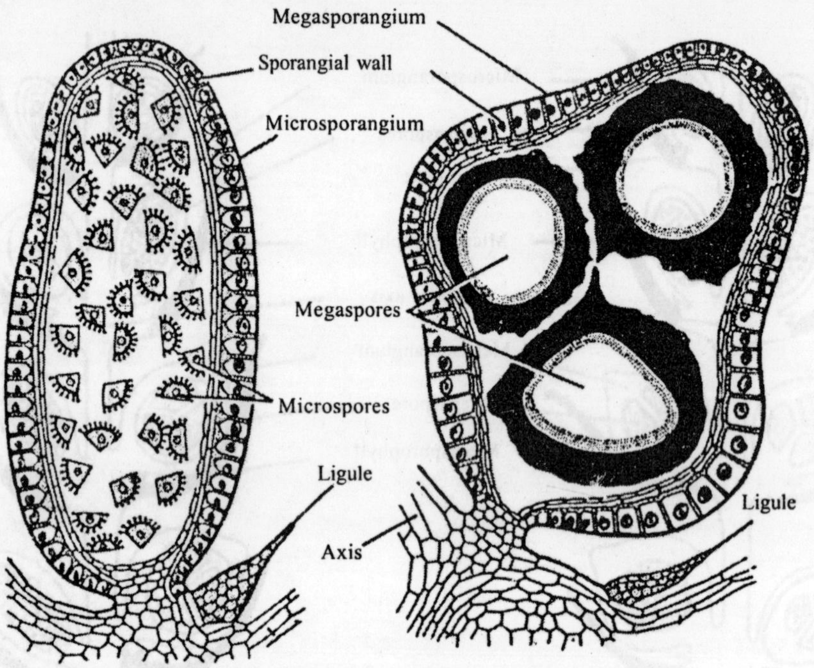

Fig. 7.28. Selaginella: Structure of a Mature Microsporangium (A) and Megasporangium (B)

6. Spores are bigger in size having a diameter of 1.5 - 5mm. Shape of the spore is triradiate or subspherical.

7. The wall of the sporangium is very thick and consists of a sculptured exine, a middle mesospore and a thin intine.

8. Cytoplasm of the spore consists of reserve food in the form of oil globules and nitrogenous matter.

Structure of a mature male gametophyte.

1. The female gametophyte is an extremely reduced structure. Its growth is confined to the spore interior.

2. A mature gametophyte has 13 cells - one prothallial cell, four primary androgonial cells and eight jacket cells.

3. The androgonial cells eventually give rise to sperms.

Structure of the female gametophyte.

1. A mature mega gametophyte has a basal free nuclear portion and an apical cushion.

2. The apical cushion is three layered thick towards the centre but one layered thick towards the margin.

3. The lower most cells of the apical cushion have a much thickened lower wall constituting the diaphragm.

4. The diaphragm separates the apical cushion from the lower free nuclear portion.

5. At later stages even the lower portion becomes cellular.

6. The cells in the apical cushion are uninucleate and polygonal.

7. The cells at the basal region are irregular in shape and sometimes multinucleate also these cells are rich in resève food material.

8. Many archegonia are found in the apical cushion.

Systematic position

LYCOPSIDA

LIGULOPSIDA

SELAGINELLALES

SELAGINELLALES

SELAGINELLACEAE

SELAGINELLA

ISOETES

Pratical work to be done

1. Study the morphology of the plant.
2. Study the internal structure of corm, root and leaf with the help of slides.
3. Mount the sporangia and observe the spores.
4. Study the morphology of the male and female gametophytes from the slides.

Morphology of the plantbody

1. The plantbody consists of a condensed, lobed structure called the axis or corm.
2. The axis or corm has a complex morphological structure.
3. The corm on its upper surface bears a number of long quill like ligulate leaves.
4. The leaves are two to several cm long and are crowded in a closed spiral forming a sort of a fascilcle.
5. In some instances the leaves may be 0.5 meter (*Isoetes coromandalianum*), long.
6. The leaves have broad spoon shaped bases and tapering tips.
7. The outermost leaves are sterile; successively within them are found megasporophylls, microsporophylls and sporophylls with imature sporangia.
8. From the lower surface of the corm are produced a number of roots.
9. The roots usually take their origin from the grooves of the lobed corm. They are arranged in vertical rows at right angles to the groove.
10. In a tow lobed corm there are four series of roots, two on each side of the corm.

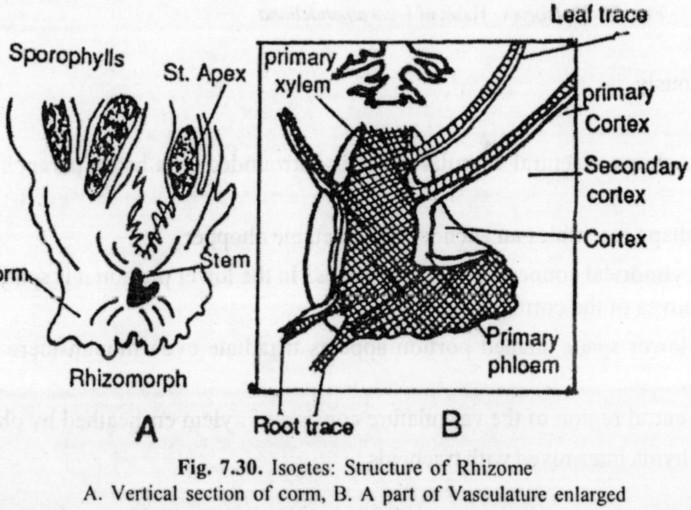

Fig. 7.30. Isoetes: Structure of Rhizome
A. Vertical section of corm, B. A part of Vasculature enlarged

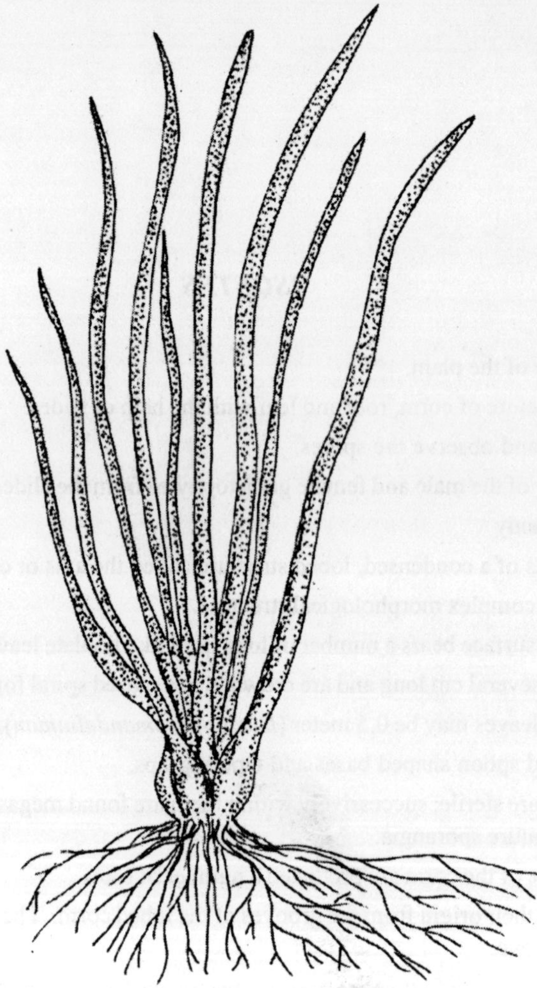

Fig. F7.29. ***Isoetes*** **: Habit of** ***I. coramandeliana***

11. The roots branch dichotomously.

Internal structure of corm.

1. A vertical section of the axis shows a central vascular cylinder surrounded by a broad parenchymatous cortex.

2. The vascular cylinder in its shape resembles an anchor or a vegetable chopper.

3. In the upper part the stele is cylindrical connected with leaf traces. In the lower portion it is spade shaped flattened in the plane of the grooves of the corm.

4. In a transverse section the lower spade shaped portion appears triradiate even though there are four radiating arms.

5. The stele is protostelic; the central region of the vasculature consists of xylem ensheathed by phloem.

6. The xylem consists of parenchyma intermixed with tracheids.

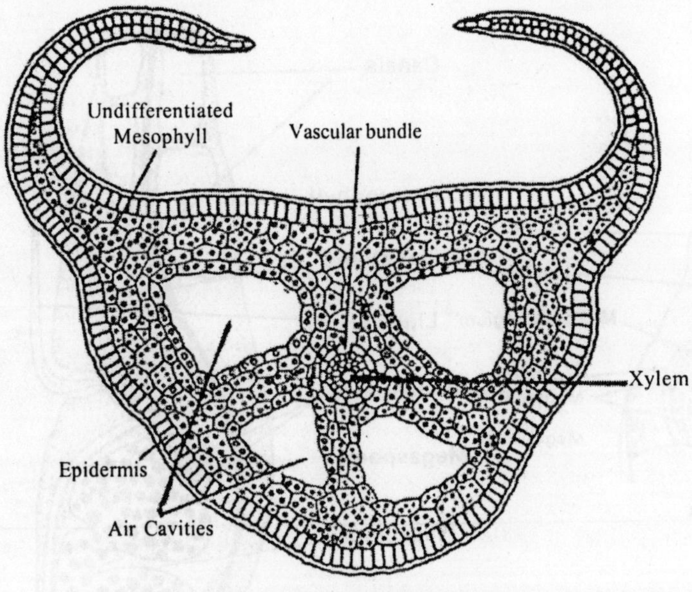

Fig. 7.31. Isoetes: T.S. of Root

7. Leaf traces depart from the stele but there are no leaf gaps.

8. Cortex is parenchymatous and has starch filled cells.

9. Secondary growth seen in *Isoetes* is of a peculiar kind.

10. A cambium is seen external to the phloem. The outer derivatives of the cambium are parenchymatous constituting the secondary cortex.

11. The inner derivatives of the cambium are called prismatic layer. Prismatic layer is a mixture of tracheids and

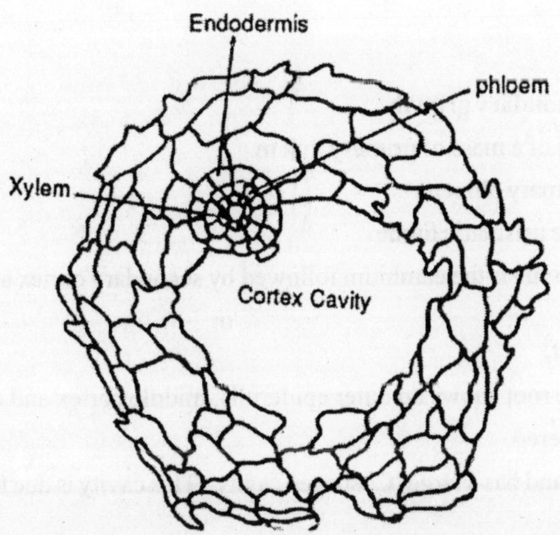

Fig. 7.32. *Isoetes* **: T.S. of Leaf**

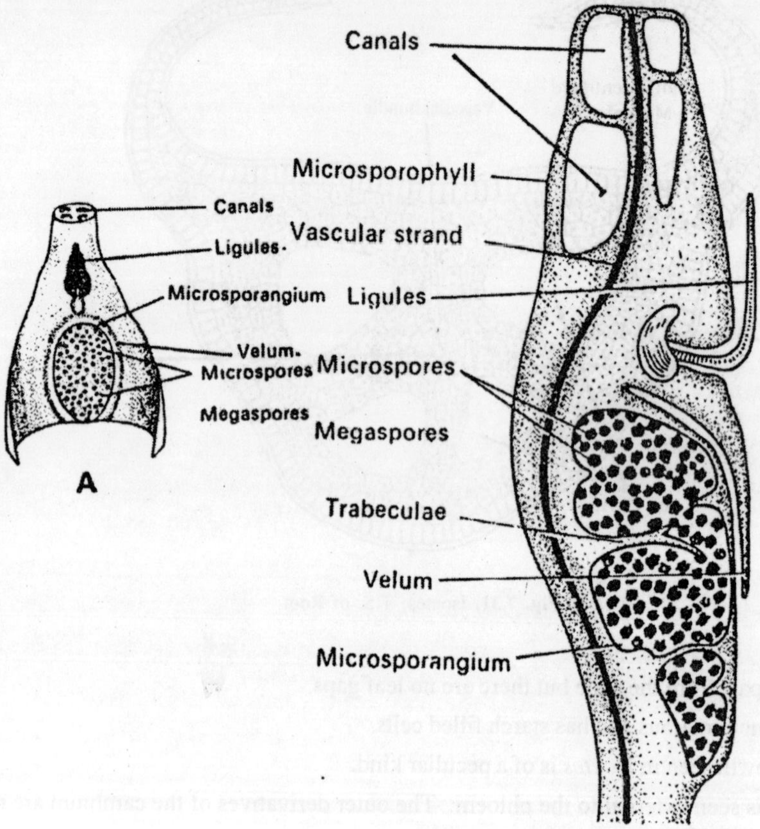

Fig. 7.33. *Isoetes* **: Diagrammatic Representation of Microsporophyll (A) and Long Sectional View of the same (B)**

seive elements.

Structure of a corm after secondary growth

1. The central region consists of a mass of primary xylem.
2. Next to the xylem is the primary phloem.
3. Outlining the phloem is the prismatic tissue.
4. External to the prismatic tissue is the cambium followed by secondary cortex and then the primary cortex, which is very broad.

Internal structure of the root.

1. A transverse section of the root shows an outer epidermis, middle cortex and central stele.
2. The epidermis is single layered.
3. Cortex is parenchymatous and has a large 'C' shaped cavity. This cavity is due to the breakdown of cortical cells.
4. The stele is a monarch protostele. The xylem and phloem are collateral and are arranged in such a way that the phloem is on the side away from the rhizome.

5. Surrounding the vasculature is a well defined endodermis.

Internal structure of the leaf

1. The leaf shows a single vascular bundle surrounded by an undifferentiated mesophyll.
2. The outermost layer is the epidermis in which stomata are found.
3. Mesophyll consists of parenchyma sometimes sclerenchyma may also be found.
4. There are four cylindrical air chambers in the mesophyll separated by parenchymatous partitions.
5. vascular bundle is collateral but may become concentric towards the tip of the leaf.

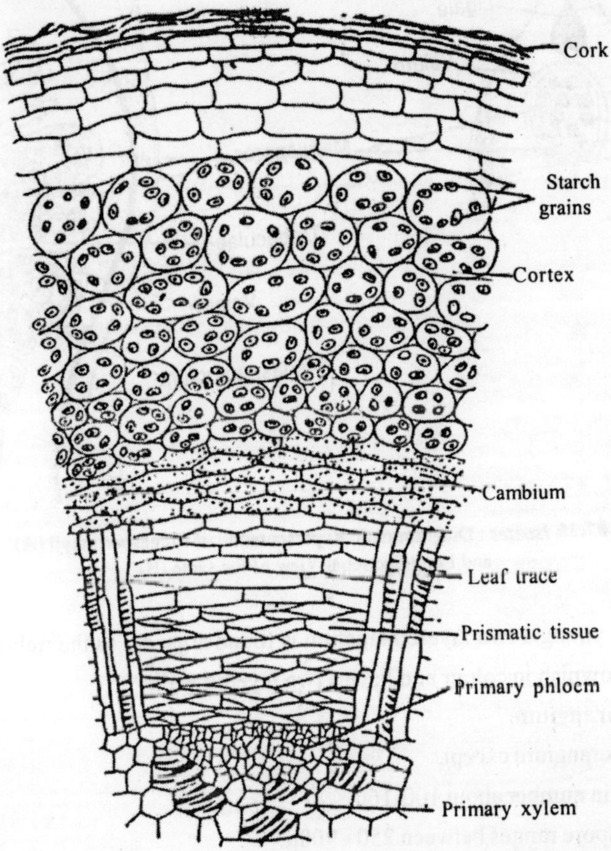

Fig. 7.34. Isoetes: Structure of Rhizome; T.S. of Corm showing Cellular details

Structure of a microsporangium

1. The sporangium has a multilayered wall.
2. Above the sporangium is situated *velum,* a projection of cells between the ligule and the sporangium.
3. Within the sporangium are found a large number of microspores.
4. The sporangium is incompletely divided by partitions called trabeculae.

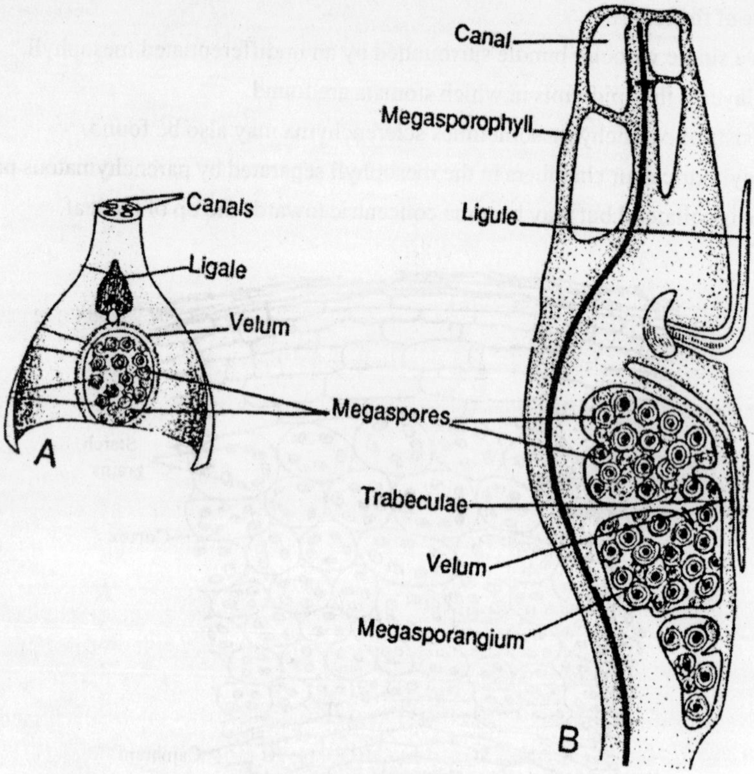

Fig. 7.35. *Isoetes* : Diagrammatic Representation of Megasporophyll (A) and Long Sectional View of the same (B)

5. When the spores are young a two layered tapetum is found adjacent to the trabeculae.
6. Spores are minute brownish in colour and have a two layered wall.

Structure of a megasporangium

Same as in microsporangium except.

1. Spores are large, less in number about 100-160.
2. The size of the megaspore ranges between 250 - 900µ.
3. Spores are white, grey or black in colour.

Systematic position

LYCOPSIDA

LIGULOPSIDA

ISOETALES

ISOETACEAE

ISOETES

Chapter 8

SPHENOPSIDA

Sphenopsida are commonly called horesetails because of the crowded secondary branches arising on stem. The group includes living as well as extinct members. Geologically the history of Sphenopsida extends to the Devonian period.

The Sphenopsida is divided into four orders - Hyeniales, Calmitales, Sphenophyllales and Equisetales. Calamitales includes *Calamites, Calamostachys* etc. *Sphenophyllales* includes Sphenophyllum, while Equisetales includes *Equisetum*.

In this book calamites, calamostachys, Sphenophyllum and Equisetum are discussed.

Sphenophyallales represent a small and compact group of sphenopsida. They have left no surviving representatives. The geological history of Sphenophyllales dates back to a time earlier than carboniferous period. Some of the genera like *Sphenophyllum* appeared during Devonian reaching maximum development during carboniferous period. A few of these Sphenophyllales survived upto Triassic.

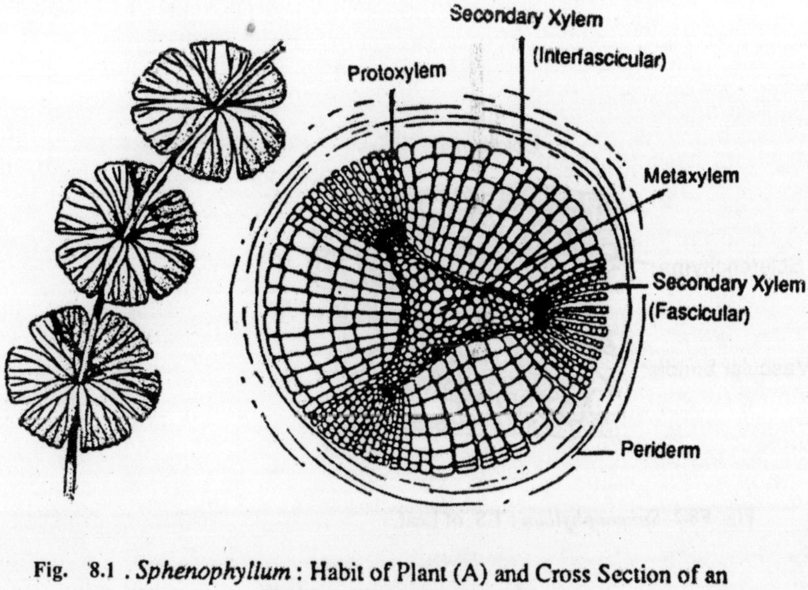

Fig. 8.1 . *Sphenophyllum* : Habit of Plant (A) and Cross Section of an Old Stem of *S. dawsonii* (B)

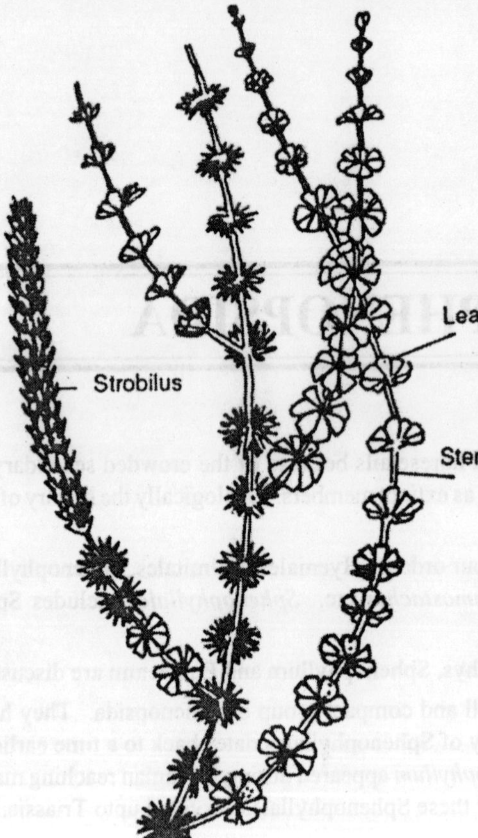

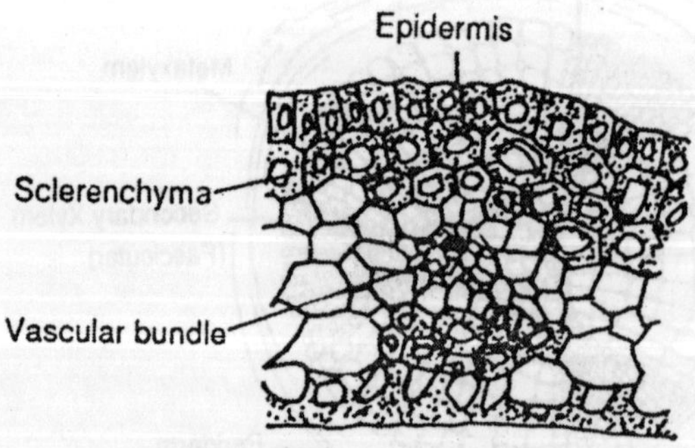

Fig. F8.2. *Sphenophyllum* : T.S. of Leaf

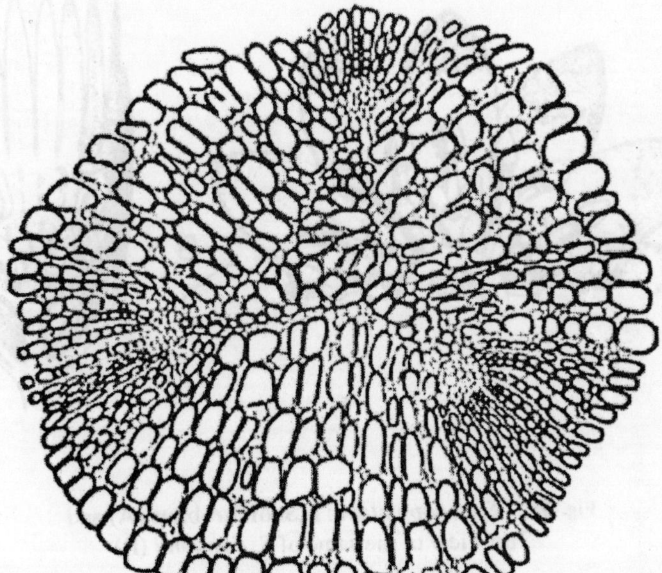

Fig. 18.3. *Sphenophyllum* : T.S. of Xylem Cylinder of *S. plurifoliatum*

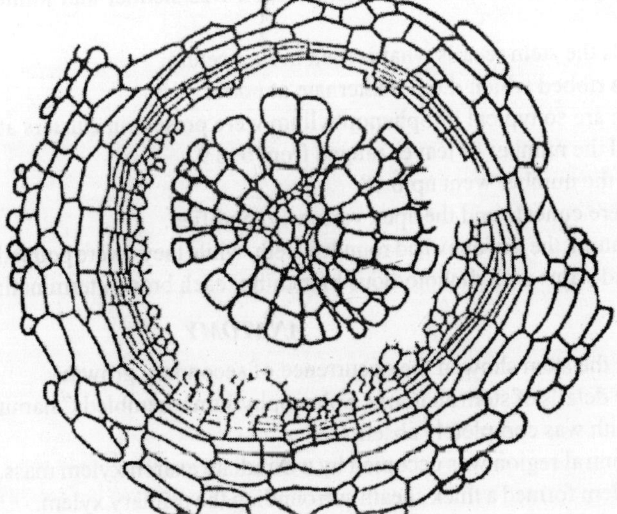

Fig. 18.4. *Sphenophyllum* : T.S. of Root

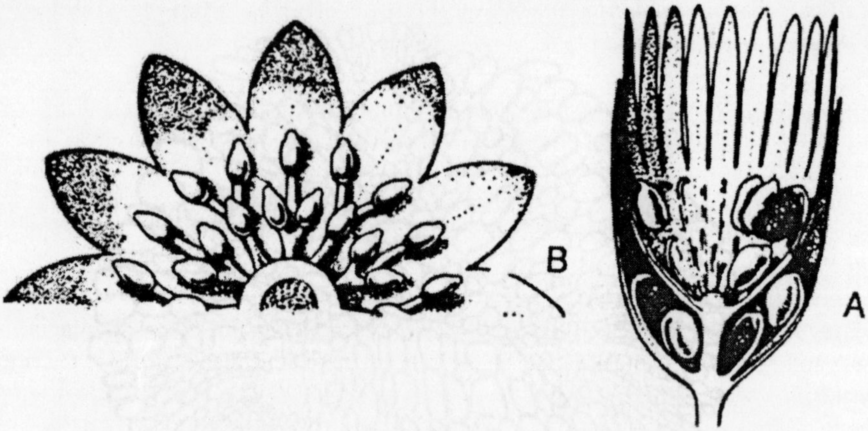

Fig. 8.5. *Sphenophyllum* : L.S. of Strobilus (A) and Top View of the same of *S. dawsonii* (B)

1. The principal genus of the order is Sphenophyllum.
2. The compressed fossils of the plant body have been found extensively scattered in coal bearing formations belonging to the carboniferous period.
3. The plant body of Sphenophyllum was small and herbaceous.
4. There is a high probability that the plant had a climbing habit.
5. While, some of the morphological features indicate an aquatic habitat, anatomical features however, point out to a terrestrial habitat.
6. The plant body consisted of a main stem, which was slender and jointed, rarely exceeding 5 mm in diameter.
7. In its externals the stem recalls what is seen in Equisetum.
8. The stem was ribbed which did not alternate at nodes.
9. Leaves which are so typical of Sphenophyllum were present in clusters at the nodal regions.
10. At each whorl the number of leaves ranged from 6 to 9.
11. Occasionally the number went upto 18.
12. The leaves were cuneate and the apex variously modified.
13. In S. emarginatum, the leaf apex had rounded teeth, while they awere pointed in S. cunifolium and S. majus.
14. The leaves had vains with dichotomous branching, each branch terminating in an apical tooth.

ANATOMY

1. Anatomically the stem showed the occurrence of secondary growth.
2. In some of the details of stem anatomy, Sphenophyllum resembled Calamites.
3. The Central pith was completely absent.
4. Instead, the central region was occupied by a triradiate exarch xylem mass.
5. Secondary xylem formed a thick sheath surrounding the primary xylem.
6. Cortical region exhibited a corky tissue indicating the activity of phellogen.
7. The phellogen probable arose deep in the primary cortex.

Roots:

1. The plant body was anchored to the soil with the help of adventitious roots that arose at the nodal

regions.

2. Anatomically the roots resembled the stems, expect that the former had a two angled xylem.
3. Secondary growth was also seen in root.

Reproduction

1. Sphenophyllum reproduced by means of sporangia borne on sporophylls.
2. The sporophylls aggregated to form long and slender cones terminating the stem apex.
3. The cones which are often given the name Bowmanites had a central axis on which were arranged many whorls of sporophylls.
4. The sporophylls of each whorl fused at their base to form a saucer like structure.
5. The sporophylls had two parts namely an adaxial fertile part and an abaxial sterile part.
6. The sporangia were borne on the fertile part either singly or in pairs.
7. In some cases the fertile lobe had two or three branches each having one to two sporangia.
8. The number and arrangement of sporangia on the lobe are taxonomic features used in the categorisation of the species.
9. The sporangia had a stalk which was highly reduced.
10. Most of the species of Sphenophyllum were homosporous.
11. Some species could have been heterosporous.

Systematic Position

SPHENOPSIDA
SPHENOPHYLLALES
SPHENOPHYLLACEAE
SPHENOPHYLLUM

CALAMITES

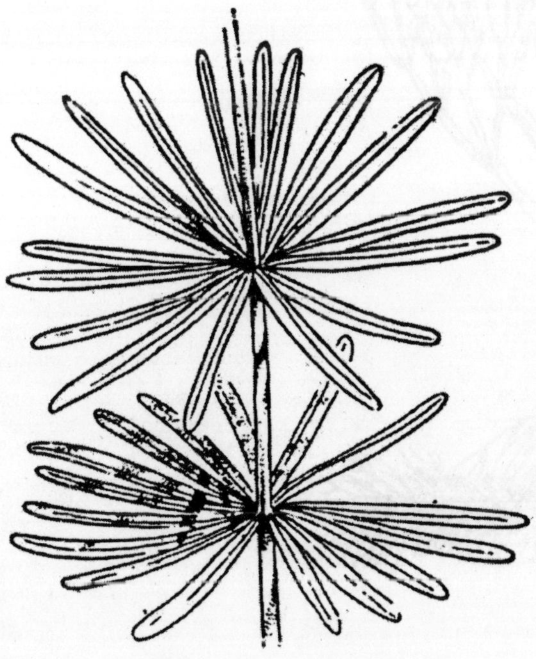

Fig. F8.6. *Calamites* : Leaves of *Annularia*

Practical work to be done

1. Study the external morphology from the reconstruction of the plantbody.

2. Study the internal structure of the stem from a slide.

Morphology of the plantbody

1. The plantbody is a small tree growing to a height of 20-30 meters.

2. The plant has an underground rhizome differentiated into nodes and internodes.

3. A whorl of adventitious roots arise at each node.

4. From the upper surface of the rhizome arise a number of aerial shoots.

5. Some of the aerial shoots produce roots at the lower node which indicates that the rhizome grew some distance below the soil.

6. Erect shoots are constricted at the point of their junction to the rhizome.

7. The aerial shoots have whorls of branches at the nodal region.

8. The surface of the stem has longitudinal ribs alternating with the grooves.

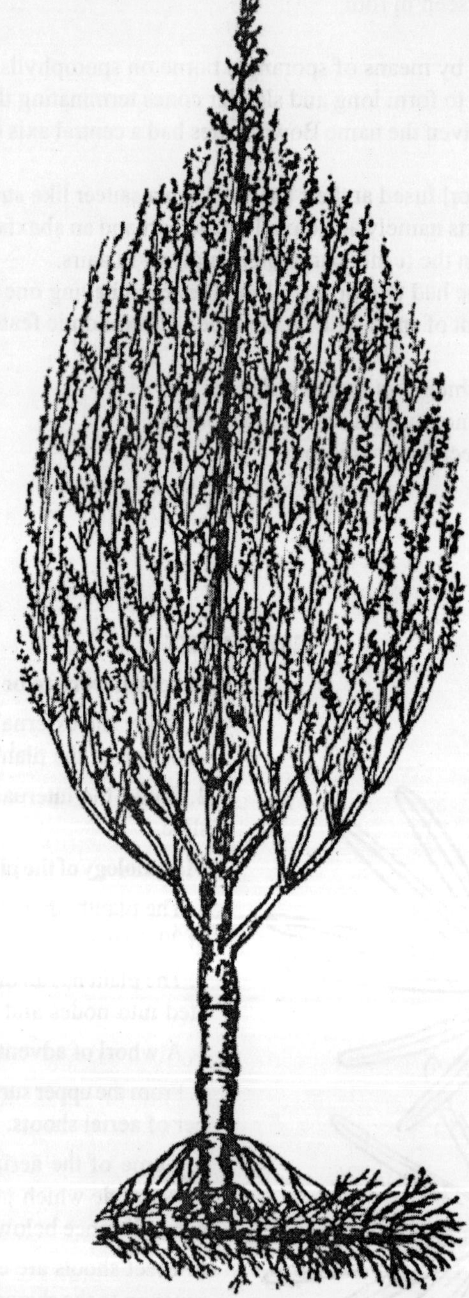

Fig. 78.7. *Calamites* : Reconstruction of Plant Body

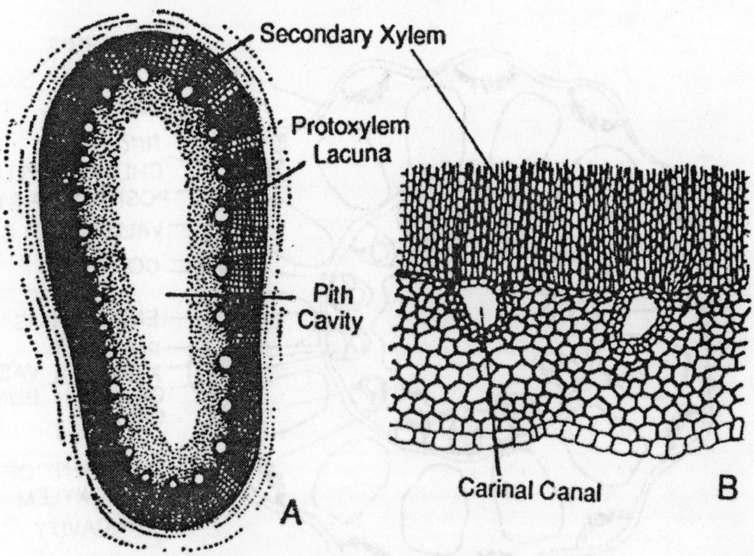

Fig. F8.8. *Calamites* : Anatomy of Stem (after secondary growth)
A. Ground Plan of Cross Section, B. A Sector enlarged

9. In some instances the ribs of successive internodes are directly above instead of alternating with the other.

Internal structure of the stem

1. A transverse section shows an epidermis, cortex and stele.
2. Epidermis is single layered.
3. The cortex has an outer sclerotic zone and an inner thin walled parenchymatous zone.
4. Stele is siphonostelic.
5. There is a pith in the nodal region in the centre but replaced by a central canal.
6. Vascular bundles have collaterally arranged xylem and phloem. The xylem is endarch.

Fig. F8.9. *Calamites* : Reproductive Structure. Comparison of Strobili of *Palaeostachys* (A) and *Calamostachys* (B)

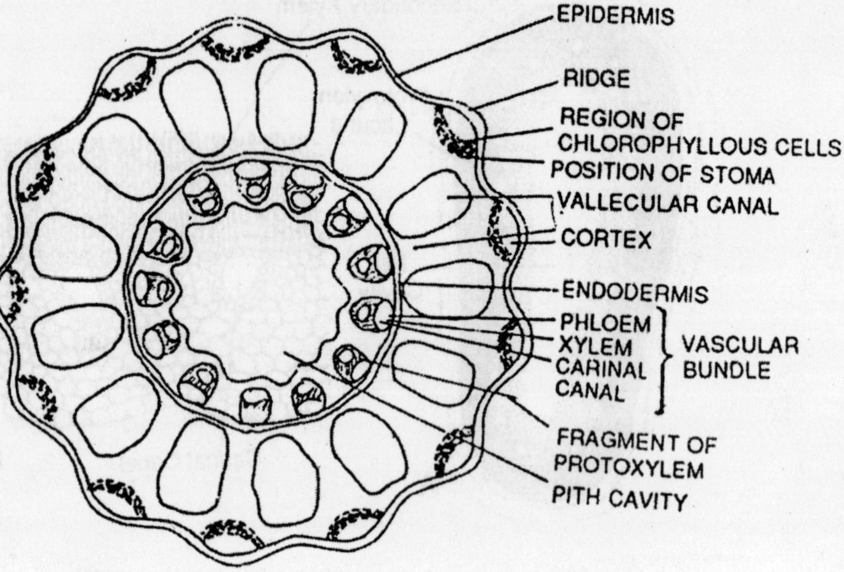

Fig. F8.10*Equisetum* : T.S. of Internode of Aerial Shoot (ground plan)

7. The tracheids had scalariform thickenings in metaxylem but annular and spiral thickenings in the protoxylem.

8. In each vascullar bundle a carinal canal is found.

9. Outside the primary vascular bundles secondary xylem is seen.

10. There is no differentiation of annual rings in the secondary wood.

Systematic position

SPHENOPSIDA

CALAMITALES

CALAMITACEAE

CALAMITES

EQUISETUM

Practical work to be done

1. Study the external morphology of the plant.
2. Cut transverse sections of

    a) Node of rhizome.

    b) Node of aerial shoot.

    c) Internode of aerial shoot.

    d) Root.

    e) Leaf.

    f) T.S. and L.S. of cone.

3. Mount a few spores in water and observe.

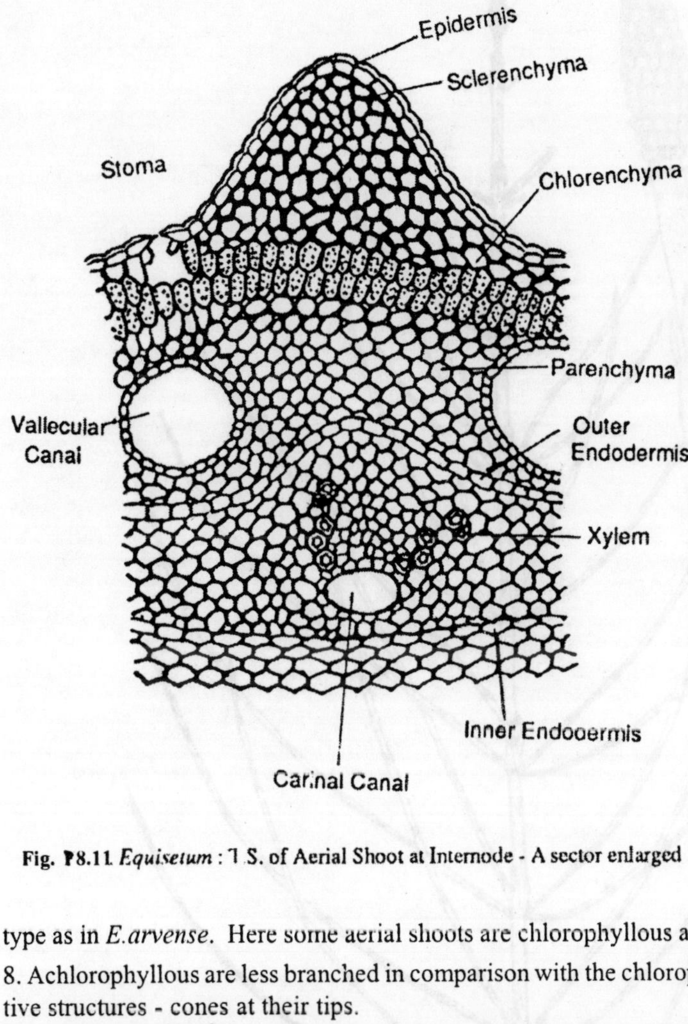

Fig. 18.11 *Equisetum* : T.S. of Aerial Shoot at Internode - A sector enlarged

4. Study the structure of the gametophyte.

External features

1. The plantbody consists of an underground prostrate rhizome from which spring up a number of aerial shoots.

2. Aerial shoots may be few cms long (*E.scirpoides*) or several meters long (*E.giganteum*).

3. The rhizome grows 1-2 feet below the soil and branches profusely. It is differentiated into nodes and internodes.

4. At each node are found a whorl of minute scale leaves. The leaves are connate at the base and form a sheath surrounding the rhizome.

5. Alternating with the leaves are found the buds and branch primordia. These buds develop either into a branch of the rhizome or the aerial shoot.

6. The aerial shoots arise from the nodal regions of the rhizome.

7. All the aerial shoots may be of the same type being chlorophyllous and much branched (*E.ramosissimum*) or may be of two type as in *E.arvense*. Here some aerial shoots are chlorophyllous and some achlorophyllous.

8. Achlorophyllous are less branched in comparison with the chlorophyllous branches. They bear reproductive structures - cones at their tips.

9. The aerial shoots produce secondary branches in whorls at the nodal regions. Usually they are 5-6 internodes in length.

10. Secondary branches arise from the branch primordia. Usually there are many of them resulting in the development of a whorl.

11. The aerial shoots are also differentiated into nodes and internodes with internodal regions having longitudinal ribs. The ribs alternate with the grooves at successive internodes.

12. Silica deposition is seen on the aerial shoot making the plantbody rough.

13. Leaves are found at the nodal regions. They are microphyllous and scaly. Phyllotaxy is whorled. Leaves are connate at the base.

14. The number of leaves at every node ranges between 3-40. Leaves have very little to do with photosynthesis. Much of the photosynthetic activity is carried out by aerial shoots.

15. Reproductive structures - cones are borne at the tip of aerial shoots.

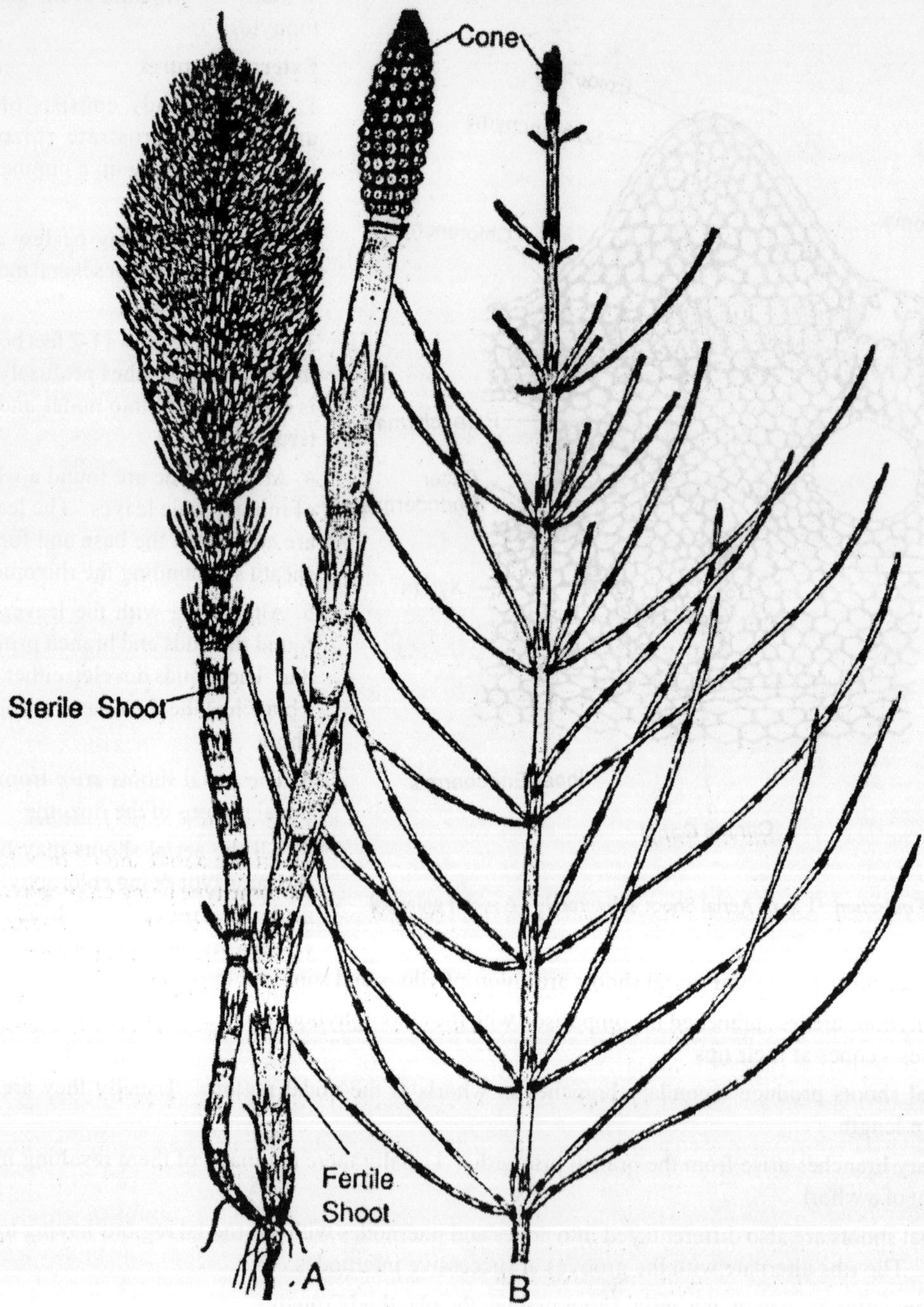

Fig. F8.12. *Equisetum* : Habit of *E. telemateia* and *E. palustrae*

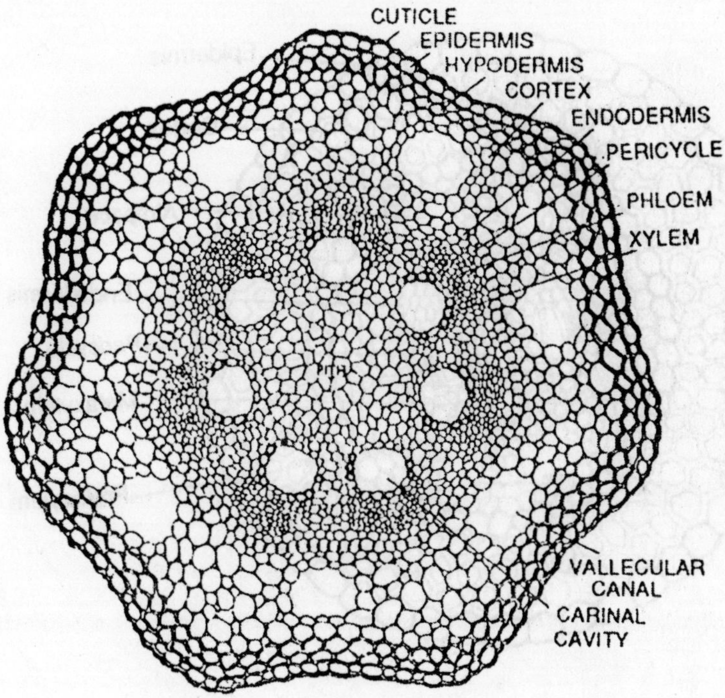

Fig. F8.13. *Equisetum* : T.S. of Rhizome

Internal structure of rhizome

1. A transverse section shows epidermis, cortex and stele. The outline of the section is wavy.

2. Epidermis is made up of a single layer of cells. There are no stomata in the epidermis.

3. Next to the epidermis is found hypodermis made up of 2-3 layers of sclerenchyma.

4. Internal to the hypodermis is the cortex. Cortex is made up of parenchyma cells. Vallecular canals are found in the cortex.

5. An endodermal layer surrounds the stele.

6. Next to the endodermis is a layer of pericycle.

7. Internal to the pericycle are seen a number of vascular bundles. These are arranged alternating with the vallecular canals.

8. Vascular bundles are collateral and endarch with poorly developed xylem. Some of the protoxylem elements degenerate to form a cavity called the carinal canal.

9. In the central region there is a parenchymatous pith.

Systematic position

SPHENOPSIDA

EQUISETALES

EQUISETACEAE

EQUISETUM

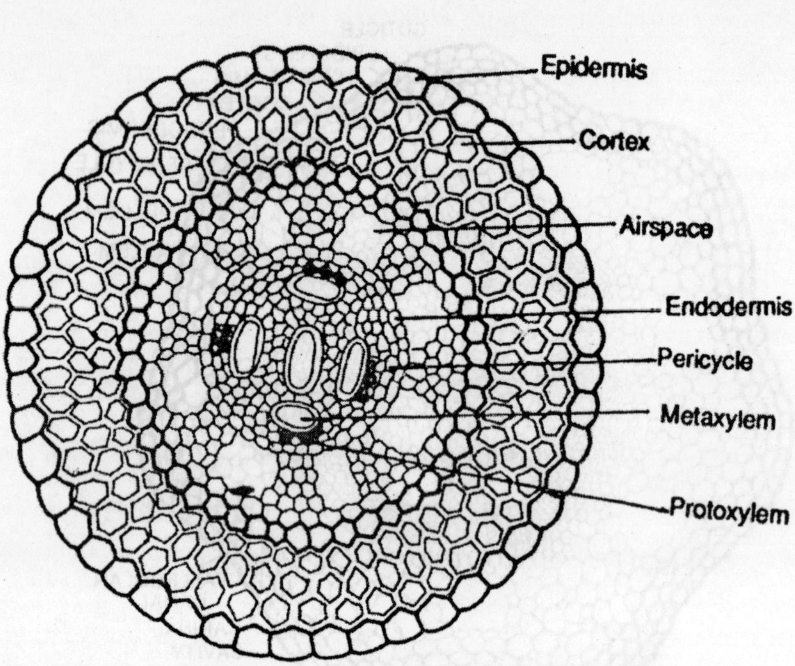

Fig. 8.14. *Equisetum* : T.S. of Root

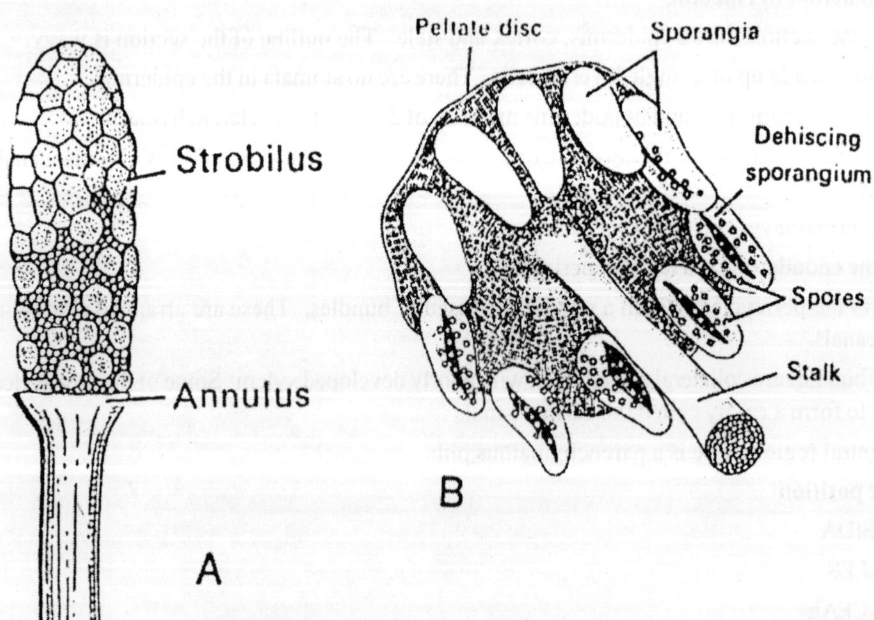

Fig. 8.15. Equisctum: Spore producing Organs
A. Cone, B. Ventral view of a Sporangiophore

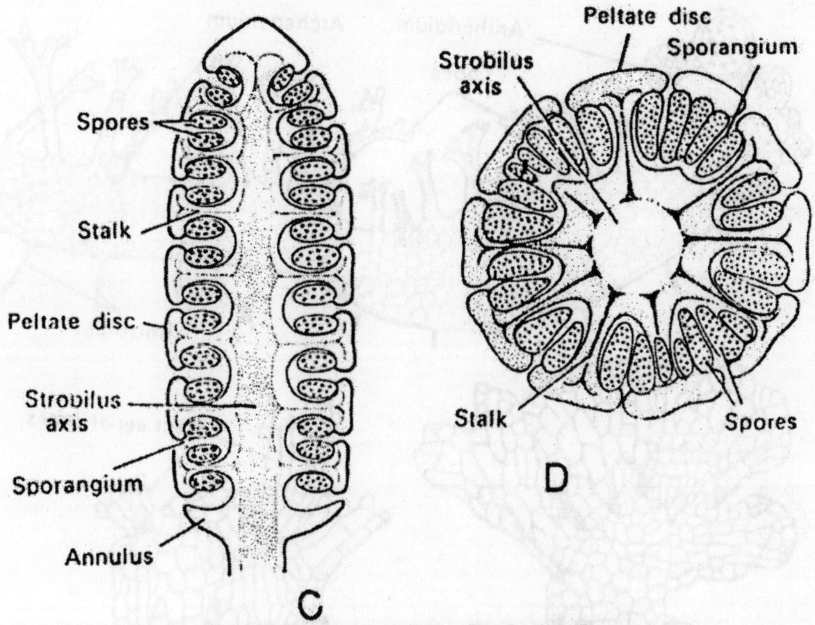

Fig. 8.16 ***Equisetum*** **: Spore Producing Organs**
C. L.S. of Cone, D. T.S. of Cone

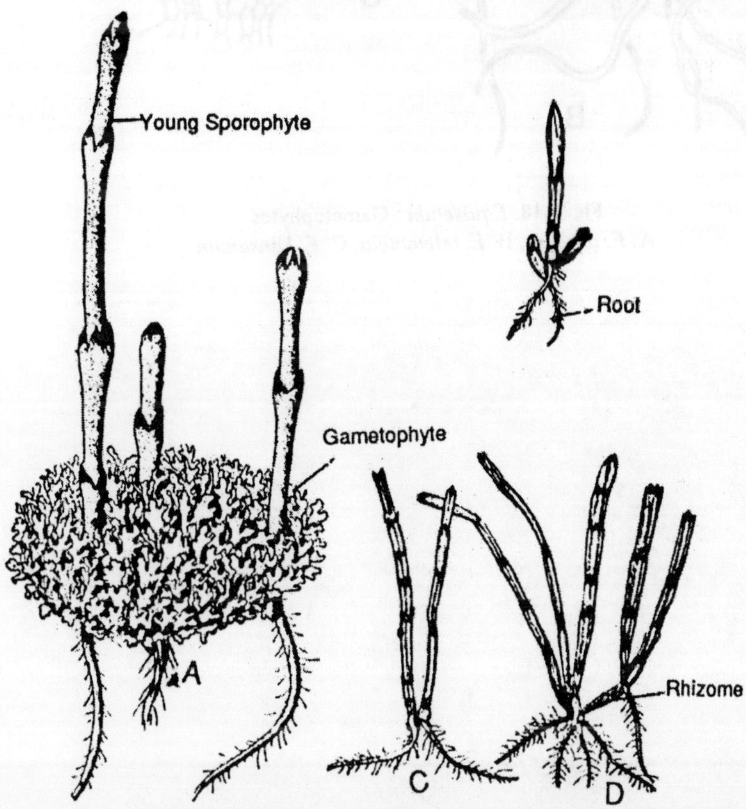

Fig. 8.17. Equisetum: Young Sporophytes (A) on the Gametophyte and Establishment of Young Sporophytes on Soil (B-D)

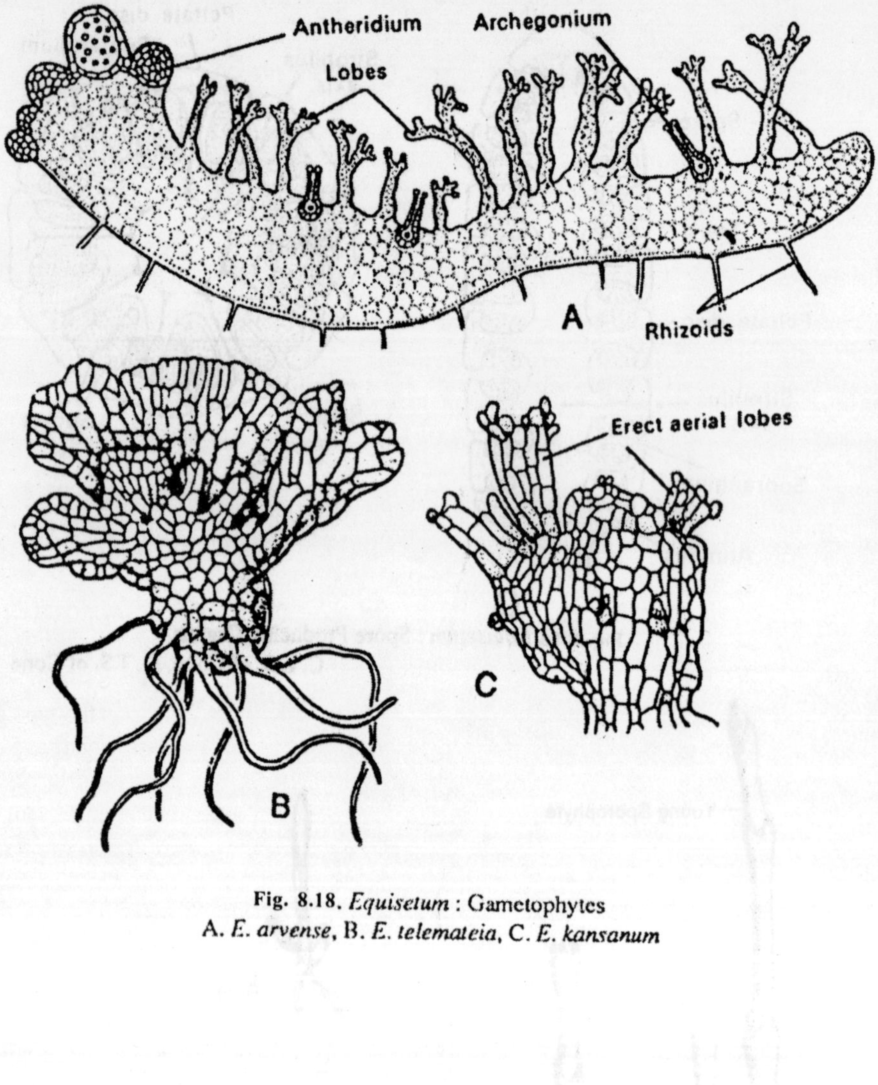

Fig. 8.18. *Equisetum* : Gametophytes
A. *E. arvense*, B. *E. telemateia*, C. *E. kansanum*

CHAPTER 9

PRIMOFILICOPSIDA

BOTRYOPTERIS

1. The genus *Botryopteris* is one of the best known among Coenopteridales.

2. It is the type genus of the family and has 5 species ranging from lower Carboniferous to the Permian.

3. The name *Botryopteris* is given to the fossil specimens of stem.

4. The stems are slender, cylindrical and few millimeters in diameter.

5. They are branched and bear spirally arranged fronds.

6. Anatomically, the main stem has a small mesarch protostele, which is surrounded by a broad cortex.

7. The cortex has a prominent band of sclerotic cells.

8. In the leaf stalk, the xylem strand is solid and has three prototoxylem points.

9. As in *B. forensis* the strand is deeply indented looking like a trident.

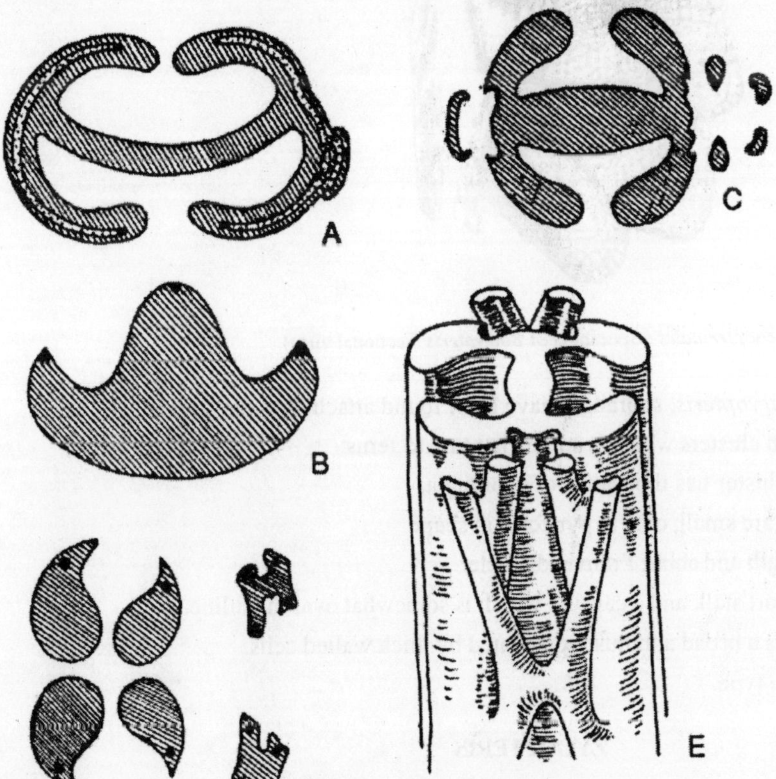

Fig. 9.1 *Coenopteridales* : Stereodiagram of Steles
A. *Ankyropteris*, B. *Botryopteris*, C. *Stauropteris*, D-E. *Etapteris*

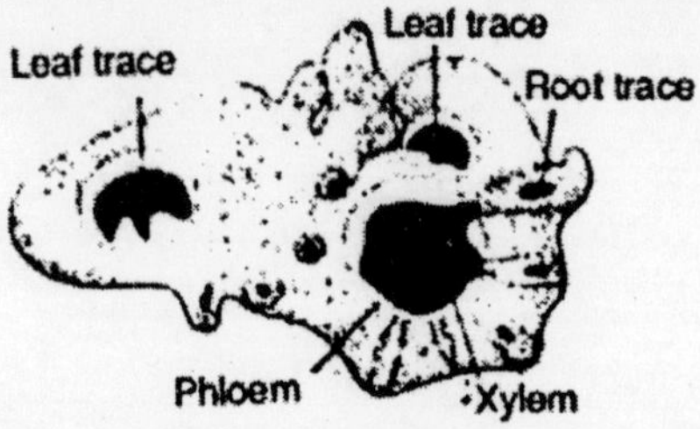

Fig. 9.2. *Coenopteridales* : Diagrammatic Transverse Section of *Botryopteris ramosa*

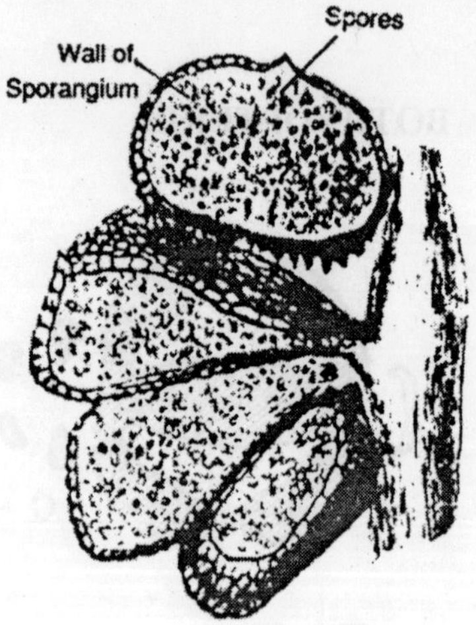

Fig. 9.3. *Coenopteridales* : Sporangia of *Botrypteris* (sectional view)

10. In very few species of *Botryopteris*, sporangia have been found attached to the fronds.
11. The sporangia are found in clusters which is somewhat rare to ferns.
12. In *B.globosa* sporangial cluster has thousands of sporangia.
13. The sporangia themselves are small, oval to pyriform in shape.
14. They measure 2 mm in length and about 1 mm in diameter.
15. Each sporangium has a short stalk and a capsule which is somewhat oval in outline.
16. The wall of the capsule has a broad annulus represented by thick walled cells.
17. The spores are of the same type.

ZYGOPTERIS

1. The ferns belonging to the familuy Zygopteridaceae are more complex and older then Botryopteridaceae.
2. The fossil specimens belong to middle Devonian and possibly have connection with Psilophytales.
3. *Zygopteris*, the type genus of the family is the best known.

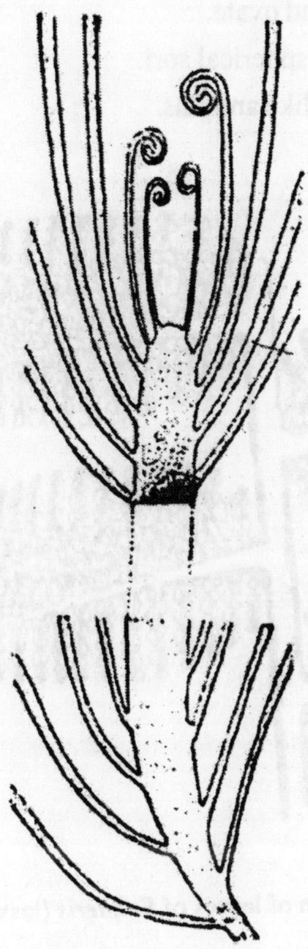

Fig. 9.4. ***Coenopteridales*** **: Reconstruction of Upper and Lower Portions of Shoots of *Zygopteris primaria* (after Sahni 1933)**

4. It has several species of which *Z.primaria* has been studied extensively.
5. The plant body of *Zygopteris* is tree like with a trunk having a diameter of 20 cm.
6. The stem as such, is only about 1.5 cm in diameter being made up of an armour of leaf stalks and adventitious roots.
7. The stem of *Zygopteris* bears an elaborately branched frond having a number of leaf stalks.
8. These are usually cylindiracal and upto 2 cm in diameter.
9. The leaf stalks have a number of pinnae.
10. Occassionally the leaf stalks are given the name *Etapteris*.
11. Anatomically the stem of *Zygopteris* has a xylem cylinder consisting of scalariform tracheids.
12. An unusual feature here is the presence of a layer of secondary wood surrounding the primary xylem.
13. This is perhaps one of the rare instances of secondary vascular tissue in a fern extinct or extant.
14. The leaf stalk internally shows a vascular strand which is H-shaped with a straight median band with some what fixed lateral arms.
15. Two small protoxylem points lie in the shallow depressions at the end of the median band.
16. Various names have been given to the fructifications of *Zygopteris*.
17. *Corynopteris* is one such fructification genus.

18. The sporangia in the fructification are large and ovate.
19. They are usually *sessile* and are grouped into spherical sori.
20. The wall of the sporangium has a broad band like annulus.

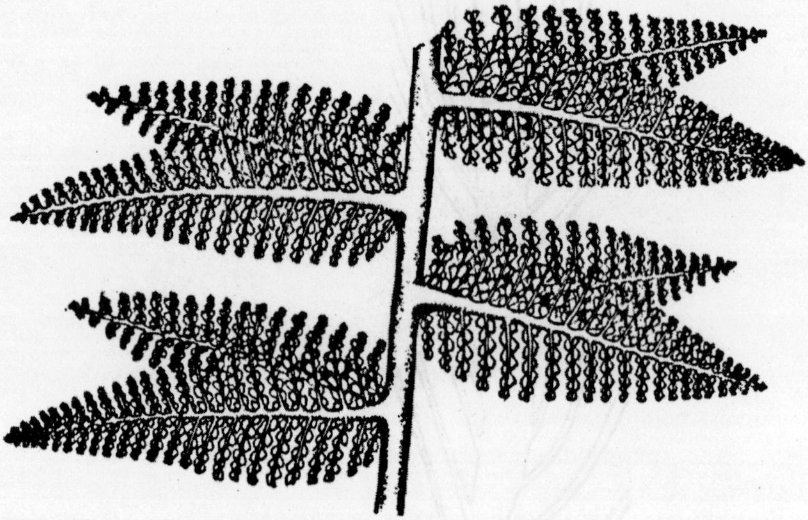

Fig. 9.5 . *Coenopteridales* : Reconstruction of leaves of *Etapteris* (leaves of *Zygopteris*)

CHAPTER 10

FILICOPSIDA

PTEROPSIDA
(Filicophyta or Pterophyta)

1. Pteropsida represent the most highly evolved group among the lower vascular plants.

2. In this group are included some of the most beautiful, and most familiar plants called *Ferns* which are the joy and pride of a gardener.

3. The delicate, varied and highly attractive foliage of ferns have made them a must in any garden.

4. Of all the pteridophytes, ferns are the most widely distributed.

5. The living ferns are represented by 305 genera and nearly 10,000 species (Holtum, 1949)

6. Geologically ferns are known from as far back as the Carboniferous period.

7. It is believed however that the group must have had its ancestors during the *Devonian* period itself.

8. A unique feature of ferns is their persistent basic characters yet sufficiently plastic to be receptive to the environmental fluctuations.

9. Pteropsida are distinct from lycopsida and sphenopsida in several characters.

10. Among the vegetative characters megaphyllous leaves with the attendant leaf gaps are most notable.

11. Among the reproductive features, (though some primitive members show some sort of a semblance to the strobilar organisation seen in the previous groups), the aggregation of sporangia on the abaxial or adaxial surface of the leaf into sori is the most significant.

12. Some of the characteristic features of ferns are given below :

PLANT BODY

1. Stem :

a) The sporophyte has an underground rhizomatous stem which may be elongated or tuberous.

b) The branching of the rhizome may not be dichotomous.

c) In some cases the rhizome is covered by hairs called "ramenta".

2. Leaves :

a) The leaves are compound and once or twice pinnate.

b) They are megaphyllous, having a dichotomous or reticulate type of venation.

c) The size of the leaves varies from few centimeters to several meters *(Angiopteris).*

d) Usually only the leaves are aerial while the rest of the plant body is subterranean.

e) Some ferns show circinate vernation in the leaves i.e., the young growing parts are coiled inwards and uncoil as they grow.

3. Roots :

Like in other pteridophytes, roots are always adventitious.

EUSPORANGIATAE

1. This sub class includes all eusporangiate ferns.
2. The sprorangial wall is more than one layered.
3. Spore output is very high.
4. There are two orders in this sub class *viz.,* Ophioglossales and Marattiales.

OPHIOGLOSSALES

1. The order include herbaceous, fleshy sporophytes with a short rhizome.
2. Sporangia are borne on a separate outgrowth *called fertile spike.*
3. This arises at the junction of the leaf blade and lamina.
4. Sporangia have a multilayered wall with a high spore output.
5. There is no special dehiscence mechanism.
6. All the members of the order are homosporous.
7. Gametophytes are tuberous and saprophytic.
8. The order has a single family ophioglossaceae, with three genera - *Ophioglossum,Botrychium* and *Helminthostachys.*
9. The genus *Ophioglossum* is discussed here.

OPHIOGLOSSUM

Occurrence and Distribution :

1. Of all the genera of the family, *Ophioglossum* is the most widely distributed being found in almost all parts of the globe.
2. The genus is divided into three sub-genera.

   (a) *Euophioglossum* E.g., O. *moluccanum,* O. *vulgatum* etc.,

   (b) *Ophioderma* E.g., *o. pendulum* etc., and

   (c) *Cheiroglossa* E.g., *O.palmatum.*

Sporophyte

Morphology of the plant :

1. In most of the species, the plant body possesses a short, erect, tuberous rhizome exhibiting radial symmetry.
2. In *O. pendulum* and O. *intermedium* the rhizome is dorsiventral. Branching of the rhizome which is infrequent, may be dichotomous or lateral.
3. The upper surface of the rhizome produces leaves in an irregular spiral.
4. The number of leaves produced per season varies.
5. In temperate species only one leaf is formed per growing season.
6. In tropical species however, the number may be 4-5 per year.
7. The growth of the leaf is extremely slow.

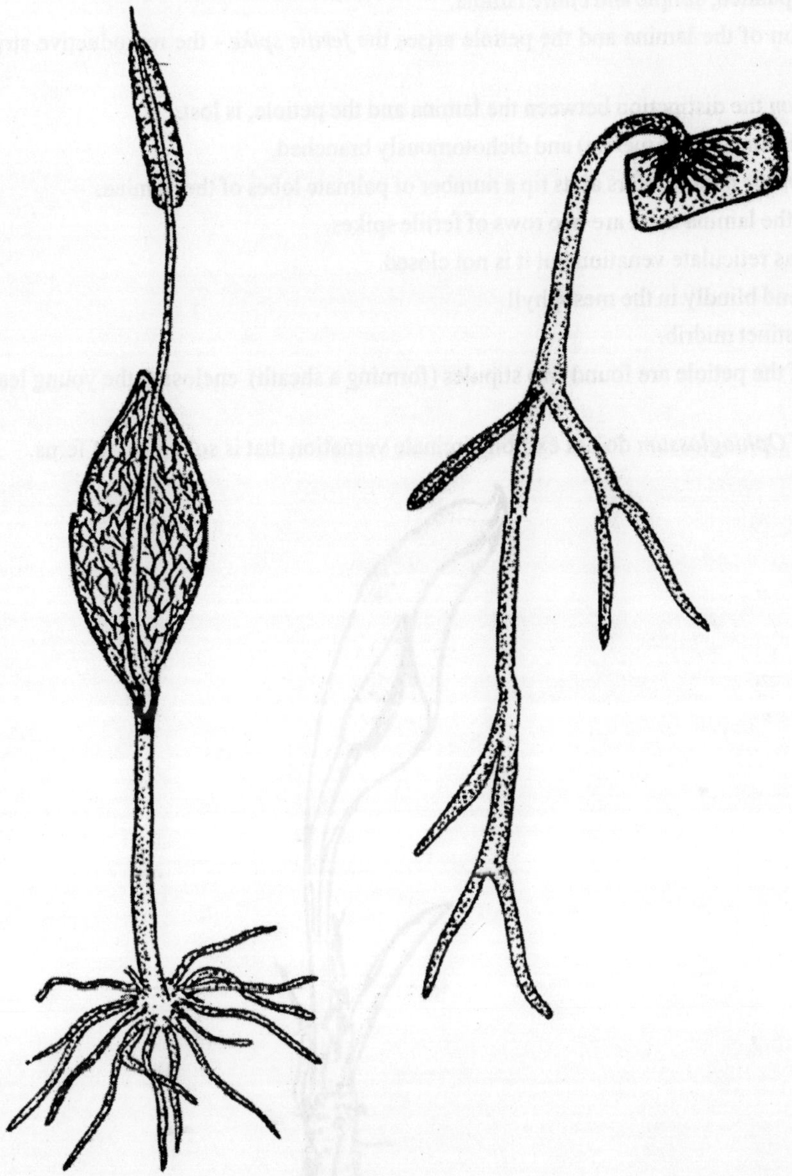

Fig. 10.1. ***Ophioglossum*** **: Habit of** ***O. vulgatum*** **(left) and** ***O. palmatum*** **(right)**

8. A leaf primordium is underground for the first three years and only in the fourth year it comes above the ground.

9. The leaves may be erect or pendant *(O. pendulum).*

10. A great amount of diversity is exhibited in the nature of leaves.

11. Generally a leaf has two parts *viz.* petiole and lamina.

12. The petiole is erect and long.

13. It bears an expanded, simple and entire lamina.

14. At the junction of the lamina and the petiole arises the *fertile spike* - the reproductive structure of the plant.

15. In *O. pendulum* the distinction between the lamina and the petiole, is lost.

16. Here the leaf is long (1-1,5 meters) and dichotomously branched.

17. In *O.palmatumtbe* petiole bears at its tip a number of palmate lobes of the .lamina.

18. In this,below the lamina there are two rows of fertile spikes.

19. The lamina has reticulate venation but it is not closed.

20. The veinlets end blindly in the mesophyll.

21. There is no distinct midrib.

22. At the base of the petiole are found two stipules (forming a sheath) enclosing the young leaf of the next season.

23. The leaves of *Ophioglossum* do not exhibit circinate vernation that is so typical of ferns.

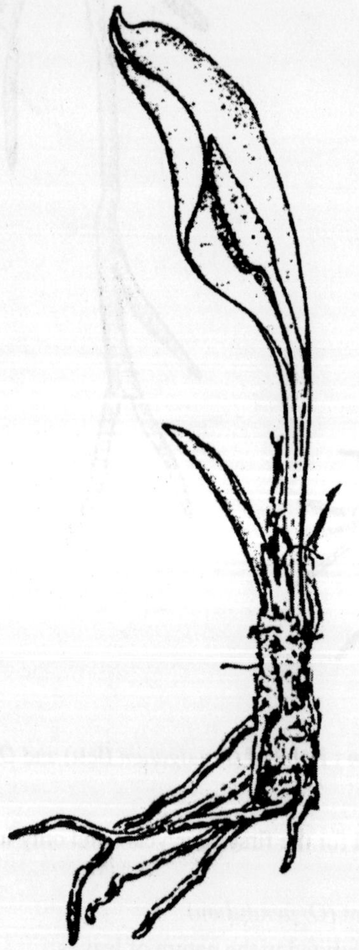

Fig. 10.2. *Ophioglossum* : Plant Body of *O. lusitanicum*

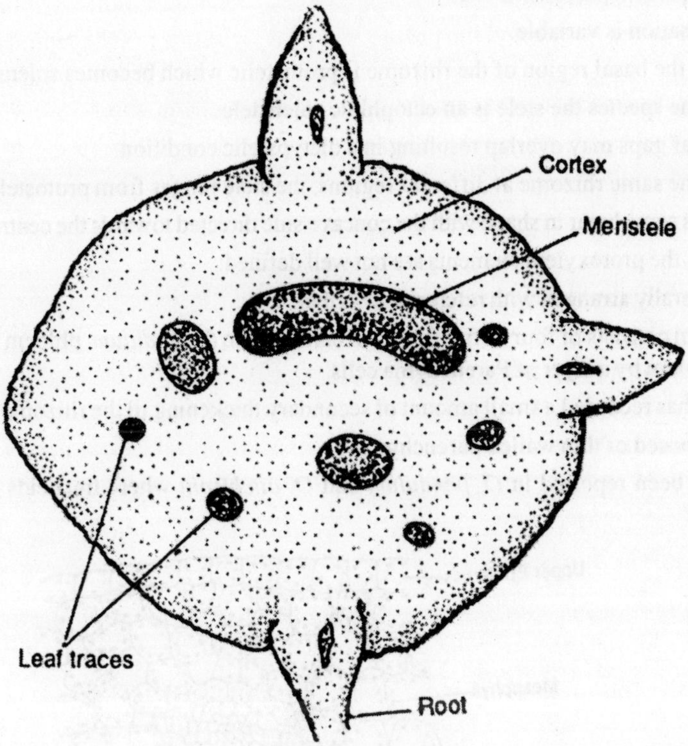

Fig. 10.3. ***Ophioglossum*** **: T.S. of Rhizome (ground plan)**

24. The rhizome is anchored to the soil with the help of numerous adventitious roots that are produced from its (rhizome) under surface.
25. The roots lack root hairs.
26. The roots may be unbranched *(Euophioglossum)*, infrequently branched *(Cheiroglassa)* or frequently branched *(Ophioderma)*.
27. Usually only one root is formed at the base of the leaf of the current season.
28. Unlike leaves, the roots are perennial.
29. The roots have a mycorrhizal association which compensates the absence of root hairs.

INTERNAL STRUCTURE

Rhizome

1. A transverse section of the rhizome shows an epidermis, Cortex and the central stele.
2. The outline of the section will be irregular and discontinuous due to the presence of roots and leaf bases.
3. The epidermis is single layered. Cortex is wholly parenchymatons-The cells are loosely arranged with large number of intercellular spaces.
4. In older rhizomes epidermis is replaced by a periderm layer. But there is no cork cambium.
5. The presence of endodermis is doubtful in older rhizomes.
6. In younger rhizomes an endodermis has been identified (*O. pendulum*).

7. Pericycle may be present (Canrobell 1939).
8. The stelar organisation is variable.
9. In *0. reticulatum* the basal region of the rhizome is protostelic which becomes solenstelic as it travels up.
10. In majority of the species the stele is an ectophloic solenstele.
11. Occasionally leaf gaps may overlap resulting in a dictyostelic condition.
12. Some times in the same rhizome at different regions, the stele ranges from protostele to a dictyostele.
13. Each meristele is semi-lunar in shape with the concave side directed towards the centre. Xylem is endarch.
14. In O. *palmatum,* the protoxylem elements are not well defined.
15. Phloem is collaterally arranged with reference to xylem.
16. Generally phoem consists of four to five layers of cells. But in *0. pendulum* phloem is single layered and is separated from xylem by a layer of Parenchyma cells.
17. Boodle C1899) has recorded a small amount of secondary thickening ill the rhizome of 0. *vulgatum.*
18. The pith is composed of thin walled parenchyma.
19. Mixed pith.has been reported in *O. pendulum* and *O. aitchisoni* where tracheids are intermixed with parenchyma cells.

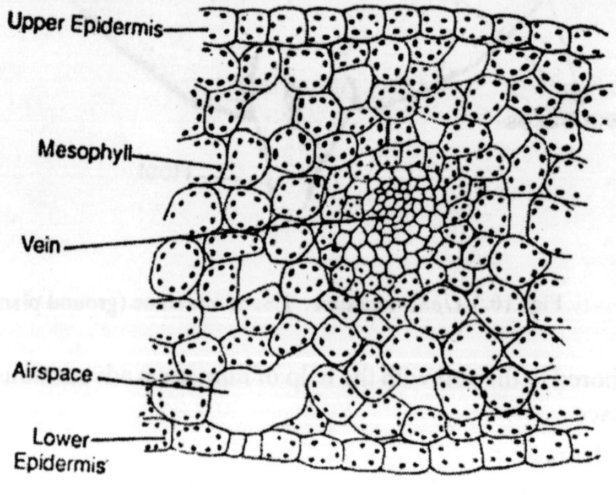

Fig. 10.4. ***Ophioglossum*** **: Venation and Anatomy of Leaf**
A. Lamina showing Venation. B. T.S. of Leaf

Leaf

1. From the rhizome a single leaf trace is given off but at the base of the petiole it divides into two strands and enters the leaf.
2. Bat in *O. palmatum* and several other species the leaf gap is associated with two leaf traces.
3. The leaf traces, further up in the petiole divide into a number of vascular bundles and are arranged in a 'C' shaped outline with the opening directed towards the adaxial surface.
4. The bundles are collateral with endarch xylem.
5. Phloem is on the outer surface of the xylem and is separated from the latter by a layer of parenchyma.
6. There is no bundle sheath.

7. Surrounding the vascular bundles is a ground tissue composed of parenchyma, a few outer layers of which have chloroplasts.

8. The ground tissue is protected by a single layered thick walled epidermis.

9. The lamina on either side is bounded by a single layered epidermis.

10. Leaves are amphistomatic, but the number of stomata in the upper and lower epidermis varies.

11. In *O. vulgatum* they are equal in number on both the surfaces; in others *(O. reticulatum)* stomata are more in number towards the lower epidermis.

12. The mesophyll is undifferentiated and is composed of thin walled isodiametric chlorophyllous cells.

13. The mesophyll consists of a number of veins (bundles) whose structure is similar to those of the petiole.

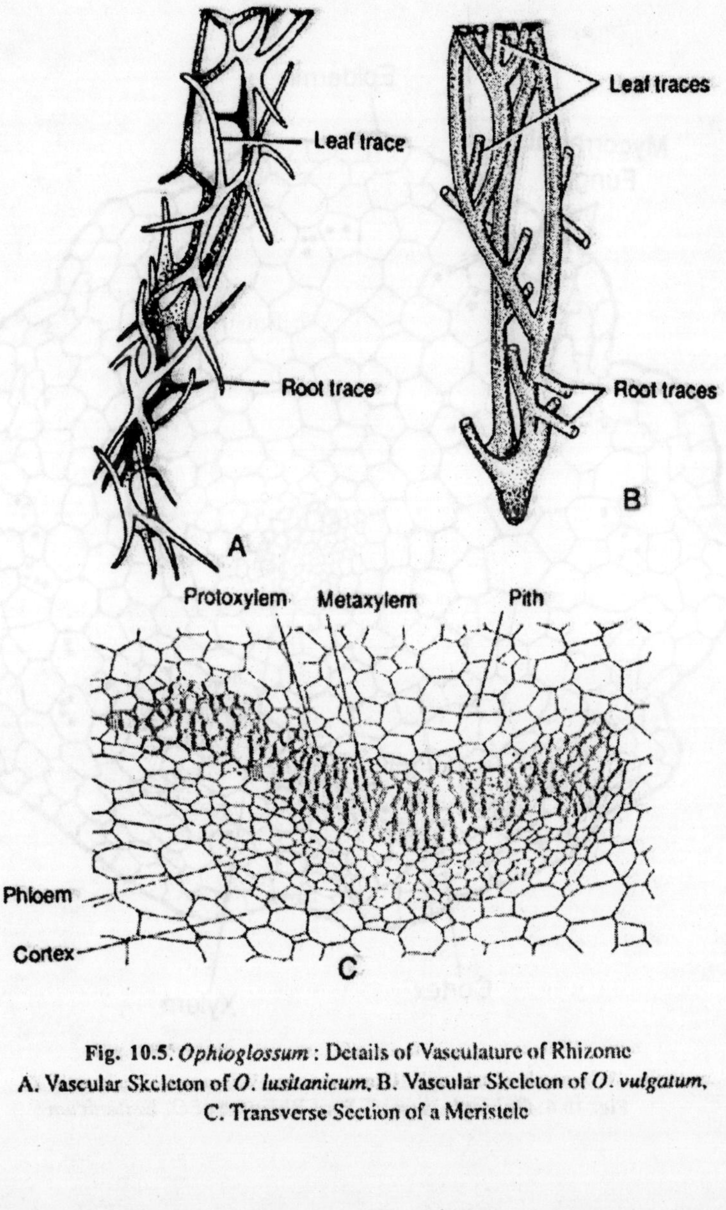

Fig. 10.5: *Ophioglossum* : Details of Vasculature of Rhizome
A. Vascular Skeleton of *O. lusitanicum*, B. Vascular Skeleton of *O. vulgatum*,
C. Transverse Section of a Meristele

Root

1. A root in a transection shows a single layered epidermis, a massive parenchymatous cortex and central stele

2. The epidermis is short lived and is replaced by exodermis *(0. fibrosum)*, or it may persist with suberization.

3. The cortex though wholly parenchymatous is divisible into two zones. In the outer zone, the cells are angular and closely packed without any intercellular spaces.

4. It is in this zone that the fungal hyphae are present.

5. The hyphae are aseptate.

6. The inner zone of cortex has loosely arranged parenchyma cells with no mycorrhizal fungus.

7. The stele of the root may be mono-, di-, tri-, tetra- or even pentarch. But the first two conditions are more frequent.

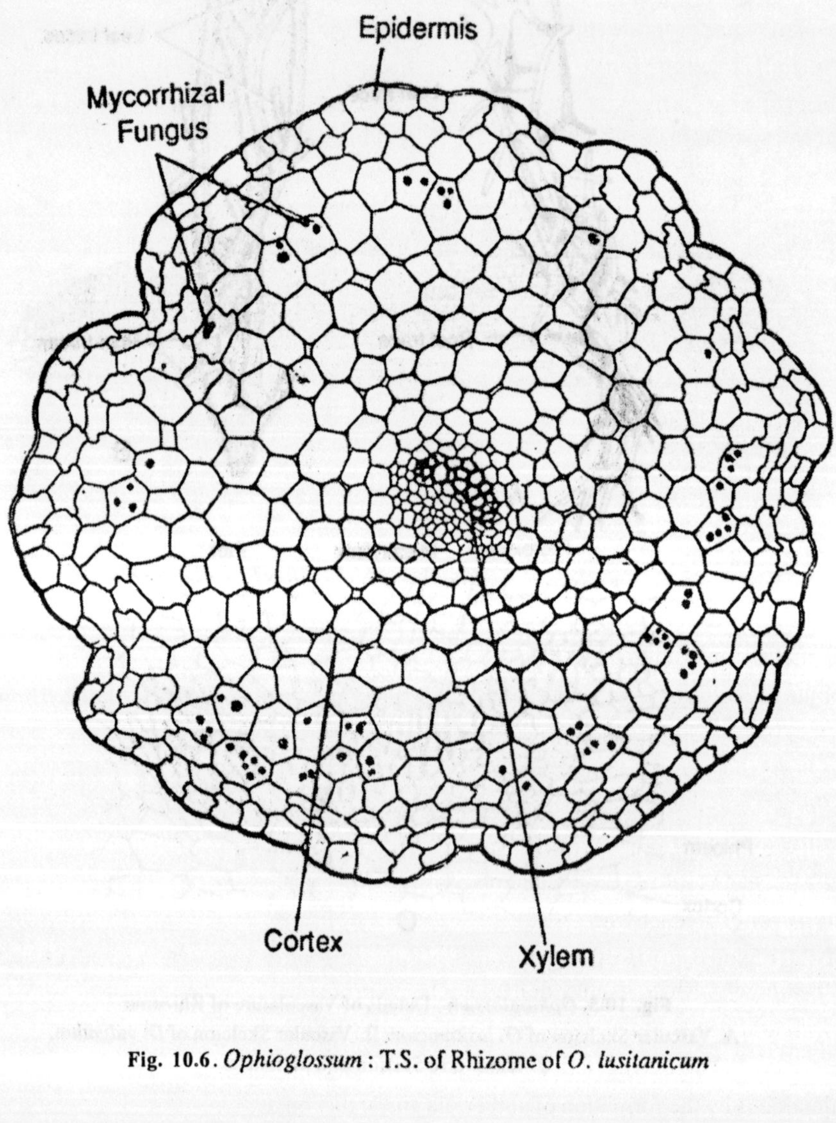

Fig. 10.6. *Ophioglossum* : T.S. of Rhizome of *O. lusitanicum*

8. In diarch roots the xylem is plate like with a protoxylem *point* at each end of the plate.

9. In triarch stele, the xylem groups are separated by parenchyma.

REPRODUCTION :

1. As usual there are two methods of reproduction.

2. Vegetative propagation is brought about by the formation of adventitious buds (root tubers). E.g., *O. pendulum, O. vulgatum* etc.

Spore producing organs :

1. The spore producing organ is generally called a fertile spike.

2. The fertile spike is very much different from the strobili of *Lycopodium* or *Selaginella* in that the sporangia are completely embedded and there is no trace of any sporophyll.

3. There is a central axis like structure and on either side of this are found two rows of sporangia, arranged parallel to *the* long axis of the fertile spike.

4. The number of sporangia per fertile spike varies from 6-20.

5. Many vascular strands run into fertile spike and they frequently anastamose.

6. The lateral strands that arise from these main strands travel in between the sporangia take a 90^0 torn and reach the base of the sporangia.

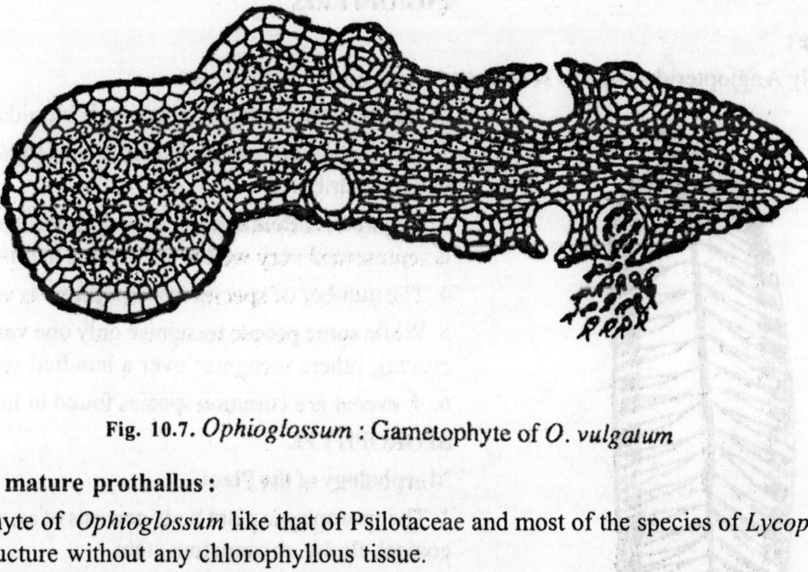

Fig. 10.7. *Ophioglossum* : Gametophyte of *O. vulgatum*

Structure of the mature prothallus :

1. The gametophyte of *Ophioglossum* like that of Psilotaceae and most of the species of *Lycopodium,* is a subterranean structure without any chlorophyllous tissue.

2. The shape of the prothallus varies in different species.

3. In *O. moluccanum* it is cylindrical and less than one cm., in length.

4. There is a slightly tuberous region at the base.

5. In *O. pendulum* the gametophyte is circular and has radiating branches.

6. The prothallus has only parenchyma.

7. The tuberous basal region consists of the mycorrhizal fungus. Prothalli may be, annual (*O. moluccanum*) or perennial *(O. pendulum)*. Rhizoids may or may not be present.

Reproduction :

1. There does not seem to be any vegetative propagation in the gametophyte.

2. Sexual reproduction is by the formation of antheridia and archegonia.

3. The prothalli are monoecious.

4. Sex organs are scattered all over the prothallus.

Systematic position

PTERROPSIDA

EUSPORANGIATAE

OPHIOGLOSSALES

OPHIOGLOSSACEAE

OPHIOGLOSSUM

MARATTIALES

1. The Marattiales include a primitive and compact group of ferns in which the sporangia are borne on the abaxial (away from the axis) surface of the leaf.
2. Sporangia are grouped into sorus.
3. Sometimes sporangia are fused laterally to form a synangium.
4. Leaves have a pair of stipules at the base of the petiole, which in not seen in the other groups of ferns.
5. All the members are homosporous.

ANGIOPTERIS

Occurrence :

1. The family Angiopteridaceae has seven genera and nearly 145 species.

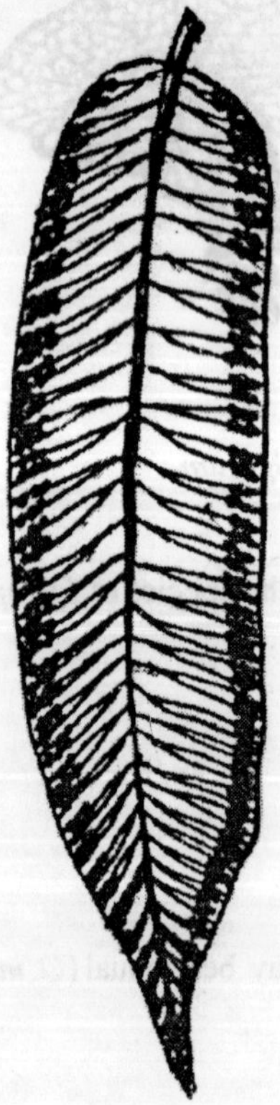

Fig. 10.8. *Angiopteris* : A Pinna enlarged

2. All are tropical plants occurring in humid forests.

2. Of the seven genera *Marattia andAngiopteris* are most widely distributed.

3. *Angiopteris* flourishes very well in the tropical forests and is represented very well in the eastern hemisphere.

4. The number of species in *Angiopteris* is variable.

5. While some people recognise only one variable species *(A. evecta),* others recognise over a hundred species.

6. *A. evecta* is a common species found in India.

SPOROPHYTE

Morphology of the Plant

1. The sporophytic plant body consists of an upright, tuberous, conical, fleshy, rhizomatous stem.

2. The stem is quite thick and in habit the plant resembles a tree fern.

3. The stem is often called a **caudex** or trunk which may be a foot or two in height and almost the same girth.

4. The upper surface of the rhizome bears a crown of graceful, stately leaves.

5. The leaves are 5-6 meters long in a luxuriously growing plant, while a petiole is as thick as a man's arm.

6. The leaves are typically bipinnately compound.

7. The venation is of the open dichotomous type.

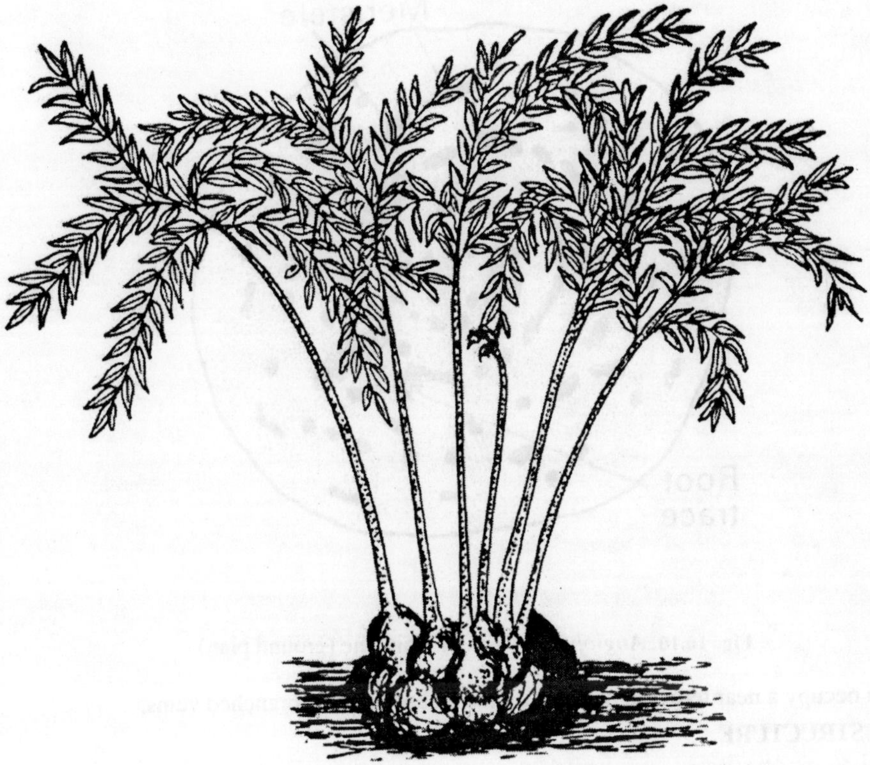

Fig. 10.9. ***Angiopteris*** **: Habit of** ***A. evecta***

8. Pinnae are glabrous (smooth).
9. It is very typical of *Angiopteris* and other marattiaceous ferns to have a pair of thick fleshy stipules at the base of the leaf.
10. When the leaves fall off, these persist with the leaf bases and form a protective armour around the stem.
11. The base of the petiole and the stipules together appear like a horse's hoof. In some species *(A. evecta)*, at the base of the pinnae and the petiole there are swollen articulations (joints) which are apparently the regions of abscission.
12. The pinnae are long (more than 4-5" long) dorsiventrally flattened and have a long drawn tip.
13. The margin is serrate.
14. Roots are produced from the undersurface of the rhizome at the base of each leaf.
15. While leaves are deciduous, roots are perennial.
16. They are thick and have a mycorrhizal association.
17. The branching of the root may or may not be profuse.
18. Roots are endogenous in origin and pierce the stipules during the course of growth.
19. Root hairs are peculiar in being multicellular.
20. At the margins of the pinnae on the abaxial surface are borne, the sori.
21. There is no distinction into foliage leaves and sporophylls.

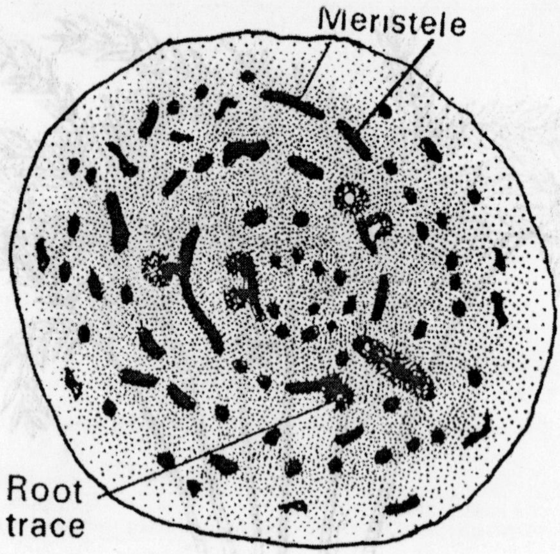

Fig. 10.10. *Angiopteris* **: T.S. of Rhizome (ground plan)**

22. The sori occupy a near terminal position on the dichotomously branched veins.

INTERNAL STRUCTURE

Stem

1. A cross section of the stem shows an outer epidermis, a broad cortex and a central region of vasculature.
2. The cortex is wholly parenchymatous.
3. Here and there mucilage ducts are seen in the cortex.
4. In young stems an endodermis may be seen at the inner face of the cortex.
5. The vascular cylinder of the first formed part of a stem is protostelic.
6. At the level of the third or fourth leaf the stele changes into a siphonostele and then to a dictyostele.
7. There are many bundles arranged in several concentric rings.
8. These concentric circles represent the funnel shaped zones of anastamosing bundles.
9. Segments of the outer zone pass into the leaves as leaf traces and the gaps thus left are filled by segments of the next inner zone.
10. According to Campbell the concentric vasculature of the stem is built up by the union of leaf traces.
11. A very weak secondary development may occur in some cases. Tracheids have scalariform pitting.

Leaf

1. The petiole shows much the same structure as the stem.
2. In the hypodermal region there is a sclerenchymatous band.
3. The leaf trace may be 'C' shaped.
4. Lamina shows a prominent midrib.
5. There is a palisade parenchyma below the upper epidermis and a loose spongy parenchyma above the lower epidermis.

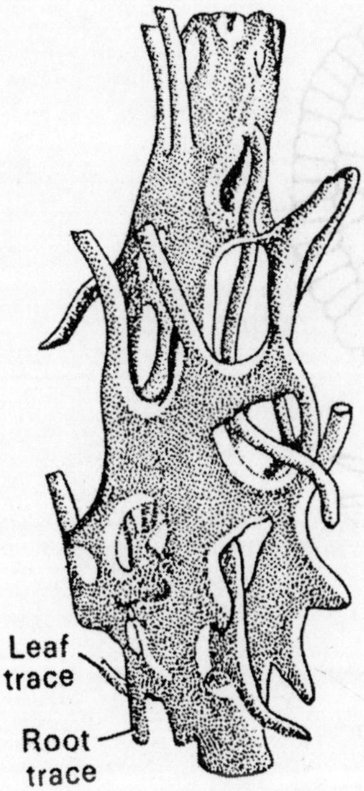

Fig. 10.11. Angiopteris: Diagrammatic Representation of Stele in the Rhizome of A. evecta

6. Leaves are generally hypostomatic.

Root

1. A transverse section shows an epidermis, cortex and stele.
2. The cortex has parenchyma as well as sclerenchyma.
3. There is a conspicuous endodermis encircling the stele.
4. Stele is an exarch, polyarch, actinostele.
5. In the young roots a mycorrhizal fungus is present.

Reproduction

1. Sporophyte reproduces vegetatively as well as by spore production.
2. Dormant buds are formed on the rhizome where the stipules join the petiole.
3. When the stipules are shed ultimately from the plant, the buds develop into new individuals.

Spore producing organs

1. Spores are produced in sporangia which aggregate to form the sori.
2. The sporangia as has already been said, lie at the back of the veins.
3. The sporangia in the sori are arranged in two linear rows.
4. The sporangia may exhibit a certain degree of lateral fusion.
5. But this is not sufficient to call it a *Synangium.*
6. The synarigial nature in *Angiopteris* is not as evident as in *Marattia.*
7. In fact many people regard the sporangia *of Angiopteris* completely independent.

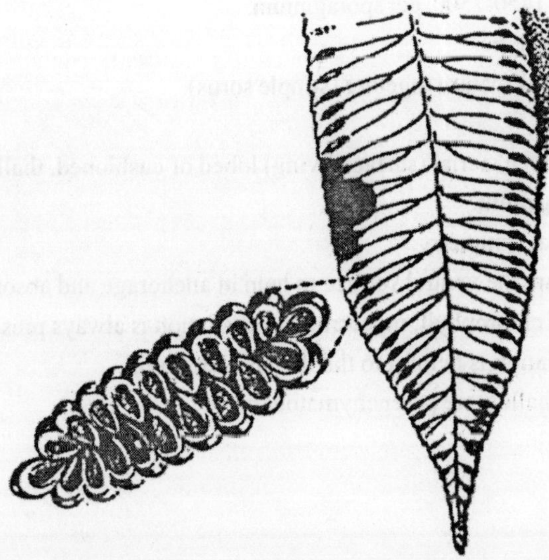

Fig. 10.12. *Angiopteris* : A Part of Pinna showing Soral Arrangement

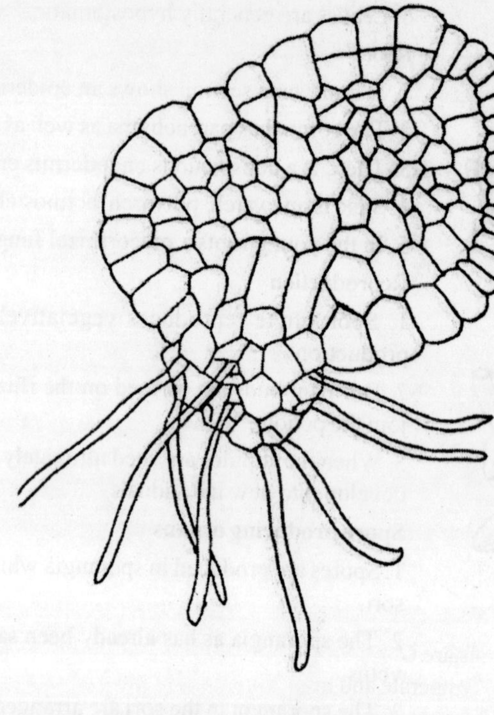

Fig. 10.13. *Angiopteris* : Gametophyte of *A. evecta*

8. There is no typical inducium.
9. Rarely inducial hairs may be present.
10. A nearly mature sporangium has a multilayered jacket enclosing a mass of spore mother cells, which undergo reduction division and produce haploid spores.
11. Spore output varies from 1450-7500, per sporaginium.
12. All spores are of the same type.
13. All the sporangia in a sorus mature together (simple sorus).

The mature prothallus :

1. The prothallus is a deep green, aerial (surface living) lobed or cushioned, thalloid structure.
2. There is a definite growing point.
3. The thallus has dorsiventral symmetry.
4. Rhizoids are produced from the ventral surface to help in anchorage and absorption.
5. In spite of the presence of chlorophyll, mycorrhizal association is always present.
6. To what extent this association is helpful to the thallus is doubtful.
7. All the cells in the prothallus are parenchymatous.

Systematic position

PTERROPSIDA

EUSPORANGIATAE

MARATTIALES

ANGIOPTERIDACEAE (Marattiaceae)

ANGIOPTERIS

OSMUNDIDAE

(Protoleptosporangiopsida)

1. Osmundidae are intermediate between eusporangiatae and leptosporangiatae in their morphological features.

2. The members are characterised by the hard, erect rhizomatous stem studded with persistent leaf bases.

3. Leaves exhibit circinate vernation typical to ferns.

4. They have winged base (as the stipular expansion) and covered with glandular hairs.

5. Stem has a complex anatomy and its cortical region is filled with 'C' shaped leaf traces.

6. Sporangia are borne on some of the pinnae of leaf which are very much crumpled and look like only aggregates of sporangia.

7. Soral organisation is not very conspicuous.

8. All the sporangia mature simultaneously.

9. Gametophytes are large and fleshy.

OSMUNDA

Occurrence and Distribution

1. *Osmunda* - the *Royal fern* is a common member of the green house.

2. It is commonly cultivated in gardens for its beautiful, attractive, lush green foliage.

3. Some of the common Indian species are *O. regalis, O. claytoniana, O. cinnamomea* etc. *Osmunda* is most widely distributed being found in the temperate and tropical regions of the southern and northern hemispheres.

4. The various species of *Osmunda* usually grow in cool, moist, shaded places such as the banks of streams, lakes, ponds etc., in the forests.

SPOROPHYTE

Morphology of the plant

1. All the species *of Osmunda* are moderately sized; a few like *O. cinnamomea* may reach a height of 2-3 metres.

2. The plant body consists of an upright rhizome which is mostly subterranean.

3. The rhizome is studded with a number of persistent, sclerotic leaf bases which are arranged in a rough spiral.

4. The rhizome is generally unbranched but occasionally a dichotomy may result.

5. From the lower surface of the rhizome are produced a number of brownish, adventitious roots.

6. Usually two roots arise at the base of each leaf.

7. The rhizome bears a crown of leaves.

8. The leaves are arranged spirally at the apex of the rhizome.

9. They are unipinnately *(O. claytoniana,* or bipinnately compound (*O. regalis)* and vary in length from 0. 30 to 3 metres.

10. Each leaf has a stout rachis with a wing like expansion (stipule) at the base covered with glandular hairs.

11. The leaves are deciduous and every year a fresh crown of leaves is produced.

12. But leaf bases are persistent and form a thick armour around the rhizome offering mechanical support.

13. The lamina has a leathery texture and is covered with brown hairs.

14. The venation is of the open dichotomous type.

15. The leaves exhibit circinate vernation characteristic of true ferns.

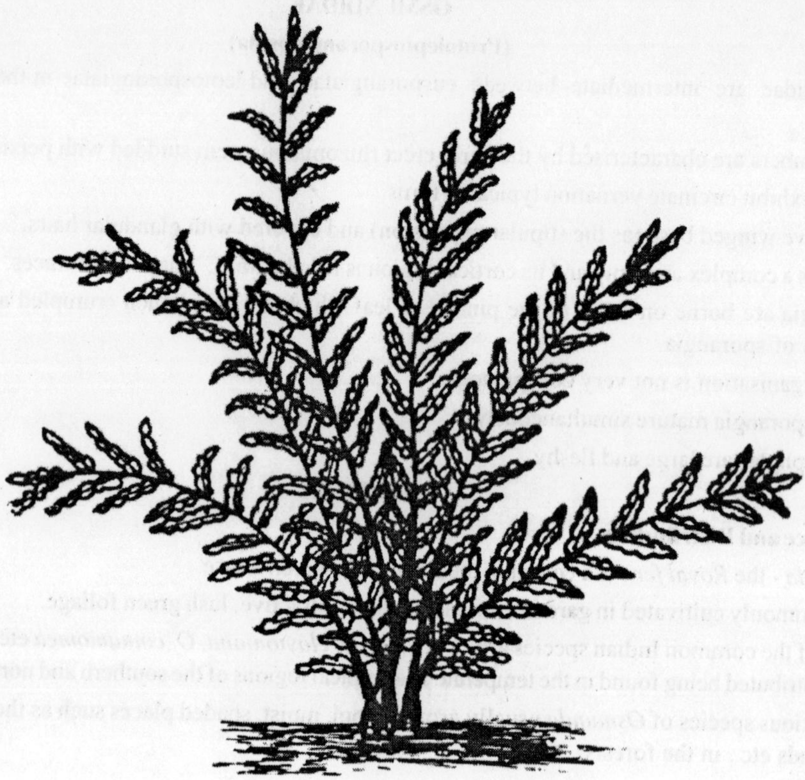

Fig. 10.14. *Osmunda* : Habit of *O. claytoniana*

16. The leaves may be **monomorphic** *(O. regalis* and *O. claytoniana)* i.e., there is no distinction into sterile and fertile leaves, or **dimorphic** *(O. cinnamomea* and *O. japanica).* i.e., fertile and sterile leaves are different.

17. In dimorphic species fertile leaves appear before the sterile ones, when a new crown of leaves is to be unfolded.

18. The chlorophyllous lamina is absent in the fertile leaves and they bear clusters of sporangia.

19. In monomorphic leaves at least the pinnae are dimorphic.

20. The fertile and sterile pinnae are variously distributed.

21. They may be at the base *(O. vachellii)*,. tip *(O. regalis)* or in the middle (*O. claytoniana* and

INTERNAL STRUCTURE

Rhizome

1. A transverse section shows the usual three regions-epidermis, cortex and stele.

2. The outline of the section will-be irregular and discontinuous due to the presence of leaf bases and roots.

3. In a section taken in the middle region of the rhizome, the cortical region shows leaf traces or roots.

4. The cortex has two well marked zones.

5. The outer zone is mainly sclerotic and extensive, while the inner zone which is narrow, is composed of thin walled parenchyma. Numerous 'C' shaped leaf traces are found in both the cortical zones.

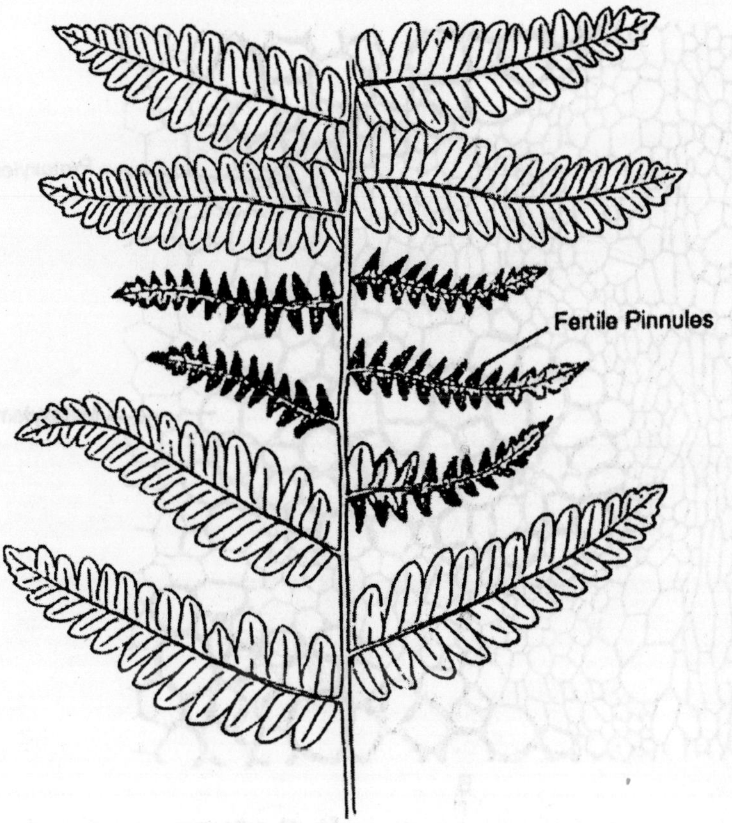

Fig. 10.15 *Osmunda* : Part of the Leaf of *O. claytoniana* enlarged to show Fertile Pinnules

6. These arise obliquely from the main vascular cylinder.
7. Each leaf trace has a well defined endodermis which is lined internally by one or two layered pericycle.
8. The xylem mass is 'horse shoe shaped with the concave side pointing towards the main stele.
9. The xylem has scalariform tracheids constituting metaxylem; the protoxylem being situated in the depressions.
10. Internal to the cortex is the endodermis which is very prominent and continuous.
11. Only when a ramular gap (branch gap) is present, the continuity of the endodermis is disturbed.
12. In the young stem (sporeling) the stele is protostelic. Subsequently i.e after the development of few leaves, it becomes medullated.
13. The stele in the adult plant is an *ectopholic solenostele.*
14. The pith occupies a large portion in the centre.
15. It may be wholly parenchymatous *(O. claytoniana)* or mixed. In *O. claytoniana* the mixed pith has sclerenchyma intermingling with parenchyma, whereas in *O. regalis* pith may have tracheids and parenchyma.
16. Outlining the pith are present the meshes of xylem forming a reticulate cylinder.
17. In a transverse section xylem appears like a circle of horse shoe shaped or oval structures separated by parenchymatous rays.
18. These rays are nothing but leaf gaps.

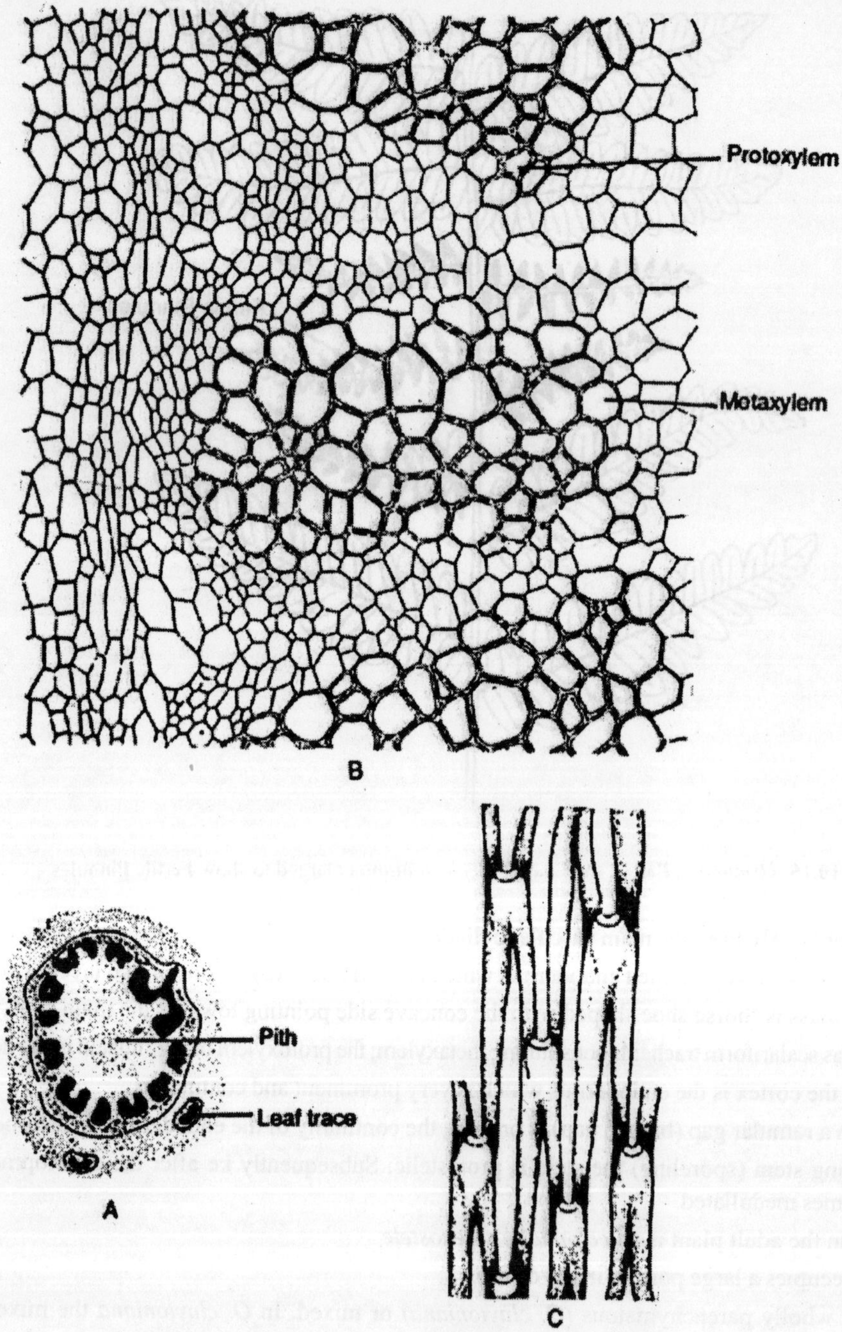

Fig. 10.16. *Osmunda* : Anatomy of Rhizome
A. Ground Plan of T.S., B. A Sector enlarged to show Steles,
C. Stereo Diagram of a part of the Stele

19. The number of xylem strands varies from 15 (*O. regalis)* to 40 *(O. claytoniana).*

20. The protoxylem lies in the sinus fonned by the horse shoe shaped xylem mass.

21. Surrounding the xylem is a sheath formed of parenchyma.

22. External to this is the phloem. In *O. cinnamomea* an internal phloem has been reported at the region of branching.

23. Phloem consists of both proto and metaphloem.

24. External to the phloem is the pericycle, 2-4 layers in thickness.

Leaf

1. The petiole has a large concentric vasculature which is crescent shaped.
2. The two arms of the crescent are incurved.
3. The opening is towards the adaxial side.
4. Xylem consists of a number of protoxylem groups embedded in a continuous mass of metaxylem.
5. The protoxylem groups are arranged on the concave face alternating in position with groups of mucilage cells.
6. Phloem completely surrounds the xylem.
7. The endodermis is not very conspicuous.
8. The vasculature is embedded in a mass of parenchymatous tissue.

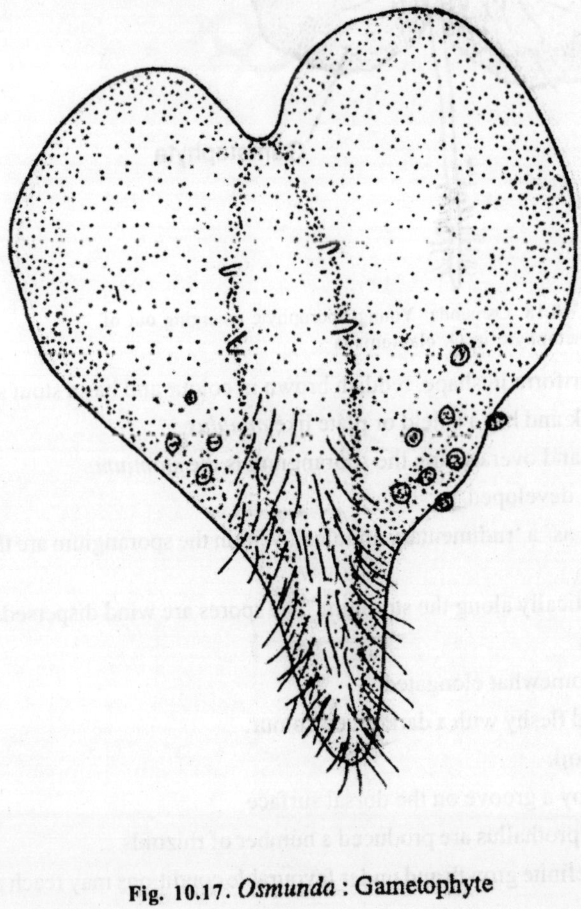

Fig. 10.17. ***Osmunda*** **: Gametophyte**

9. Some of the outer layers are thick walled.
10. The pinna anatomically shows bifacial arrangement.
11. The mesophyll is differentiated, but palisade parenchyma is not distinct.
12. Bounding the mesophyll on either side is the epidermis.
13. The lamina is hypostomatic.

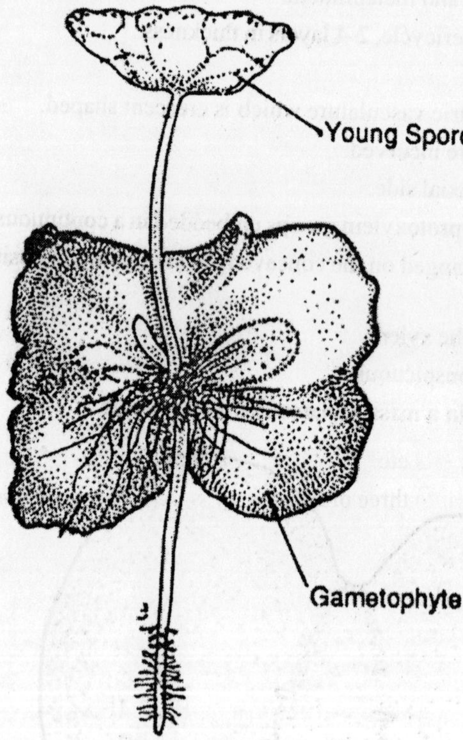

Fig. 10.18. Osmunda: Young Sporophyte emerging out of Gametophyte in O. claytoniana

14. The mature sporangium is pyriform in shape, reddish brown in colour and has a stout stalk.
15. The jacket is one layered thick and has a shield or plate like *annulus*.
16. Perpendicular to the annulus and overarching the sporangium is the *stomium*.
17. The annulus is not very well developed.
18. Hence it is often referred to as a 'rudimentary annulus'. Within the sporangium are the spores ranging in number from 256-512.
19. The sporangium dehisces vertically along the stomium. The spores are wind dispersed.

Structure of a mature prothallus

1. The protballus is cordate but somewhat elongated.
2. It is dorsiventrally flattened and fleshy with a dark green colour.
3. It grows on the surface of the soil.
4. There is a mid rib represented by a groove on the dorsal surface.
5. From the ventral surface of the prothallus are produced a number of rhizoids.
6. The prothalli are capable of indefinite growth and under favourable conditions may reach a size of 2-3 cms.

7. The prothallus is wholly parenchymatous.

8. The margin of the prothallus is thin; but towards the centre it may be 16-18 cells in thickness.

9. Nutrition is autotrophic.

10. The gametophytes are monoecious, but underdeveloped ones tend to produce only male sex organs.

11. Archegonia occur near the mid rib on the ventral surface and antheridia occur towards the fringes of the prothallus.

Systematic position

PTERROPSIDA

OSMUNDIDAE

OSMUNDALES

OSMUNDACEAE

OSMUNDA

LEPTOSPORANGIATAE

1. Among the ferns, leptosporangiatae represent a highly successful evolutionary line.

2. As the name indicates they have their sporangia traceable to a single initial cell.

3. Some of the other typical characters of leptosporangiatae are the single layered wall of the sporangium, definite number of spores per sporangium, a definite dehiscence mecnanism for the sporangium, a regular soral organisation, simple antheridia with relatively less number of antherozoids, early differentiation (at the quadrant stage) of the embryonal parts etc.

4. Leptosporangiatae are classified into three orders *viz.*, Filicales, Marsileales and Salviniales.

FILICALES

1. This is the largest order of leptosporangiatae and comprises all the homosporous members.

2. Bower has classified Filicales into three groups *viz.*, *simplices* (simple *Sorus*), *gradate* (gradate sorus) and *mixtae* (mixed sorus).

3. Simplices includes a varying number of families like Schizaeaceae. Gleicheniaceae etc. Gradatae includes two families *viz.*, Hymenophyllaceae and Cyatheaceae.

4. In this book *Hymenophyllum, Adiantum* and *Pteris* are discussed.

HYMENOPHYLLUM

Members of Hymenophyllaceae are commonly called "Filmy ferns". Genera in the family is variable. Usually there are three well defined genera *viz.*, *Hymenophyllum, Trichomanes* and *Cardiomanes.*

Occurrence and distribution

1. *Hymenophyllum* is distributed in humid, tropical and warm temperate regions of the world.

2. It is characteristically found in tropical rain forests.

3. In the west land forests of New Zealand it is especially abundant.

4. It usually grows on the forest floor or on the wet branches of trees as epiphytes.

5. In Karnataka State *Hymenophyllum* is available in Shiradi ghats near Sakleshpur (Hassan Dist.) where they grow on wet trunks of trees.

6. The number of species in the genus is variable.

7. It has about 200- 230 species.

8. Some of the common species are *H. australe, H. multifidum* and *H. recurvum* etc.

Fig. 10.19. *Hymenophyllum* : Habit of *H. australe*

SPOROPHYTE

Morphology of the plant

1. The plant body rarely exceeds 3-4" in size But in *H. pulcherimum* the pendant fronds may be 2-3 feet in length.

2. It has a rhizome which is very slender and extensive branching of the rhizome is axillary.

3. Roots are adventitious (lacking in some smaller forms) and are produced in pairs at the base of each leaf.

4. But some times roots may be seen between the leaves.

5. This is due to the failure of a leaf primordium to develop above the region where the roots are produced.

6. The leaves of *Hymenophyllum* are very characteristic-true to the name filmy ferns and are objects of beauty.

7. The translucent leaves when growing in close mats may easily be mistaken for bryophytic thalli.

8. Leaves are produced in acropetalous succession and show circinate vernation.

9. But it is not uncommon to find some times the unfolding of a young leaf between two mature leaves.

10. This shows that many leaf primordia may remain dormant and may resume growth at a later stage.

11. Usually in *H. australe* one out of six primordia grows further.

12. The leaf primordia are formed in double series on the rhizome.

13. The leaves have a flattened petiole.

14. The lamina may be simple *(H.cruenta)* or dissected into unequal dichotomies as in the majority of species.

15. It is mostly one celled thick and has an open dichotomous venation.

16. In each shank of dichotomy there is a single vein.

17. In *H. cruenta* there are a series of veins reaching the margin of the simple lamina.

18. Sori are borne along the margin or at the apex of each dichotomy.

19. The leaves are pale green in colour and together with the roots they also seem to take part in absorption.

INTERNAL STRUCTURE

Rhizome

1. A transver.se section shows a relatively narrow cortex surrounding a central stele and being surrounded by an epidermis.

2. The cortex may be wholly sclerotic or partly thin walled (outer cortex) and partly thick walled (inner cortex).

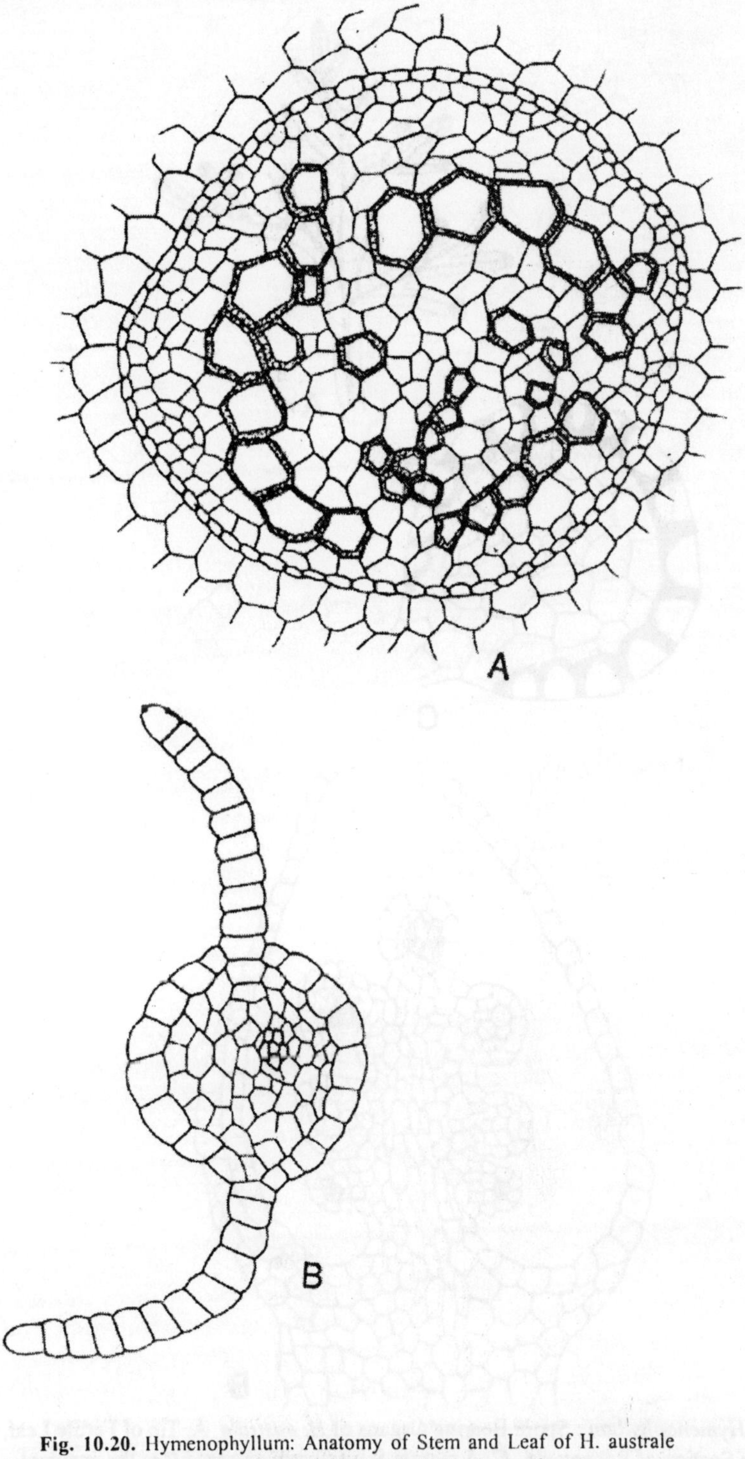

Fig. 10.20. Hymenophyllum: Anatomy of Stem and Leaf of H. australe
A. T.S. of Stele of Stem, B. T.S. of a Leaf

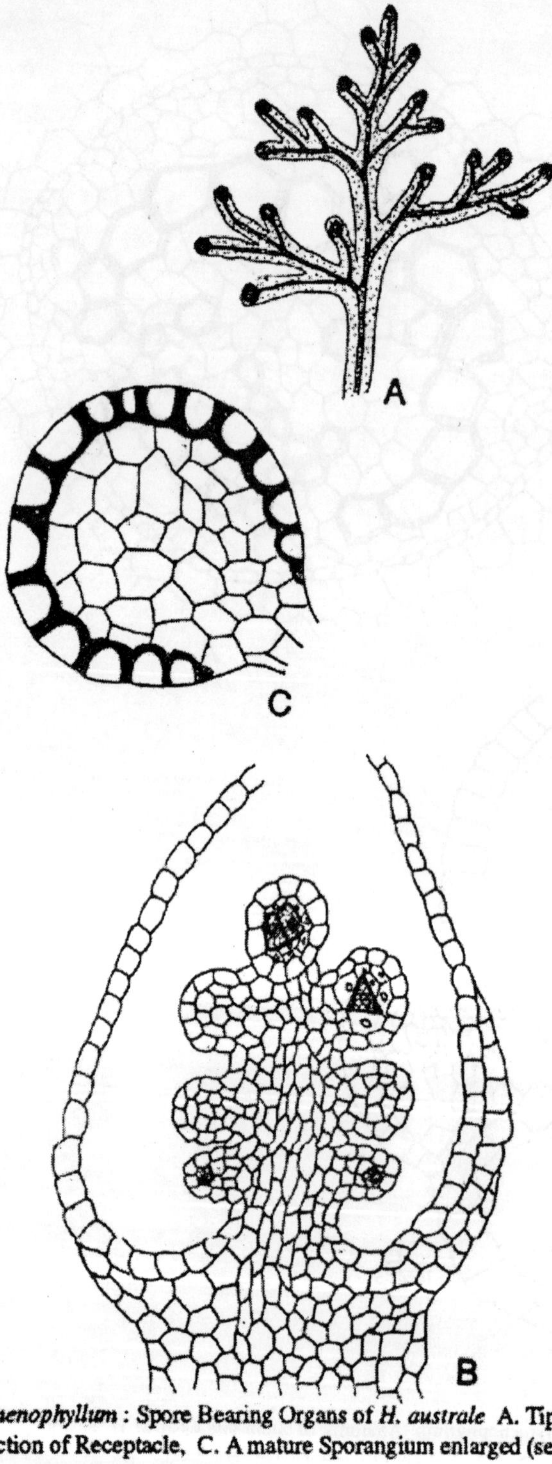

Fig. 10.21. *Hymenophyllum* : Spore Bearing Organs of *H. australe* A. Tip of Fertile Leaf, B. Long Section of Receptacle, C. A mature Sporangium enlarged (see the annulus)

3. Vascular cylinder is protostelic even in the mature plant, a feature unusual to ferns, but not unusual in the background of its habitat.
4. There is an outer endodermis surrounding the vasculature.
5. Internal to the endodermis is the pericycle one to many cells thick.
6. Phloem surrounds the xylem but the two are separated by conjunctive parenchyma.
7. The architecture of the xylem is variable in different species; in *H. demissum* there is a ring of metaxylem surrounding a parenchymatous region in which is embedded in the protoxylem, in *H. scabrum* the metaxylem ring is broken at two points to form two arcs, in *H. cruenta* one of the arcs (ventral) may be absent.
8. Leaf traces that depart from the vascular cylinder do not leave any gap.
9. Branch traces are connected to the leaf traces or more often arise from it.

Root

1. There is no root cap (Sadeback, 1898), the vascular cylinder is protostelic as in other leptosporangiatae.
2. Xylem may be monarch or diarch.

Leaf

1. Leaf blade is one cell in thickness towards the margin but many celled thick towards the central vein.
2. In *H. dilatatum* leaf blade also is many celled thick. The vein consists of poorly developed xylem and phloem.

REPRODUCTION

Vegetative propagation may be brought about by the fragmentation of the rhizome when each portion grows into a new individual.

Spore producing organs

1. The characteristic method of reproduction of the sporophyte is by the spore formation.
2. Sori are developed singly at the tip of each dichotomy.
3. In a dissected leaf blade sori are terminal and in an entire blade they are marginal.
4. The sorus has a fertile tissue called the receptacle from which sporangia are produced.
5. Receptacle development on the lamina is associated with the development of two flaps of tissue from the adaxial and abaxial surface of the leaf.
6. This is the two lipped inducium.
7. It is a true inducium as it does not represent the incurving margin of the lamina.
8. The inducium overarches the-sorus, offering protection to it.
9. The sporangia develop in a basipetalous succession true to the gradate nature of the sorus.

GAMETOPHYTE

1. A mature gametophyte of *Hymenophyllum* is filamentous, ribbon like and dichotomously branched or lobed.
2. Rhizoids are produced from the ventral surface.
3. Some of the branches of the gametophyte may grow erect.
4. Sex organs are mostly formed on the ventral side.
5. Antheridia are borne near the margins of the branches.
6. They *are* scattered all over the thallus.

Systematic position

PTERROPSIDA

LEPTOSPORANGIATAE

FILICALES - GRADATAE
HYMENOPHYLLACEAE
HYMENOPHYLLUM

ADIANTUM

Occurrence and distribution

1. *Adiantum* is popularly called *Maiden hair fern* because of the shiny black rachis of the leaves.
2. It is one of the most widely distributed genera of the family growing luxuriantly in both tropical and sub tropical regions of the world.
3. It grows ubiquitously wherever nature offers moist, shaded locality.

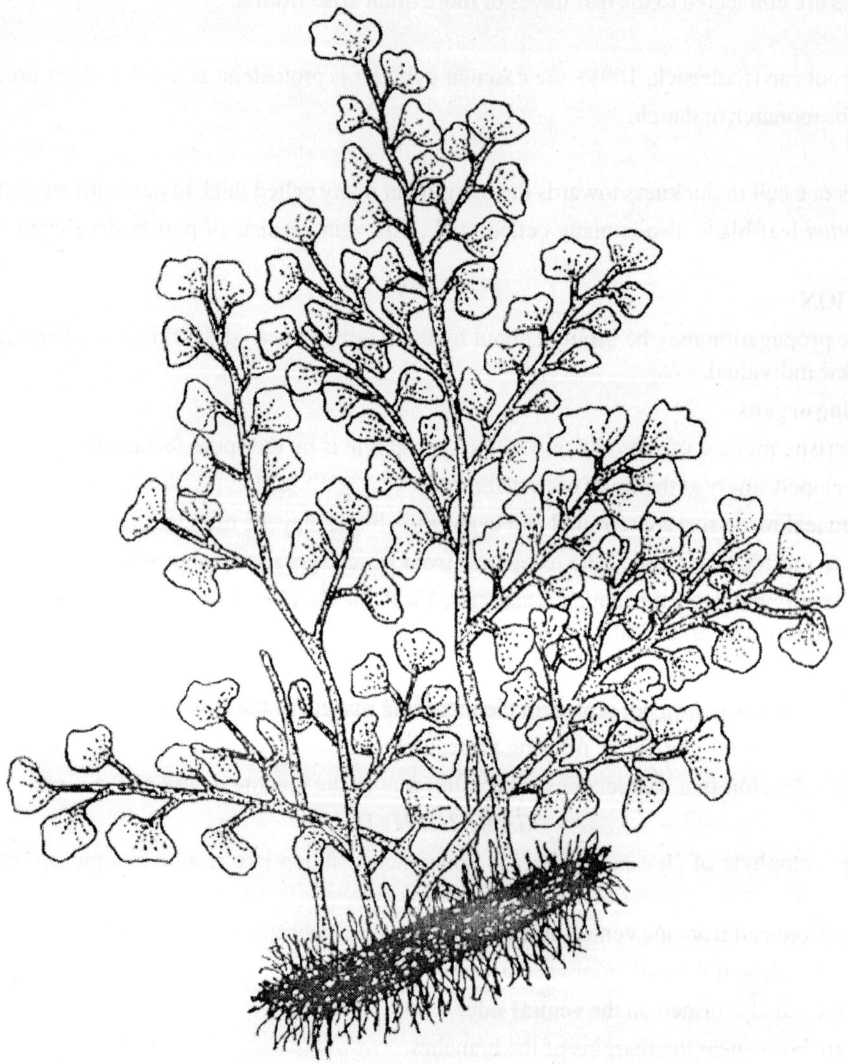

Fig. 10.22. *Adiantum* : Habit of *A. capillus-veneris*

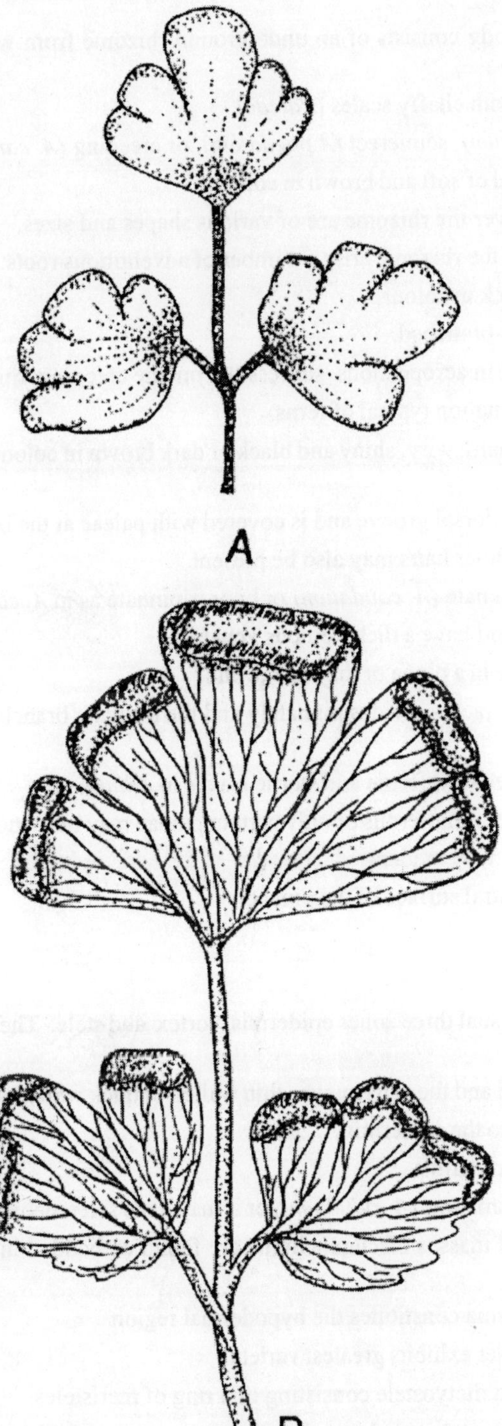

Fig. 10.23. *Adiantum* : Sterile (A) and Fertile (B) Leaflets of *A. capillus veneris*

SPOROPHYTE

Morphology of the plant

1. The sporophytic plant body consists of an underground rhizome from which are produced leaves and roots.
2. The rhizome is covered with chaffy scales *(Paleae).*
3. It may be erect *(A.caudatum),* semierect *(A.paedatum),* or creeping *(A. capillus - veneris).*
4. The rhizomes may be hard or soft and brown in colour.
5. The chaffy scales that cover the rhizome are of various shapes and sizes.
6. From the undersurface of the rhizome arise a number of adventitious roots.
7. The roots are stiff and black in colour.
8. Occasionally they may be branched.
9. The leaves. are produced in acropetalous succession on the creeping rhizome.
10. They show circinate vernation typical of ferns.
11. The rachis of the leaf is hard, wiry, shiny and black or dark brown in colour thus giving the name maiden hair fern.
12. The rachis has a median dorsal groove and is covered with paleae at the basal region.
13. In addition to this, glandular hairs may also be present.
14. The leaves may be unipinnate *(A. caudatum)* or bi or tripinnate as in *A. capillus-veneris.*
15. The pinnae are stalked and have a dichotomous venation.
16. The rachis may terminate in a pinna or may bear a bud.
17. In *A.capillus-veneris* the rachis divides pinnately and the ultimate branches bear pinnae in an alternate fashion.
18. There is no distinction between fertile and sterile leaves in *Adiantum.*
19. The whole leaf may be sporangiferous or only certain pinnae may bear sporangia.
20. The soral organisation is very evident.
21. Sori are borne on the ventral surface of the pinnae.

INTERNAL STRUCTURE

Rhizome

1. A transection reveals the usual three zones epidermis, cortex and stele. The outlines of the section would be wavy.
2. Epidermis is single layered and the cells may be thin walled or thick walled.
3. There is a cuticle external to the epidermis.
4. Cortex lies internal to the epidermis.
5. It may be wholly parenchymatous *(A. rubellum),* or it may have sclerenchyma and parenchyma.
6. In *A.pectinatum,* scattered masses ofsclerenchyma are found embedded in the parenchymatous ground tissue.
7. In *A. caudatum* sclerenchyma constitutes the hypodermal region.
8. The central vascular cylinder exhibits greatest variety.
9. *In A.capillus veneris,* it is a dictyostele consisting of a ring of meristeles.
10. In the young condition the stele may be asolenostele.

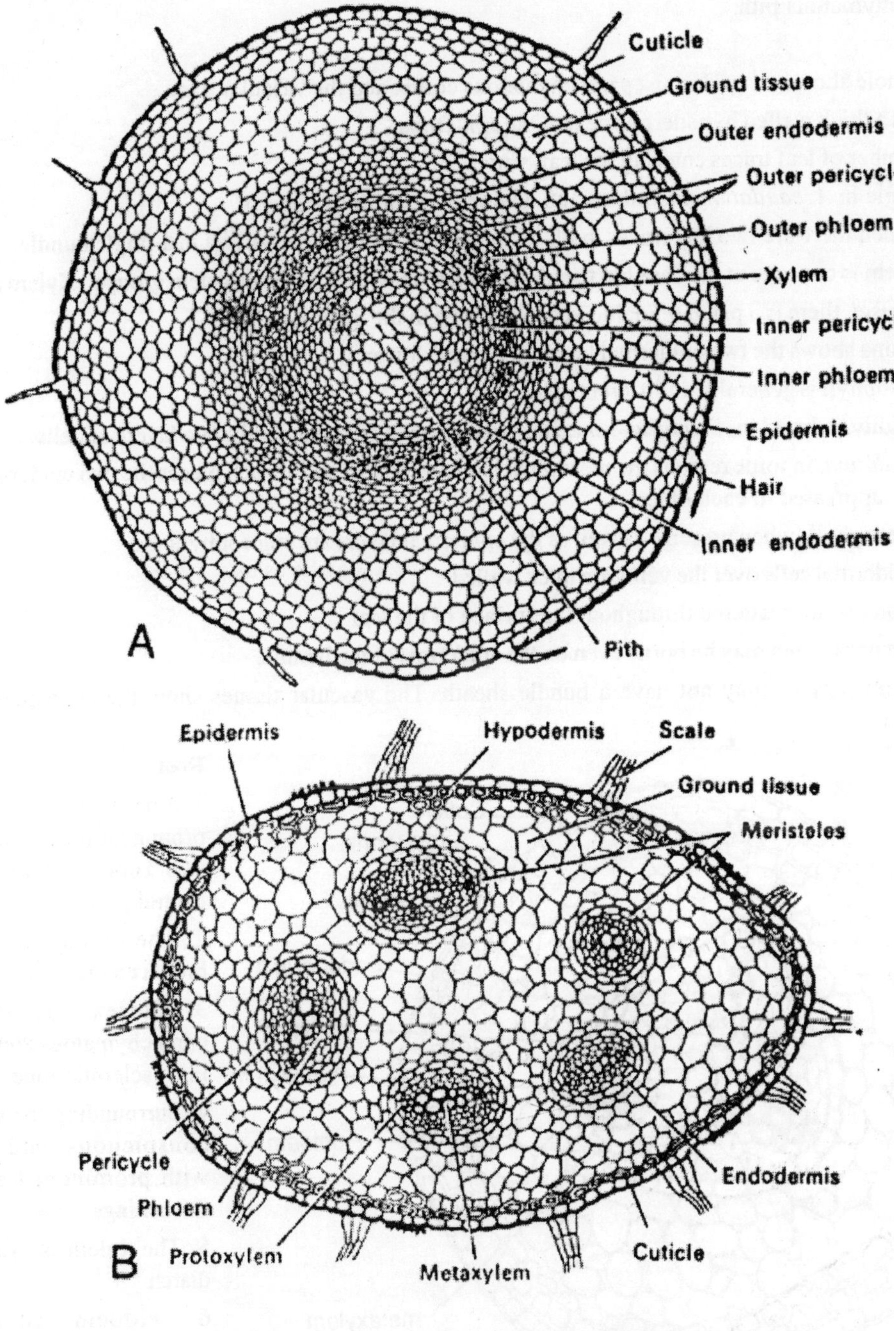

Fig. 10.24. *Adiantum* : Anatomy of Rhizome
A. Amphiphloic Siphonostele in *A. rubellum*, B. Dictyostele in *A. capillus veneris*

11. *In A. rubellum* the stele is a typical amphipbolic solenostele, with the characteristic features such as outer endodermis, outer pericycle, outer phloem, xylem, inner phloem, inner pericycle and inner endodemns lining the parenchymatous pith.

Leaf

1. The petiole shows an epidermis, parenchymatous cortex and the vascular trace.
2. There is a thick walled hypodermis next to the epidermis.
3. The number of leaf traces entering the leaf, varies.
4. It is single in *A. caudatum* and others and double in *A. capillus-veneris.*
5. Even when there are two leaf traces, both of them unite further up resulting in a single bundle.
6. The xylem is concave at the base but triradiate higher up with three protoxylem groups. Xylem is exarch.
7. In *A. bausei,* there is a patch of included parenchyma in the xylem.
8. The lamina shows the two epidermal layers upper and lower.
9. The mesophyll is generally undifferentiated.
10. It is highly reduced in *A.capillus-veneris, A. pedatum,* etc., having only two layers of cells.
11. In *A.pedatum,* in some regions the mesophyll is totally absent and at such places the two epidermal layers are closely appressed to each other.
12. The mesophyll (when present) as well as the epidermal layers are chlorophyllous.
13. The epidermal cells over the veins are thick walled.
14. The stomata are scattered throughout the surface of the leaf.
15. Palaeae or ramenta may be borne even on the epidermis of the lamina.
16. The vein may or may not have a bundle sheath. The vascular tissues show the characteristic X/P arrangement.

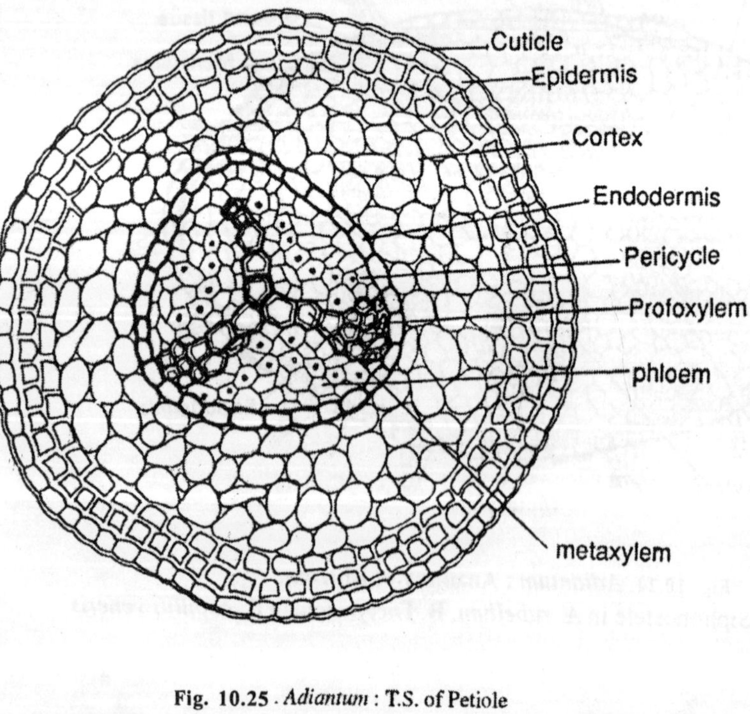

Fig. 10.25 . *Adiantum* : T.S. of Petiole

Root

1. A transection shows a very prominent piliferous layer, a two zoned cortex and the central protostele.
2. The piliferous layer has brown colored cell walls.
3. Cortex has an outer parenchymatous zone and an inner sclerotic zone.
4. Surrounding the stele is a conspicuous endodermis with prominent Casparian thickenings.
5. The xylem is exarch and diarch.
6. Phloem completely surrounds the xylem.
7. External to phloem is a single layered pericycle.

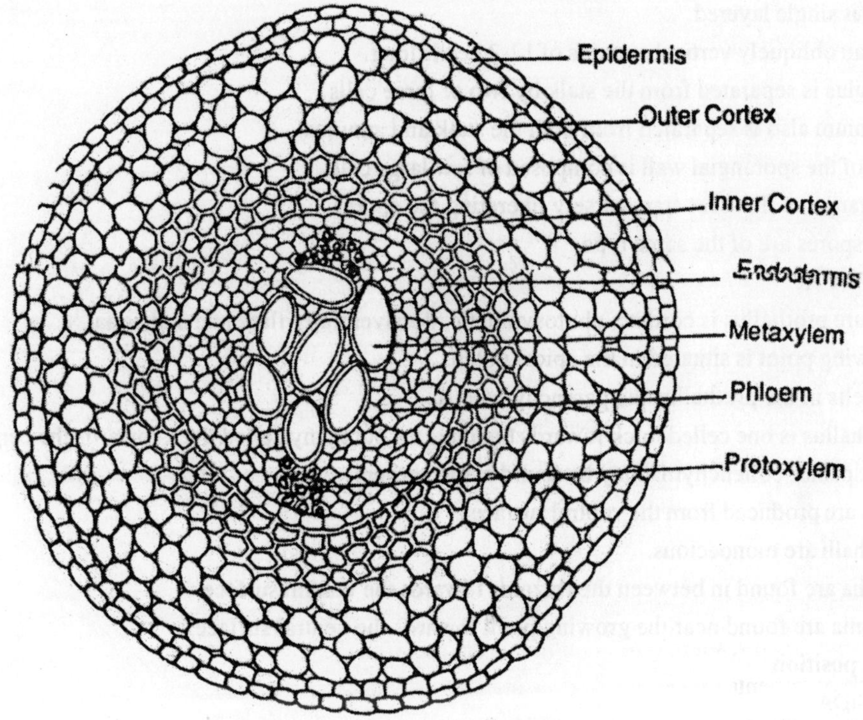

Fig. 10.26. *Adiantum* : T.S. of Root

REPRODUCTION

1. Vegetative propagation is brought about by buds produced at the leaf tips.
2. The buds enter the ground when the leaf bends and touches the soil.
3. There, they develop into a new individual.
4. This, in turn may repeat the process leading to the "Walking habit".
5. Walking habit is seen in *A. caudatum.*

Spore producing organs

1. As has already been said there is no distinction into fertile and sterile leaves.
2. The sori are borne on the distal end of the pinnae.
3. But the sori are not exactly marginal.
4. They are borne a little behind the tip of the veins.
5. The sorus bearing margin of the leaf incurls and forms the false inducium.
6. In some cases sporangia may develop at the distal ends of the veins *(A. philippense).*
7. In the sori paraphyses may be present in between the sporangia as in *A. mbellum, A. tenerum,* etc.
8. The sorus is of the mixed type.

Development and structure of the sporangium

1. The development is the similar to what is seen in *Pteris.*
2. A mature sporangium has a stalk made up of three rows of cells.

3. The stalk terminates in a globose or biconvex capsule.
4. The wall is single layered.
5. There is an obliquely vertical annulus of 12-24 cells long.
6. The annulus is separated from the stalk by two or three cells.
7. The stomium also is separated from both the stalk and annulus.
8. The rest of the sporangial wall is composed of few large cells.
9. The sporangium dehisces transversely liberating the spores.
10. All the spores are of the same type.

GAMETOPHYTE

1. The mature prothallus is cordate, photosynthetic, dorsiventrally flattened and aerial.
2. The growing point is situated in the apical notch.
3. All the cells in the prothallus the parenchymatous.
4. The prothallus is one celled thick towards the margins but many celled thick towards the centre.
5. In some species collenchyma may be found at the corners.
6. Rhizoids are produced from the ventral surface.
7. The prothalli are monoecious.
8. Antheridia are found in between the rhizoids towards the ventral surface.
9. Archegonia are found near the growing point towards the ventral surface.

Systematic position

PTERROPSIDA
LEPTOSPORANGIATAE
FILICALES - MIXTAE
ADIANTACEAE
ADIANTUM

PTERIS

Occurrence and distribution

1. This is a cosmopolitan fern being distributed in almost all geographical regions.
2. *Pteris* however prefers tropical and sub tropical climates.
3. Plants usually grow in well drained places or in the crevices of rocks.
4. They are very common along the slopes of hills and can be seen even at 1200 metres above sea level.
5. There are about 250-280 species reported for the genus.
6. Some of the common Indian species are *P. quadriaurita, P. critica, P. vittata, P. pellucida, P. wallichiana, P. stenophylla, P. biaurita,* etc.

SPOROPHYTE

Morphology of the plant

1. The plant body has a rhizomatous stem that produces roots and leaves.
2. The rhizome may be creeping *(P.grandiflora)* or compact and erect as in *P. critica* and *P. longifolia.*
3. The rhizome may or may not show branching and is covered with scales.
4. Roots arise at the base of the leaf or all over the rhizome.
5. The growing point of the rhizome is covered with ramenta.

Fig. 10.27. *Pteris* : Habit

6. The leaves arise from the upper surface of the rhizome and have a long rachis.

7. They are unipinnately compound, decompound or multipinnately compound.

8. The dissections of the pinnae are not as deep as in *Pteridium.* Venation if of open dichotomous type.

9. The pinnae are small near the base, large towards the middle and once again small towards the apex as in *P.vittata.*

10. The leaves show circinate vernation that is typical to true ferns.

11. Very often pinnae are coriaceous (leathery).

12. The leaves bear the reproductive structures - sori along the ventral margin of the pinnae.

13. The sorus.is continuous along the margin but avoiding the apices of the segments and usually in the sinuses between them.

INTERNAL STRUCTURE

Rhyzome

1. The rhizome in a transection exhibits quite a good amount of diversity particularly in the stelar organisation.
2. It is solenostelic *(P.grandiflora, P. vittata),* or Dictyostelic.
3. In *P. biaurita* the rhizome shows a mixed protostele in the lower region.
4. It becomes siphonostelic a little higher up and near the apex it shows a polycyclic dictyostelic organisation.
5. The epidermis is single layered and it encloses a broad cortex.
6. There is a hypodermal sclerenchyma.
7. The bulk of the cortex however is made up of parenchyma.
8. The stele consists of a number of meristeles.
9. Usually there are two rings of meristeles.
10. The inner ring consist of 2 or 3 large meristeles whereas the outer ring consists of a number of small meristeles.
11. Each meristele has a band or plate like xylem mass.
12. Sometimes it is angular.
13. One or two protoxylem groups are embedded in the metaxylem (mesarch).
14. Surrounding the xylem is the phloem.
15. Each meristele may have its own endodermis.

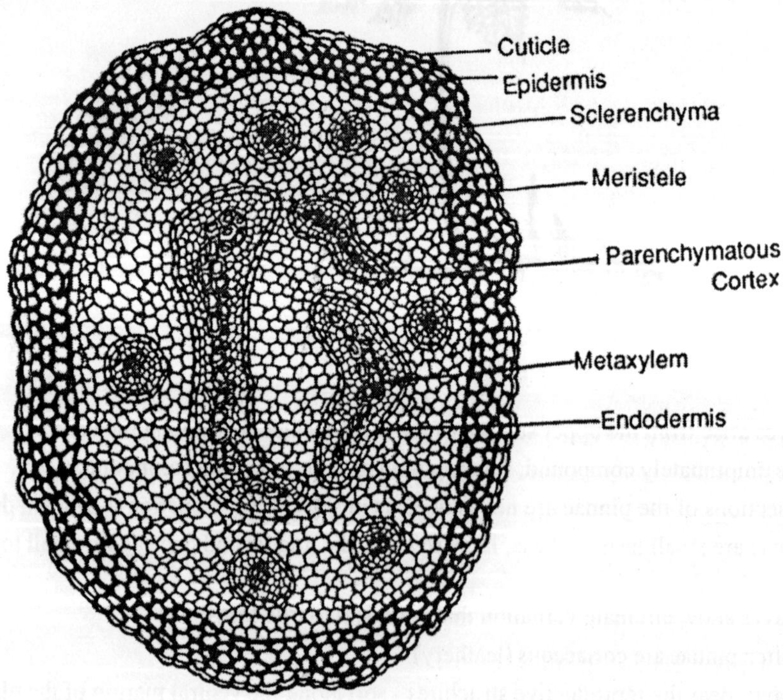

Fig. 10.28. *Pteris* **: Anatomy of Rhizome**

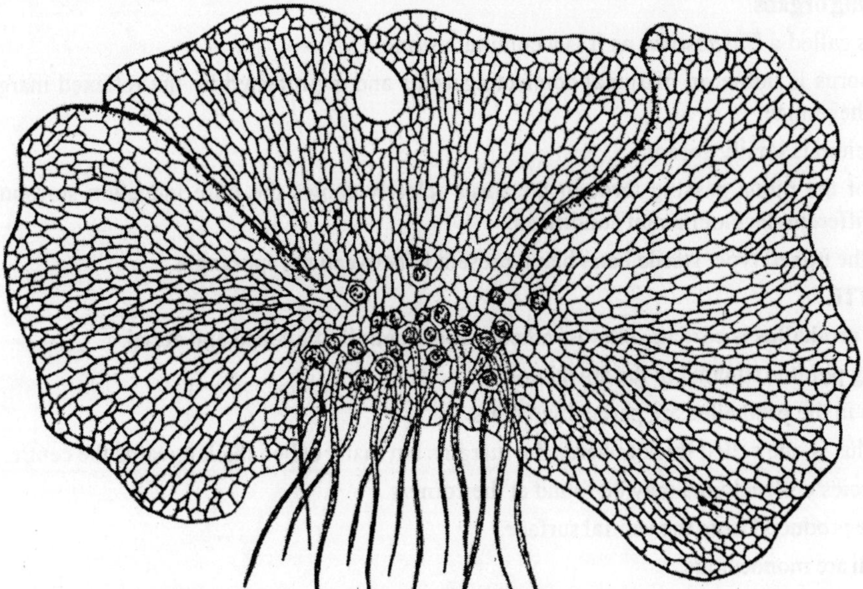

Fig. 10.29. *Pteris* : Gametophyte

Leaf

1. The rachis is. traversed by a single leaf trace which is variable in shape. 'C' shaped leaf traces are found in *P. vittata.*

2. In *P. biaurita* the leaf trace is 'U"shaped while entering the leaf base, but further up it becomes 'V shaped.

3. The xylem strand appears hooked.

4. As usual xylem is surrounded by phloem, pericycle and endodermis.

5. The cortical region has an inner parenchymatous zone and an outer sclerotic zone.

6. Epidermis is single layered with a deposition of cuticle over it.

7. Ramenta arise from some of the epidermal cells.

8. The lamina appears bifacial.

9. It may or may not have a differentiated mesophyll.

10. Leaves are generally hypostomatic.

11. The midrib has a concentric vascular bundle *(P. vittata)* with a distinct endodermis (bundle sheath).

Root

1. Anatomically the root shows an outer piliferous layer and a cortex divided into an outer parenchymatous zone and an inner sclerotic zone.

2. The stele is protostelic and has an exarch, diarch xylem.

REPRODUCTION

This is brought about by spores.

Spore producing organs

1. The sorus is called a *Coenosorus* as it is a continuous one.

2. The coenosorus is borne on marginal connecting veins and is protected by the reflexed margin (false inducium) of the pinnae.

3. A true inducium is totally absent.

4. A section of the pinna passing through the sorus clearly shows the false inducium surrounding the sporangia in different developmental conditions.

5. Sorus is of the mixed type; sterile hairs being present in between the sporangia.

GAMETOPHYTE

1. The mature prothallus is cordate, photosynthetic, dorsiventrally flattened and aerial.

2. The growing point is situated in the apical notch.

3. All the cells in the prothallus are parenchymatous.

4. The prothallus is one celled thick towards the margins but many celled thick towards the centre.

5. In some species collenchyma may be found at the corners.

6. Rhizoids are produced from the ventral surface.

7. The prothalli are monoecious.

8. Antheridia are found in between the rhizoids towards the ventral surface.

9. Archegonia are found near the growing point towards the ventral surface.

Systematic position

PTERROPSIDA

LEPTOSPORANGIATAE

FILICALES - MIXTAE

DENNSTAEDTIACEAE

PTERIDOIDEAE

PTERIS

MARSILLEALES

1. These are heterosporous, leptosprangiate, aquatic or semiaquatic ferns.

2. Spore bearing organs are encased in a special bean shaped structure called the sporocarp.

3. The order comprises a single family Marseliaceae with three genera *viz., Marsilea, Pilularia* and *Regnellidint. Marsilea* is discussed here.

MARSILEA

Occurrence and distribution

1. It is one of the most widely distributed among ferns.

2. But the genus has a preference to the warmer parts of the world.

3. The plants grow in aquatic *(M. quadrifolia)* or semiterrestrial *(M. condensata)* habitats.

4. A few species are known to grow in dry soil.

5. The genus consists of about 53 species (Reed, 1954; Gupta 1962) of which about nine are recorded from India, some of the common species are *M. quadrifolia, M. hirsuta, M minuta, M. vestita, M. aegyptiaca* etc.

SPOROPHYTE

Morphology of tbe plant

1. The plant body consists of a creeping cylindrical rhizome which is partly or completely subterranean.

2. The rhizome has in definite length and under luxuriant growth conditions may reach a length of 25 metres.
3. Branching in the rhizome is quite frequent and it may be axillary or lateral.
4. The rhizome is distinguished into nodes and internodes.
5. The nodal regions are slightly swollen.
6. The internodes are longer in aquatic species *(M. quadrifolia)* and shorter in terrestrial species.
7. The leaves and roots are generally borne at the nodes.
8. The leaves have a long and slender petiole.
9. In aquatic species only the lamina floats on the surface of water.
10. The leaves are arranged alternately in two rows on the upper surface of the rhizome.
11. The leaves are compound and have a varying number of pinnae.
12. The number of pinnae per leaf may be 3, 5, 6, 7 or 8. Frequently it is four as see in n *M. quadrifolia*.
13. The leaves show circinate vernation.

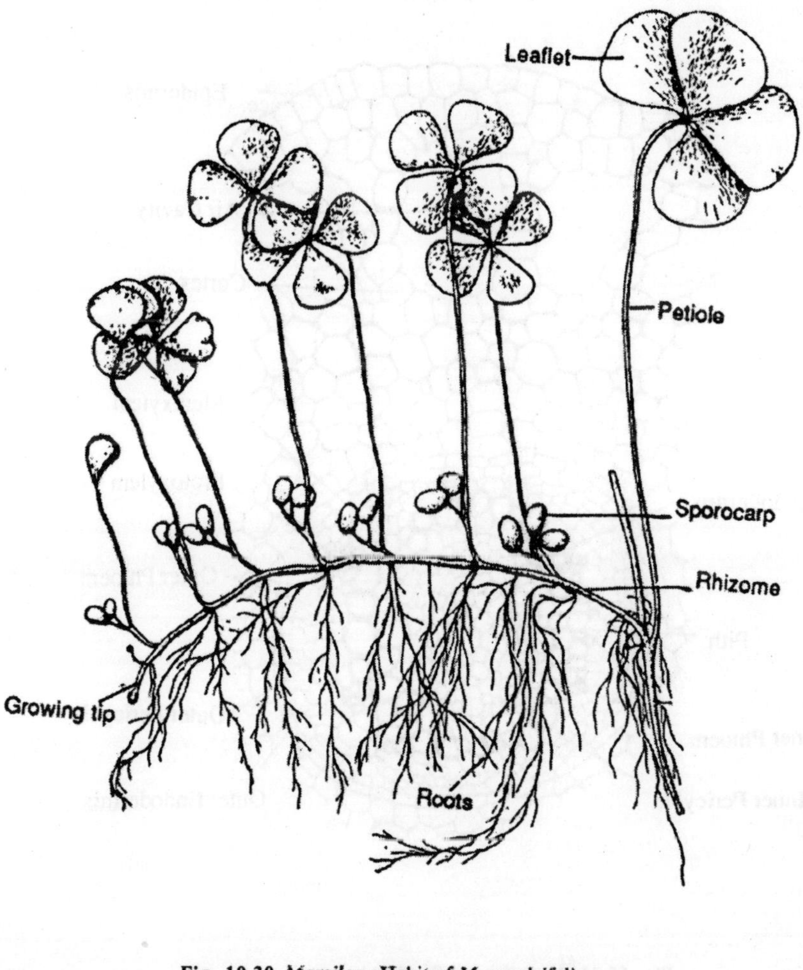

Fig. 10.30. ***Marsilea*** **: Habit of** ***M. quadrifolia***

14. The pinnae vary in their shape.
15. They may be obovate, obcuneate or elliptical.
16. They may or may not be lobed.
17. The margin of the pinnae may be smooth or dentate (*M. minuta*).
18. The venation is of the closed reticulate type.
19. The pinnae exhibit sleeping movements.
20. Roots are borne on the undersurface of the rhizome at the nodal region.
21. But it is not uncommon to find roots emerging from the internodal region.
22. Roots develop in acropetalous succession (youngest near the growing tip of the rhizome).
23. The plant bears reproductive structures which are very special to Marsileales.
24. These are called the *sporocarps*.
25. The sporocarps have a stalk called the peduncle or the pedicel.
26. The peduncle may be attached to the stem or the petiole.
27. The sporocarps are bean shaped and bisporangiate.

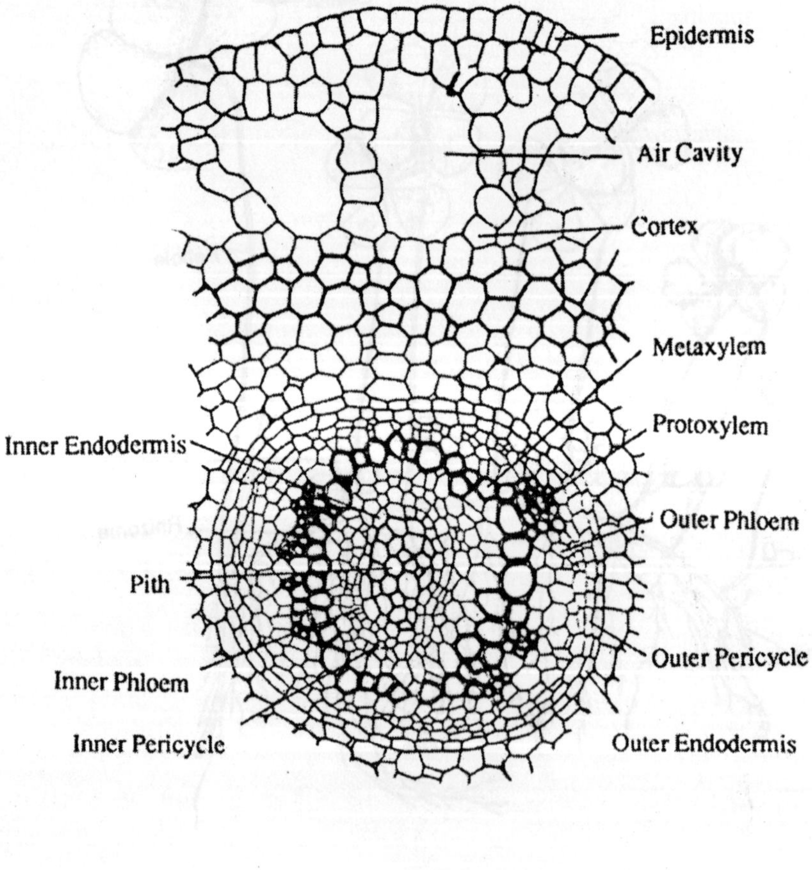

Fig. 10.31. *Marsilea* : T.S. of Rhizome

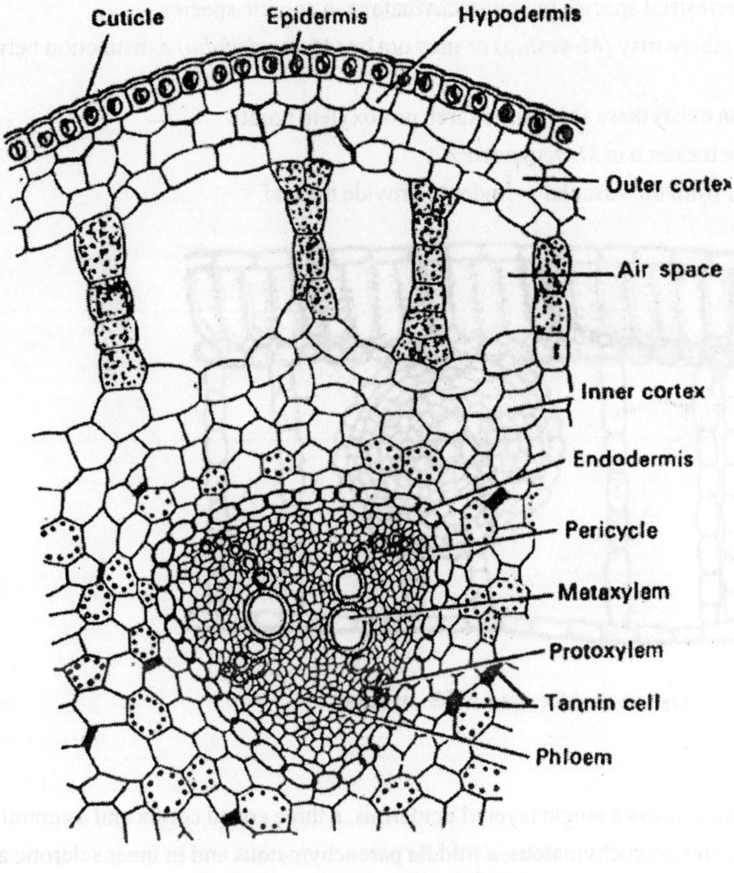

Fig. 10.32 . *Marsilea* : T.S. of Petiole

INTERNAL STRUCTURE

Rhizome

1. A transection shows the usual three regions *viz.*, epidermis, cortex and central stele.
2. Epidermis is single layered, continuous and lacks stomata.
3. The cortex has many regions.
4. The outermost cortex consists of compactly arranged parenchyma cells.
5. Next to this is the aerenchymatous cortex.
6. This consists of broad airspaces separated by parenchymatous portions.
7. Internal to the arenchyma is the sclerotic zone having one or two layers of thick walled cells.
8. The innermost region of the cortex is again parenchymatous.
9. In this region may be found a few tannin cells.
10. The central vasculature is an *amphiphloic solenostele.*
11. As is typical to such steles there is an outer endodermis, outer pericycle, outer phloem, xylem, inner phloem, inner pericycle and inner endodermis.

12. The composition of the pith varies with the habitat.
13. It is scelrotic in terrestrial species but parenchymatous in aquatic species.
14. In the xylem ring, there may *(M. vestita)* or may not be (*M. quadrifolia)* a distinction between proto and metaxylem.
15. When a distinction exists there are many exarch protoxylem points.
16. The xylem may be mesarch in *M. aegypticaca*.
17. Leaf traces depart from the vascular cylinder to provide the leaf.

Fig. 10.33. *Marsilea* : T.S. of Leaf

Root

1. Anatomically the root shows a single layered epidermis, a three zoned cortex and a central protostele.
2. The cortex has an outer aerenchymatous, a middle parenchymatous and in inner sclerotic zone.
3. Stele is bounded by an endodermis and pericycle.
4. Xylem consists of two large metaxylem tracheids and two protoxylem. tracheids.
5. It is exarch and diarch.
6. Vessels have been reported in *M. quadrifolia, M. drummnndii* and *M. hirstlta*.

Leaf

1. The petiole has an outermost epidermis.
2. The cortex has two zones *viz.,* an outer aerenchymatous and an inner parenchymatous zone.
3. The inner zone is rich in starch and tannin.
4. The vasculature consists of a 'V' shaped xylem with the two arms slightly bent outwards.
5. Some times the two arms may be separated at the base also.
6. The xylem is surrounded by phloem, pericycle and endodermis.
7. The lamina is bifacial.
8. It has an upper and lower epidermis bounding the mesophyll.
9. Leaves may be epistomatic (aquatic) or amphistomatic (terrestrial).
10. The mesophyll shows clearly defined palisade and spongy parenchyma.
11. Vascular bundles are concentric with the centrally situated xylem surrounded by phloem.

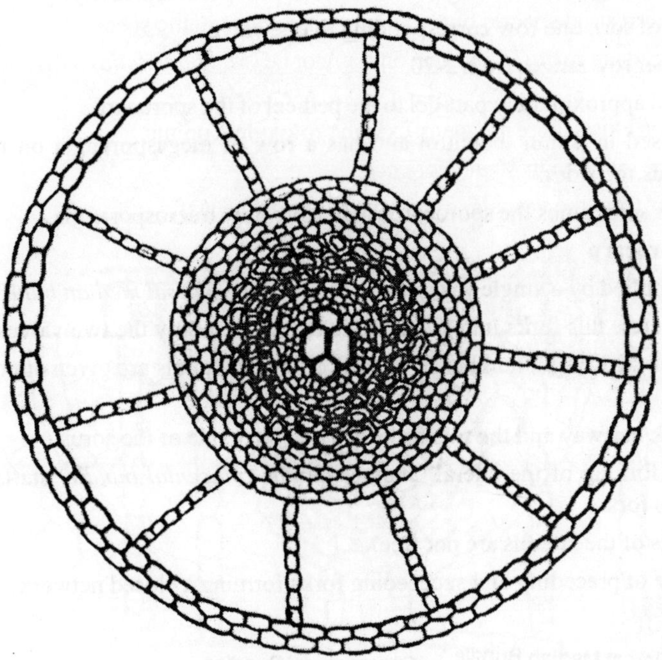

Fig. 10.34. Marsilea: T.S. of Root

REPRODUCTION

1. The sporophyte reproduces by vegetative propagation as well as by spore production.
2. Vegetative propagation is brought about *by tubers (M. hirsuta).*
3. These are small, condensed side branches having leaf primordia.
4. The tubers persist even after a plant dies during unfavourable conditions and on the availability of favourable conditions, develop into a new plant.

Spore producing organs

1. As has already been pointed out, the spore bearing structures, are called sporocarps.
2. Each sporocarp has a short stalk - the pedicel
3. The mode of attachment of the pedicel to the plant is varied.
4. In *M.polycarpa* and *M. subangulata* the pedicels are directly attached to the petiole in a linear sequence.
5. In *M. quadrifolia* the pedicels of many sporocarps join together and have a common stalk joined to the petiole.
6. In *M. vestita* a single pedicel is attached to the base of the petiole.
7. The sporocarp is bisporangiate and is usually light green in colour.
8. At the point of attachment of the pedicel to the body of the sporocarp there are one to two teeth like projections called tubercles.

Structure of the mature sporocarp

1. The sporocarp is made up of two valves which can be split open along the dorsiventral plane so as to reveal the internal structure.
2. The wall of the sporocarp is very thick.

3. Adhering to the wall is a *gelatinous* ring that extends in the sporocarp cavity.

4. There are two rows of sori, one row corresponding to one valve.

5. The number of sori per row ranges from 2-20.

6. The sori are arranged approximately parallel to the pedicel of the sporocarp.

7. Each sorus is encased in a thin indicium and has a row of megasporangia on top and two rows of microsporangia towards the sides.

8. In *M. rajasthanensis* sometimes the sporocarps may have only microsporangia.

Vasculature in the sporocarp

1. The sporocarp is supplied by a single vascular trace called the *Dorsal median bundle* (DMB).

2. At the region of tubercle this forks into two and the branches supply the two valves.

3. From each branch of the DMB a number of laterals or commissurals are given out one corresponding to each sorus.

4. The laterals also fork halfway and the two forks run on either side of the sorus.

5. From the point of forking of the lateral, another bundle *(placental bundle)* starts and enters into the receptacle where it also forks.

6. The ends of the forks of the laterals are not free.

7. They fuse with those of preceding and succeeding forks forming a closed network.

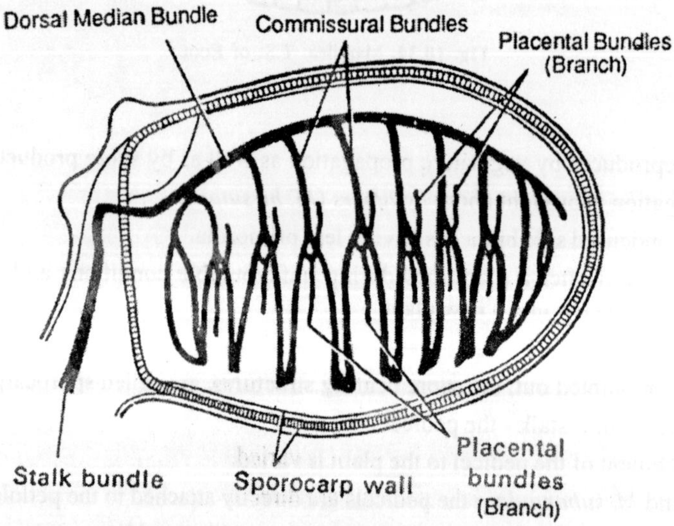

Fig. 10.35. *Marsilea* : Vascular Supply of the Sporocarp

Detailed internal structure of the sporocarp

1 Sections cut at different planes clearly reveal the internal structure.

2. The following are the three planes of section:

a) Vertical longitudinal section or section cut parallel to the flat surface of the sporocarp.

b) Horizontal section or section perpendicular to the pedicel.

c) Vertical transverse section or section perpendicular to the flat surface of the sporocarp.

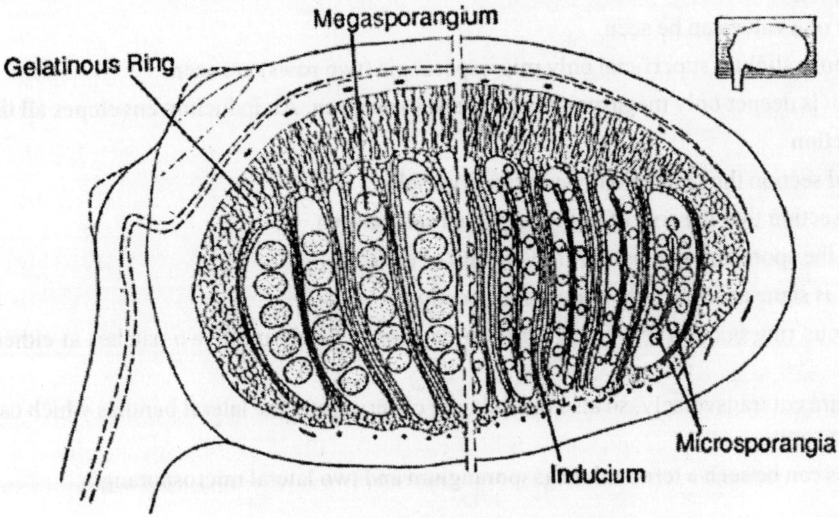

Fig. 10.36. ***Marsilea*** **: Internal Structure of the Sporocarp (vertical long section parallel to the surface)**

Vertical longitudinal section

1. The section shows a wall made up of three layers of cells.
2. The outermost is epidermis which has stomata and two hypodermal layers.
3. The hypodermal cells are columnar.
4. Next to the wall is the continuous gelatinous ring.

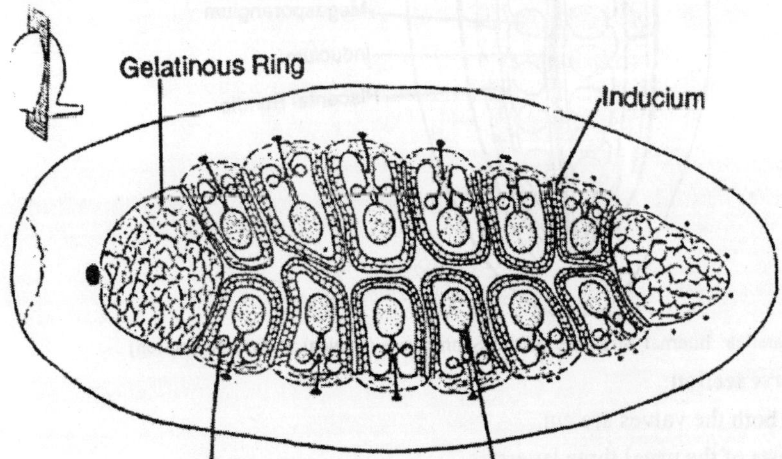

Fig. 10.37. ***Marsilea*** **: Internal Structure of the Sporocarp (Horizontal Section)**

5. Sori are cut lengthwise.
6. Sori of only one valve can be seen.
7. If the section is slightly superficial only microsporangia (two rows) are seen.
8. If the section is deeper only megasporangia (one row) are seen. An inducium envelopes all the sori.

Horizontal section

1. In horizontal section the pedicel will be cut transversely.
2. It is in this section that the sori of both the valves can be seen.
3. The wall of the sporocarp consists of three layers of cells.
4. Its structure is same as in vertical longitudinal section.
5. The gelatinous ring is cut transversely and can be seen in the form of two patches at either end of the section.
6. All the sori are cut transversely, so also the inducial covering and the lateral bundles which can be seen at the base of the sorus.
7. In each sorus can be seen a terminal megasporangium and two lateral microsporangia.

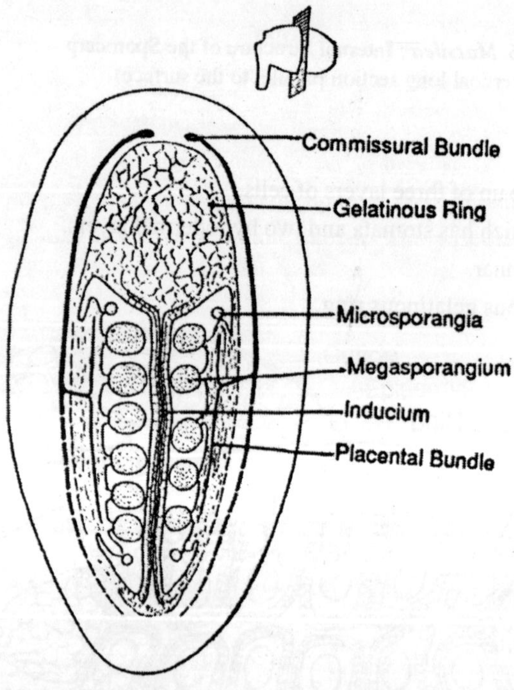

Fig. 10.38. Marsilea: Internal Structure of the Sporocorup (vertical transverse section)

Vertical transverse section

1. In this section both the valves are cut.
2. The wall consists of the usual three layers.
3. The gelatinous ring is cut transversely and is present in the form of two patches at either end of the section.
4. Sori are cut lengthwise and each sorus represents a valve.
5. Inducia are attached to this.
6. A slightly superficial section shows only microsporangia and a deeper section shows a row of megasporangia with one or two microsporangia at either end of the sours.

Systematic position
PTERROPSIDA
LEPTOSPORANGIATAE
MARSILEALES
MARSILEACEAE
MARSILEA

SALVINIALES

1. The order Salviniales belongs to sub class Leptosporangiate.
2. The order is closely related to Marsileales.
3. The members of the order are commonly referred to as 'water ferns'.
4. The orderis similar to Marsileales in having the reproductive structures encased within a bean shaped sporocarp.
5. The sporocarp in Salviniales is fundamentally different from that of Marsileales in that, in the latter the sporocarp represents a single sorus.
6. The order includes two families namely Salviniaceae and Azollaceae.
7. The family Salviniaceae has a single genus *Salvinia* with about 12 species and family Azollaceae includes one genus *Azolla* with 6 species.

SALVINIA

Occurrence And Distribution
1. The genus *Salvinia* has 12 species most of which are found to be distributed in the African continent.
2. The genus is represented in India by two species namely *S.natans* and *S.auriculata*. *S.natans* is found in Kashmir.

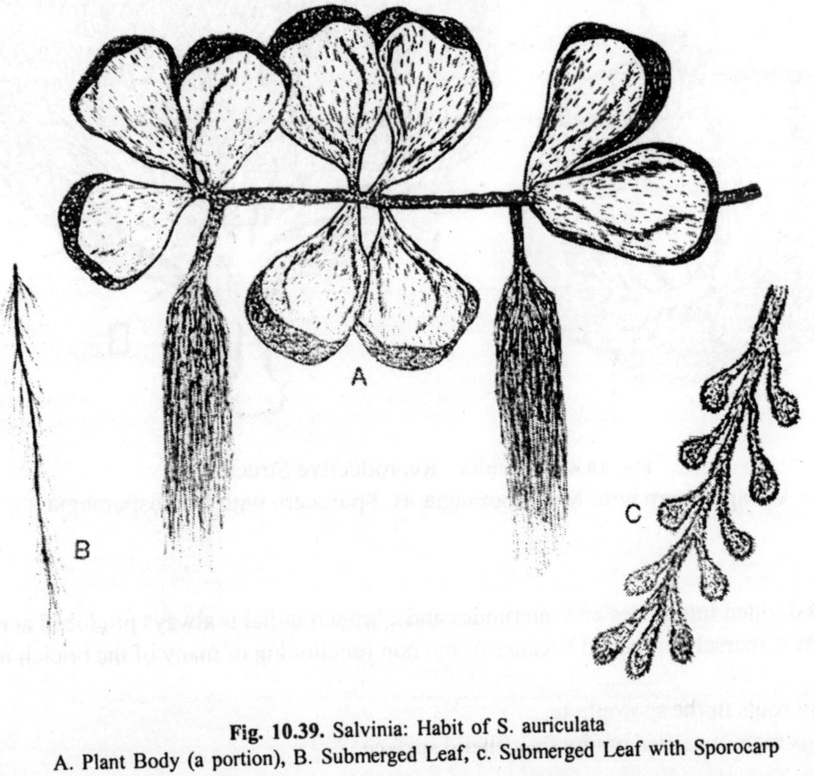

Fig. 10.39. Salvinia: Habit of S. auriculata
A. Plant Body (a portion), B. Submerged Leaf, c. Submerged Leaf with Sporocarp

3. Of the several specied of *Salvinia*, *S.natans* is the only annual.

SPOROPHYTE

Morphology of the Plant

1. The sporophytic plantbody consists of a long slender floating rhizome which grows horizonatally.
2. The rhizome is completely covered by the whorls of leaves.

Leaves

1. Leaves are oblong or sometimes hemispherical.
2. They usually arise in groups of three.
3. Of the three leaves, two lateral ones float on the surface of water while the third one is submerged.
4. There is a distinct morphological differences between the submerged and floating leaves.
5. The submerged leaves are highly dissected, long and filiform.
6. They very much resemble the roots.
7. But is it not certain whether they also function like roots.
8. It is very often suggested that the function of filliform leaves is
   (i) to proctect the sporocarp and
   (ii) to act as a ballast or a stabilizer in preventing undue drift of the plant.
9. The floating leaves are green and ovate or oblong.
10. They are covered withstiff hairs that prevent wetting.
11. Besides the hairs, papillate projections are also found.
12. The members of the leaf whorl alternate at successive nodes, so that there are actually six rows of leaves though the arrangement apparently suggests three rows.

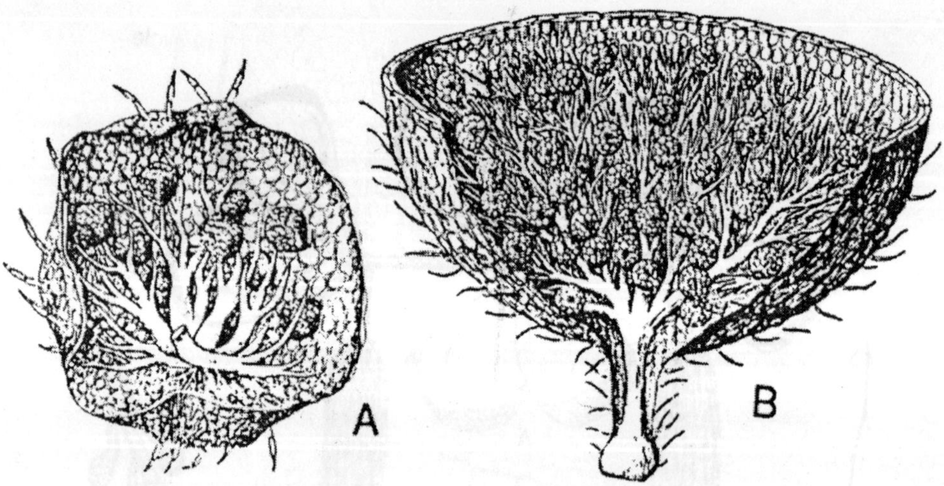

Fig. 10.40. *Salvinia* : Reproductive Structures
A. Sporocarp with Megasporangia, B. Sporocarp with Microsporangia

Stem

1. The stem is divided into nodes and internodes and a branch initial is always produced at every node.
2. But the stem is sparsely branched because of the non functioning of many of the branch initials.

Roots

1. There are no roots in the sporophyte.
2. Absorption is mainly carried out by the general surface.

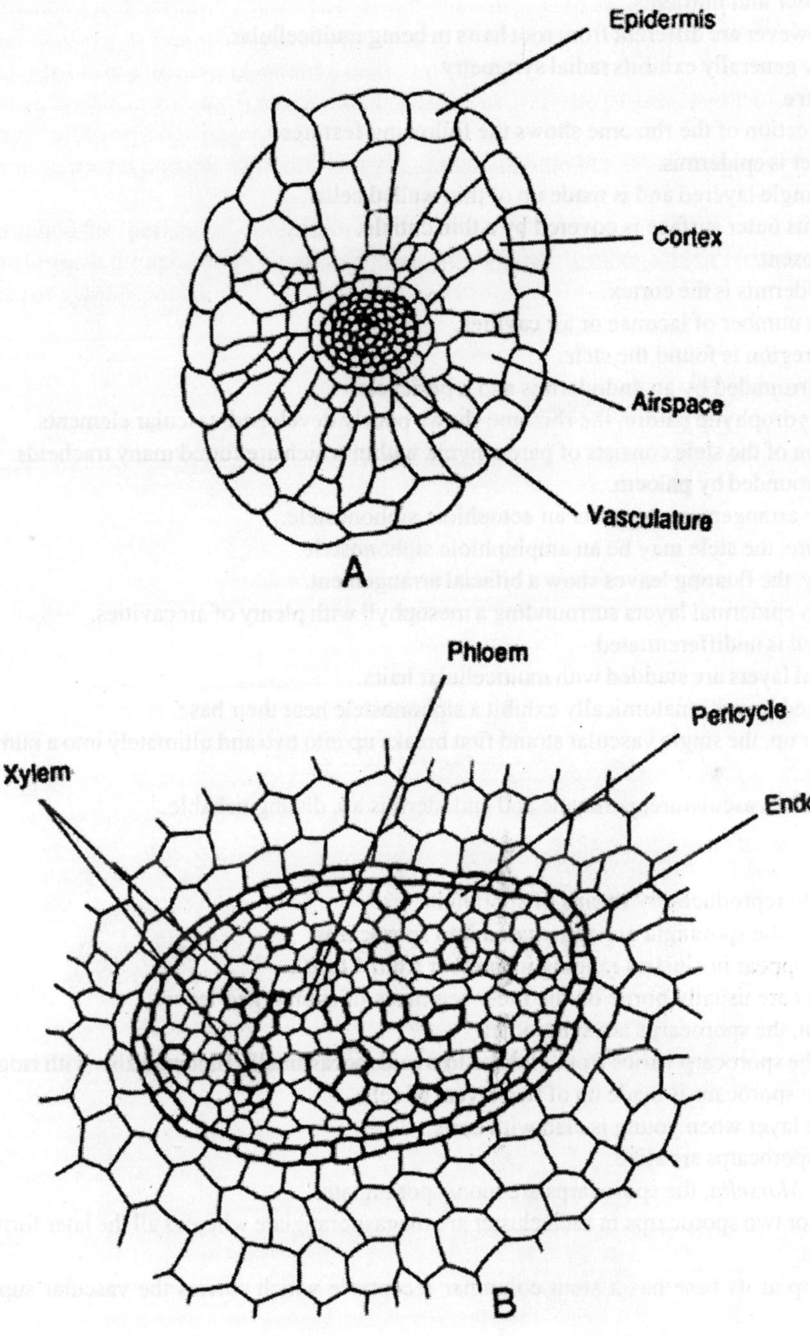

Fig. 10.41. ***Salvinia*** **: Anatomy of Stems**
A. T.S. of Stem (entire), B. T.S. of Stem (Stelar portion enlarged)

3. It is suggested that the hair like structures found on the submerged filiform leaves may help in the absorption of water and nutrients.
4. These hairs however are different from root hairs in being multicellular.
5. The plant body generally exhibits radial symmetry.

Internal Structure

1. A transverse section of the rhizome shows the following features.
2. Outermost layer is epidermis.
3. It is distinct, single layered and is made up of thin walled cells.
4. Epidermis on its outer surface is covered by a thin cuticle.
5. Stomata are absent.
6. Next to the epidermis is the cortex.
7. It consists of a number of lacunae or air cavities.
8. In the central region is found the stele.
9. The stele is surrounded by an endodermis and a pericycle.
10. True to the hydrophytic nature, the rhizome shows poorly developed vascular elements.
11. Central region of the stele consists of parenchyma within which are found many tracheids.
12. Xylem is surrounded by phloem.
13. The vascular arrangement suggests an ectophloic siphonostele.
14. In *S.auriculate*, the stele may be an amphiphloic siphonostele.
15. Anatomically, the floating leaves show a bifacial arrangement.
16. There are two epidermal layers surrounding a mesophyll with plenty of air cavities.
17. The mesophyll is undifferentiated.
18. The epidermal layers are studded with multicellular hairs.
19. The submerged leaves, anatomically exhibit a siphonostele near their base.
20. A little higher up, the single vascular strand first breaks up into two and ultimately into a number of small ectophloic steles.
21. Surrounding the vasculature, pericycle and endodermis are distinguishable.

Reproduction

1. The sporophyte reproduces by means of sporangia.
2. As in *Marselia*, the sporangia are aggregated into sporocarps.
3. The sporcraps appear in clusters ranging in number from 4 to 20.
4. The sporocarps are usually borne on the inner segments of submerged leaves.
5. In arrangement, the sporocarps are sympodial.
6. The shape of the sporocarp varies from globose to ovoid, occasionally flattened also with ridged surfaces.
7. The wall of the sporocarp is made up of two layers of cells.
8. The outermost layer when young is clad with hairs.
9. Externally all sporocarps are alike.
10. But unlike in *Marselia*, the sporocarps are monosporangiate.
11. The first one or two sporocarps in each cluster are megasporangiate whereas all the later formed ones are microsporangiate.
12. The sporocarp at its base has a stout columnar receptacle which carries the vascular supply into the sporocarp.
13. The receptacle is unbranched in the megasporangiate sporocarps of *S.natans*.

Structure of a Mature Sporocarp

1. A sectional view of mature sporocarp shows a well developed wall enclosing the, sporangia.
2. In a megasporangiate sporocarp, only about 25 megasporangia are seen whereas in a microsporangiate sporocarp, the number is much larger because of the branching of the receptacle.

Structure of Megasporangium

1. The megasporangium consists of a short stalk and a generally ovoid capsule.
2. The capsule is made up of a single layer of cells.
3. Internal to the wall of capsule is a layer of tapetum.
4. Enclosed by the tapetum are found 8 megasporocytes which divide meiotically to produce 32 megaspores.
5. Of these spores, all except one degenerate.
6. The degenerating spores and the tapetum harden and surround the functional megaspores in the form of a thick layer.
7. This is often called the 'perispore' or 'epispore'.
8. The perispore towards the apex of the sporocarp forms a traingular chamber, which very much simulates the pollen chambers of gymnosperms.

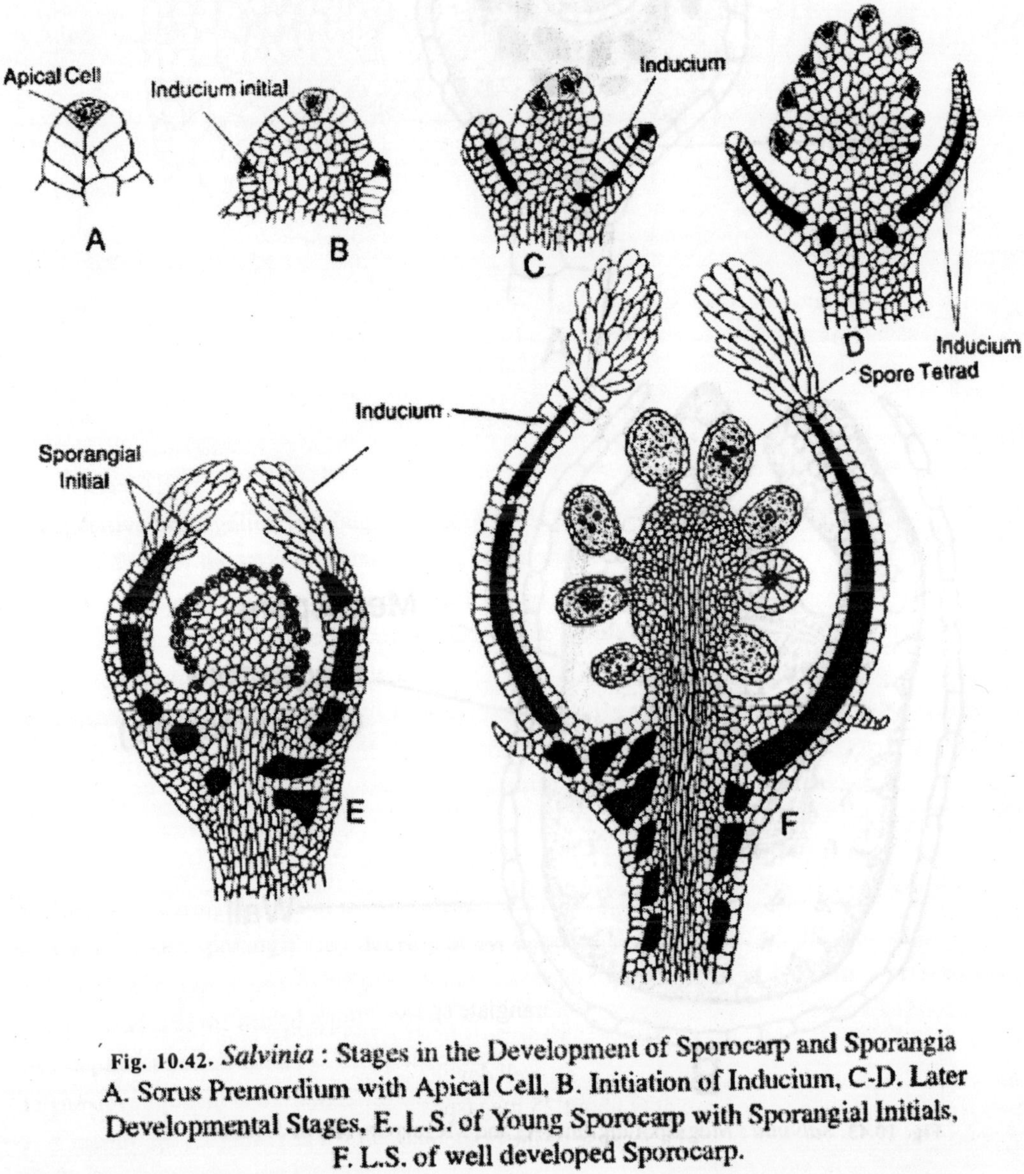

Fig. 10.42. *Salvinia* : Stages in the Development of Sporocarp and Sporangia A. Sorus Premordium with Apical Cell, B. Initiation of Inducium, C-D. Later Developmental Stages, E. L.S. of Young Sporocarp with Sporangial Initials, F. L.S. of well developed Sporocarp.

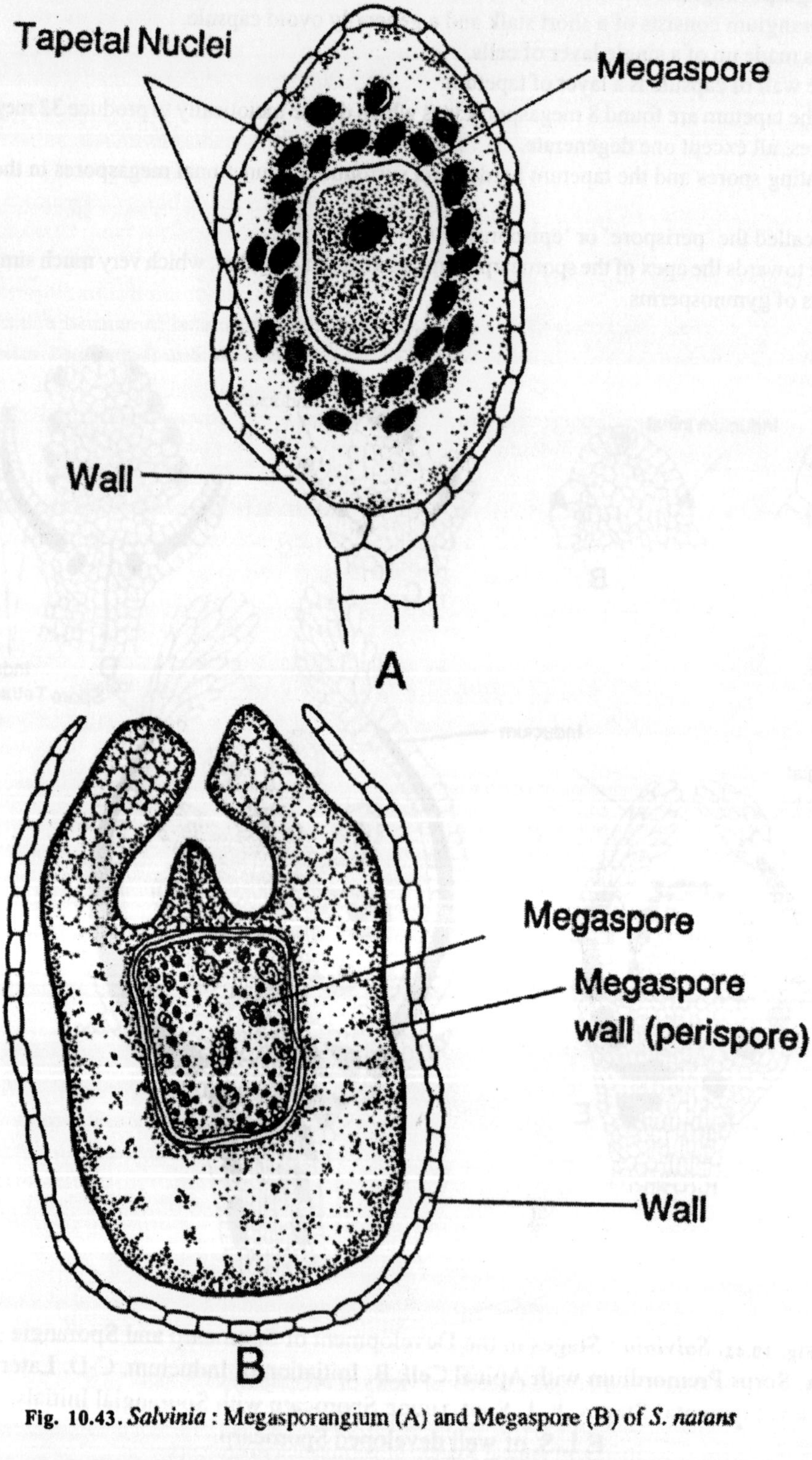

Fig. 10.43. *Salvinia* : Megasporangium (A) and Megaspore (B) of *S. natans*

9. Arising from the floor of the pollen chamber is a central mound of tissue with three flaps.
10. These three flaps get separated during the germination of megaspore.

Structure of Microsporangium

1. The microsporangia are basically similar to megasporangia in their construction.
2. Enclosed in the sporangium are 16 microspore mother cells and consequently 64 microspores are formed.
3. All the spores attain maturity and survive.
4. The microspores are embedded in the cytoplasmic fluid derived from the tapetum.
5. As the spores mature, the tapetal cytoplasm encompassing the spores gets hardened and forms a mass called 'massula'.
6. The microspores are extermely smali, triradiate in appearance and have two wall layers.

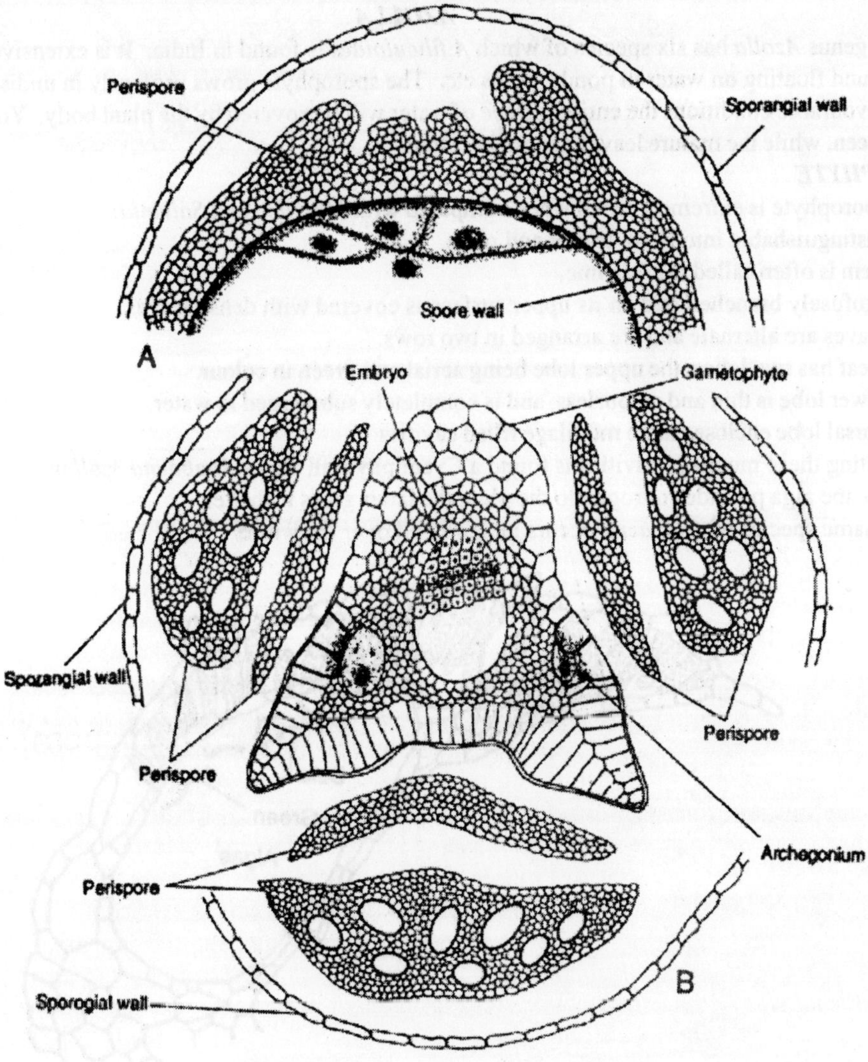

Fig. 10.44. *Salvinia* : Sectional views of Megaspore of *S. natans*
A. Median L.S. through apex of Megaspore,
B. T.S. of Apical Region of germinated Megaspore

7. There is a single nucleus in each microspore.
8. Occasionally the number of microspores per microsporangia might vary.
9. In *S.auriculata*, there are only eight microspore mother cells and consequently only 32 microspores are formed.

Systematic position

PTERROPSIDA

LEPTOSPORANGIATAE

SALVINIALES

SALVINIACEAE

SALVINIA

AZOLLA

The genus *Azolla* has six species of which *A.filiculoides* is found in India. It is extensively distributed and is found floating on water in ponds, pools etc. The sporophyte grows profusely in undistrubed waters. Under favourable conditions the entire surface of water will be covered by the plant body. Young plants are bluish green, while the mature leaves are red in colour.

SPOROPHYTE

1. The sporophyte is extremely small when compared with *Marsilea* and *Salvinia*.
2. It is distinguishable into stem, leaves and roots.
3. The stem is often called the rhizome.
4. It is profusely branched and on its upper surface is covered with dense leaves.
5. The leaves are alternate and are arranged in two rows.
6. Each leaf has two lobes, the upper lobe being aerial and green in colour.
7. The lower lobe is thin and colourless, and is completely submerged in water.
8. The dorsal lobe encloses large mugilage filled cavities.
9. Inhabiting these mucilage cavities is found a Cyanophycean alga - *Anabaena azollae*.
10. While the alga provides nitrogen to the plant the latter gives it shelter.
11. The same species of *Anabaena* occurs in *Azolla* all over the world.

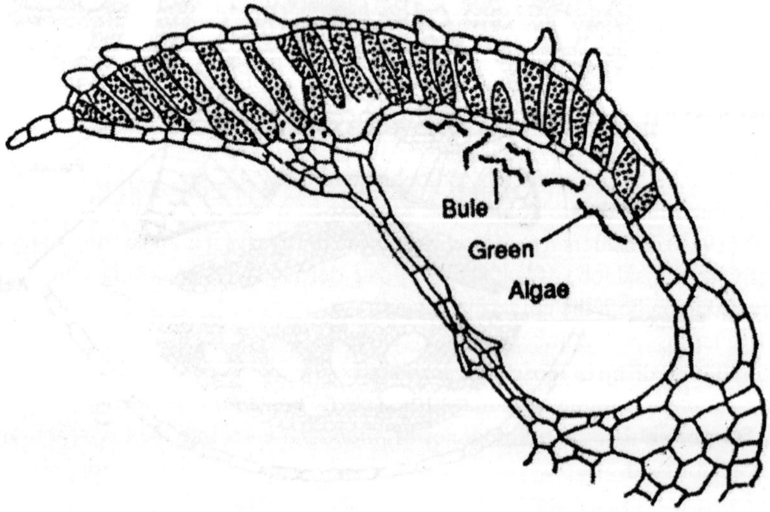

Fig. 10.45. *Azolla* : Anatomy of Leaf of *A. filiculoides* showing endophytic Blue Green Algae

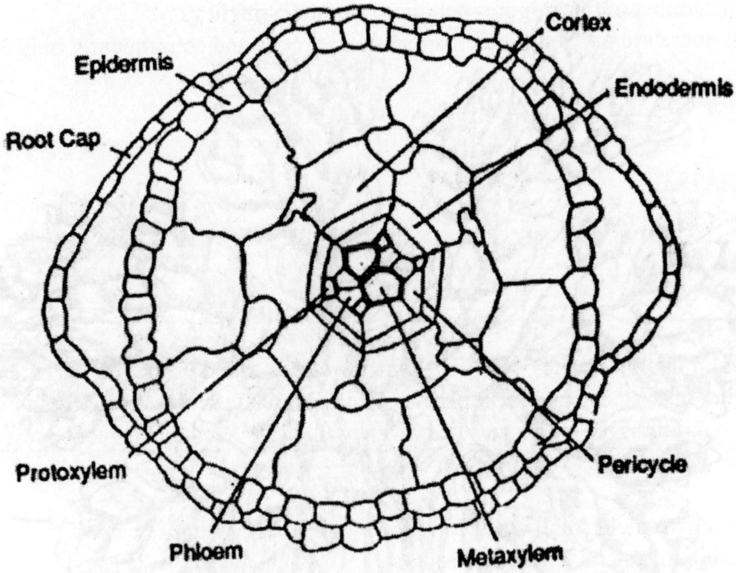

Fig. 10.46. *Azolla* : Anatomy of Root of *A. filiculoides*

12. The rhizome on its lower surface produces simple roots either singly or in clusters.
13. These roots help in stabilising the plants in water.

Anatomy

1. The rhizome, anatomically, resembles other ferns.
2. A cross section shows an epidermis, a middle cortex and a central stele.
3. Epidermis is single layered and it encloses a cortex which is 4-8 cells broad.
4. There are no air cavities in the cortex as in *Salvinia*.
5. The stele is surrounded by an indistinct endodermis.
6. Internal to the endodermis is the pericycle.
7. The vascular elements are greatly reduced.
8. There are a few tracheids surrounded by phloem elemnts.
9. It is difficult to determine the stelar type.
10. While Smith (1955) considers it protostelic, Eames (1936) says it is apparently siphonostelic.
11. Anatomically, the dorsal lobe of leaf shows two epidermal layers enclosing a thin mesophyll.
12. Stomata are found on both the epidermal layers.
13. The upper epidermis has a number of one to two celled layers.
14. The mesophyll is made up of loosely arranged cells.
15. Within the mesophyll is found a mucilage filled cavity containing *Anabaena*.
16. It has been suggested that the hormogones of *Anabaena* enter into the cavity through a small pore and get themselves established.

Reproduction

1. The main method of reproduction in *Azolla* seems to be vegetative fragmentation.
2. The lateral branches get separated and develop into new individuals.

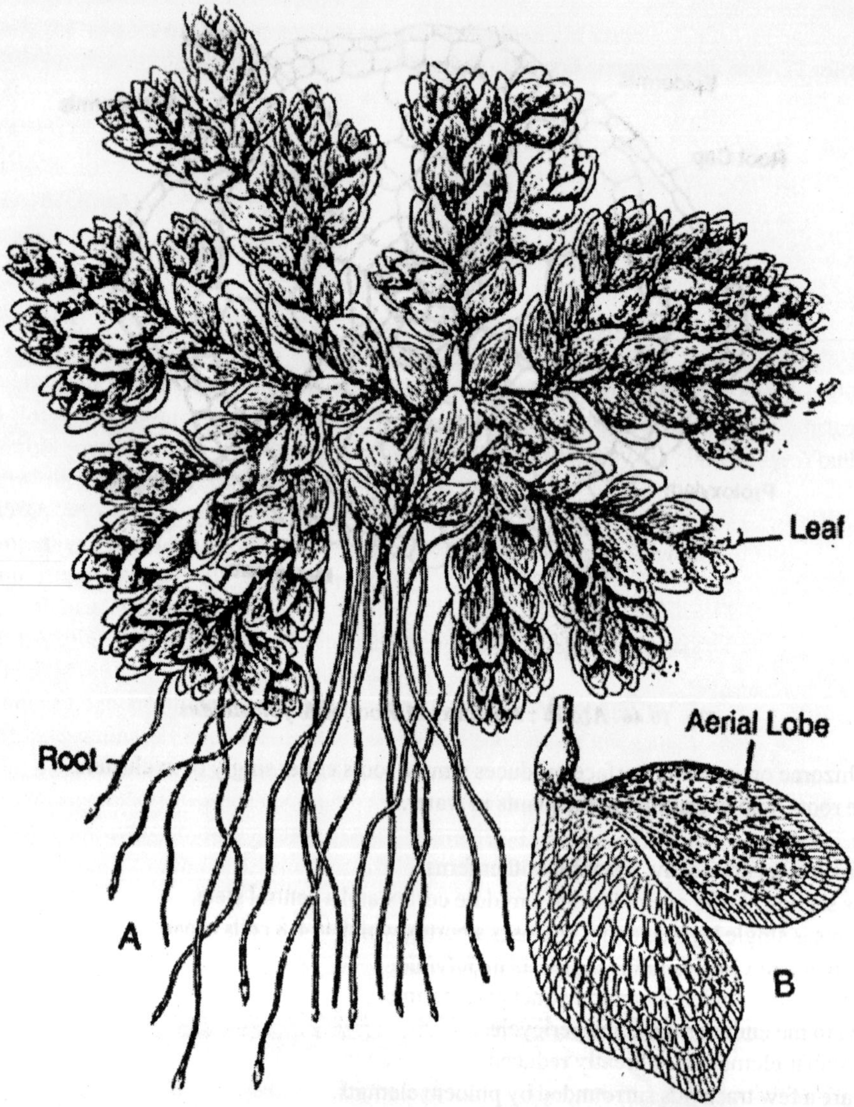

Fig. 10.47. *Azolla* : Habit of *A. microphylla* (A) and a Leaf Enlarged (B)

Spore Production

1. The spores in *Azolla* are produced in sporangia which in turn are enclosed in sporocarps as in *Marsilea* and *Salvinia*.
2. The sporocarps are usually borne on the first leaf of the lateral branch.
3. In fertile leaves, the submerged lobe is usually divided only twice, and on each a sporocarp is produced terminally.
4. The upper lobe of the leaf form a marginal flap covering the sporocarp.
5. The sporocarps are monosporangiate.
6. They have either microsporangia or megasporangia.
7. There is a size difference also between megasporocarps and microsporocarps.

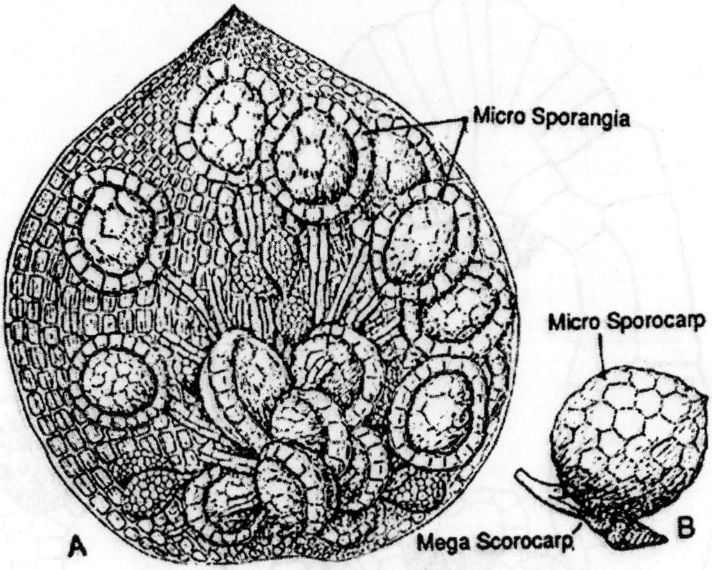

Fig. 10.48. *Azolla* : Sporocarps of *A. microphylla*
A. Sporocarp Split open to show Microsporangia.
B. Leaf Segment with Macro and Microsporocarps

8. The former are small and have only one megasporangium, while the latter are large and have a number of microsporangia.
9. The wall of sporocarp is two layered.
10. In a megasporocarp, the megasporangium arises on a small receptacle at the base.
11. The sporangium is covered by a two layered inducium.
12. In a microsporocarp there is a central cushion like receptacle which gives rise to a number of microsporangia.

Systematic position

PTERROPSIDA

LEPTOSPORANGIATAE

SALVINIALES

AZOLLACEAE (SALVINIACEAE)

AZOLLA

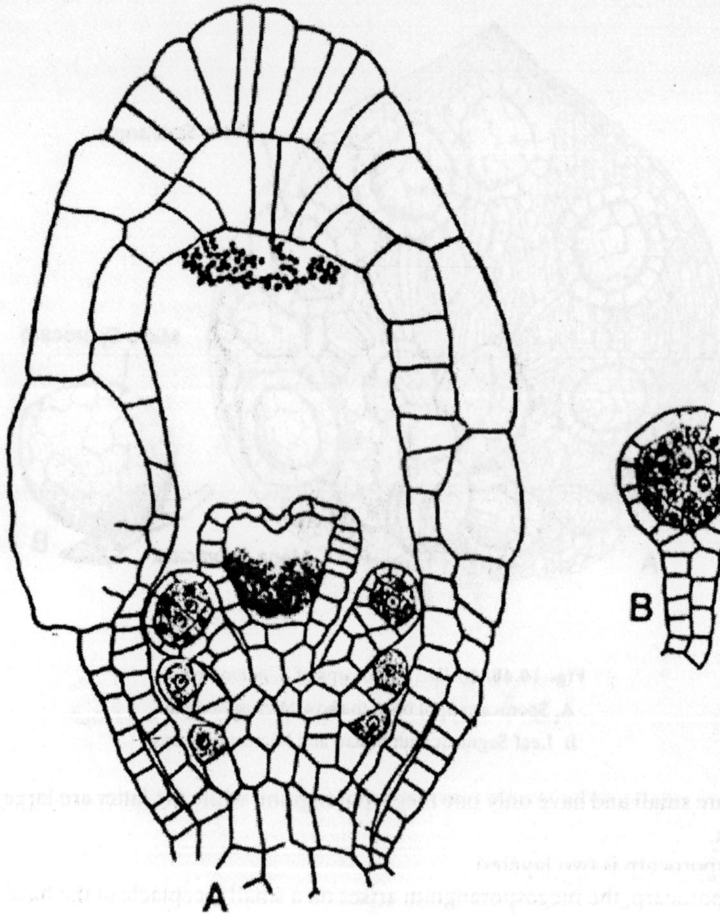

Fig. 10.49. *Azolla* : Structure of Microsporocarp in long Section (A) and a single Microsporangium (B)

PART TWO
GYMNOSPERMS

CHAPTER 11

GENERAL ACCOUNT AND CLASSIFICATION

The spermatophytes (seed bearing plants) are classified into two main groups *viz..*, *Gymnosperms* and *Angiospeims*. In Gymnosperms (Gr.gymnos = naked, *spermos* = seed) the seeds are exposed and do not have any covering. In angiospernis (Gr. *angio* = sac; *Spermos* = seed) the seeds are enclosed by the ovary wall, which (ovary) after fertilization becomes the fruit.

1. The first use of the word gymnosperms was by Theophrastus in 300 B.C.

2. In his book "Enquiry into plants" Theoprastus included all the plants with unprotected seeds under gymnosperms.

3. Geologically gymnosperms represent the most ancient among the seed plants.

4. Fossils of gymnosperms have been obtained from rock strata belonging to the late paleozoic era.

5. During mesozoic era there is an abundance of gymnospermous fossils indicating that they reached the climax of development.

6. In the present day flora however, gymnosperms - form a little part of the vegetation except perhaps for the conifers.

7. The life cycle of a typical gymnospermous plant comprises of a heteromorphic alternation of generations.

8. In the establishment of the sporophyte and in the reduction of the gametophyte, gymnosperms represent a further advancement over pteridophytes.

9. In gymnosperms, not only the sporophyte is the predominant generation, but the gametophytes have become microscopic and parasitic on the sporophyte due to extreme reduction.

Occurrence and distribution

1. Gymnosperms are found, both in tropical and temperate regions.

2. The same plants however are not found in both the regions.

3. Some are found even in arctic regions.

4. The living gymnosperms comprise approximately 70 genera and 725 species.

Morphology of the plant

1. The sporophytic plant body of gymnosperms is bushy or mostly a tall tree.

2. Some of the tallest and oldest trees are gymnospermous. *Sequoiadendron giganteum,* growing in the red wood park of California is 4000 years old and reaches a height of nearly 125 metres, with a girth of 30 metres.

3. The stem is anchored to the soil with the help of a tap root system.

4. Some of the secondary roots may have symbiotic association either with fungi (mycorrhizal.roots of *Pinus)* or blue green algae (coralloid roots of *Cycas).*

5. The stems are aerial, erect or climbing *(Gnetum),* branched or unbranched.

6. The stem is often surrounded by an armour of persistent leaf bases.

7. The branches of the stem are usually of two types *viz..,* branches of limited growth and branches of unlimited growth.

8. The former are short and usually bear the foliage leaves as in *Pinus;* the latter resemble the main shoot and in turn bear the branches of limited growth.

9. The leaves of gymnosperms are generally of two types - the green foliage leaves and the scaly leaves.

10. The green foliage leaves are evergreen and may be flattened or needle like.

11. The leaves may be simple or pinnately compound. Both microphyllous and megaphyllous leaves are seen in gymnosperms.

INTERNAL STRUCTURE

Stem

1. Anatomically the stem shows an epidermis, cortex and stele.

2. The epidermis is single layered and very soon replaced by a periderm. Cortex is parenchymatous.

3. Based on the relative quantity of the soft tissues and mechanical tissues gymnosperms are classified into *manoxylic* and *pycnoxylic* types.

4. In manoxylic forms (Cycads) the wood is not compact and has parenchyma.

5. In pycnoxylic forms wood is compact and relatively less amount of soft tissues is present.

6. The stele is surrounded by a typical endodermis followed by a layer of pericycle.

7. The vasculature consists of a number of vascular bundles arranged in the form of a ring surrounding a central pith.

8. The vascular bundles are collateral, endarch and open.

9. A stele of this type is given the name *Eustele.*

10. The xylem consists of mostly tracheids.

11. Vessels are found in the secondary wood as in *Gnetum.* Phloem consists of only sieve tubes and phloem parenchyma.

12. There are no companion cells.

13. Secondary growth is seen in all the members.

14. It is usually normal but anomalous secondary growth may also be found as in *Cycas* and *Gnetum.*

15. Normal secondary growth takes place due to the activity of a fascicular cambium, which produces the typical annual rings.

16. Occupying the central region of the stem is found the parenchymatous pith.

17. In some cases mucilage ducts or resin canals are found both in the pith and cortex.

Leaf

1. Anatomically the leaf exhibits xerophytic characters.

2. The epidermis is always covered by a cuticle.

3. The stomata are sunken and are often present in pits.

4. The mesophyll may be differentiated with palisade and spongy parenchyma *(Cycas)* or undifferentiated with arm parenchyma *(Pinus)*.

5. In some members *(Cycas)* with a single veined leaf, a *transfusion tissue* is present to help in lateral conduction.

6. Transfusion tissue may be regarded as a *paravascular tissue,* since it performs the function of a vascular tissue without having all the structural attributes of the latter.

7. The vascular bundles are collateral with a mesarch xylelm.

8. Vasculature shows the characteristic X/P arrangement:

Root

1. Most of the gymnospermous roots show a typical dicotyledonous root structure.

2. The xylem is either diarch or triarch.

3. Cortex is mostly parenchymatous.

4. As in the stem, secondary growth is seen both in the cortex and the stele.

REPRODUCTION

1. All the gymnosperms are heterosporons.

2. Their typical mode of reproduction is by spore formation.

3. Vegetative propagation is very infrequent and is seen in some Cycads where it is brought about by the formation of *bulbils.*

4. The reproductive organs are generally grouped into strobili, except perhaps in the female reproductive structures of *Cycas.* Plants maybe monoecious or dioecious with reference to strobili.

5. Very rarely bisporangiate strobili may be found.

6. The male strobilus consists of a central axis on which are arranged spirally the microsporophylls.

7. These on their abaxial surface bear the microsporangia or the pollen sacs.

8. The number of pollen sacs per microsporophyll ranges from two *(Pinus)* to many *(Cycas).*

9. The pollen sacs produce the pollen grains.

10. Within the pollen grains are found the male gametophytes.

11. The male gametophytes are incompletely developed when they come out of the pollen sacs.

12. Further development generally takes place within the ovule, after pollination.

13. The female strobilus consists of a number of megasporophylls arranged spirally on an axis.

14. In *Cycas* however the megasporophyus are not compact.

15. Each megasporophyll bears a varying number of ovules (megasporangia).

16. The ovules have a thick envelope (integument) which is generally three layered *viz.* .,outer fleshy, middle stony, and inner fleshy layers.

17. Internal to this integument, lies the *nucellus* which is the seat of development of the female gametophyte.

18. The female gametophyte arises by the development of one of the hypodermal cells of the nucellus.

19. Its development is either monosporic or tetrasporic.

20. The development of female gametophyte bears a resemblance to that of heterosporous pteridophytes.

21. Archegonia are highly reduced or they may be totally eliminated as in *Gnetum.*

22. Pollination is mostly anemophilous.

23. Only a part of the zygote develops into embryo. Such a development is said to be meroblastic.
24. Polyembrgony is of frequent occurrence.
25. The embryo usually has a suspensor.
26. The number of cotyledons varies from two to many.
27. Seed has a thick seed coat.
28. Germination is epigeal or hypogeal.
29. Growth of the seedling is slow.

GYMNOSPERMS

A. CYCADOPSIDA

1. PTERDIDOSPERMALES*
2. BENNETTITALES*
3. PENTOXYLALES*
4. CYCADALES

* Indicates fossil groups.

B. CONIFEROPSIDA

1. CORDIATALES*
2. CONIFERALES*
3. TAXALES
4. GINKGOALES

C. GNETOPSIDA

In this book the system of classification proposed by Sporne is followed.

AFFINITIES OF GYMNOSPERMS

Gymnosperms exhibit resemblances with pteridophytes as well as with angiosperms.

Resemblances with Pteridophytes

1. Life cycle shows heteromorphic alternation of generations with a reduced gametophyte.
2. Some of the gymnosperms (*Cycas*) exhibit circinate vernation.
3. There is no companion cell in phloem : occasionally vessels are present.
4. Development of the female gametophyte resembles that of heterosporous pteridophytes.
5. Motile sperms present in cycads recall what is seen in pteridophytes.
6. Gametophytes are endosporic as in heterosporous pteridophytes.

Differences with Pteridophytes

1. The sporophyte is much more complex.
2. Root system is always a tap root (adventitious in pteridophytes)
3. Stele is eustelic.
4. Seeds are produced.
5. Regular occurrence of secondary growth.
6. Permanent retention of female gametophyte in the megasporangium. Development of integument in the megasporangium.

Resemblances with Angiosperms

1. Tap root system
2. Possession of eustelic vascular cylinder.
3. Development of seed.
4. Permanent retention of the female gametophyte the ovule.
5. Siphonogamous fertilization.
6. Highly reduced male gametophytes.

7. Non motile male gametes (except in *Cycas).*

Difliefences with angiospenns :

1. Gymnosperms are mostly bushy or arborescent while it is not so in Angiospenns.
2. Phloem lacks the companion cells.
3. Flowers are bisexual or unisexual in Angiospenns, while strobili are mostly unisiexual in Gymnosperms.
4. Pollination is always anemophilous in Gymnosperms.
5. The ovules do not have a carpellary covering as is seen in angiosperms.
6. Archegonia are present in the female gymetophyte except in *Gnetum.*
7. There is no postfertilization development of endosperm. The female gametophyte itself is tbe endosperm. There is also no double fertilization.

CHAPTER 12

PTERIDOSPERMALES

The members of the order are commonly called seed bearing ferns or *Cycadofilicales.* Tbe order comprises many heterogeneous plants which appeared in the upper Devonian and extended through the Carboniferous and Permian to the Mesozoic.

The group was first discovered in 1877, by Grand ury. In India the group occurs in lower, middle and upper Gondwana systems belonging to Paleozoic and Mesozoic eras.

The group is subdivided into seven families *viz., Lyginopteridaceae, Medullosaceae, Calamopityaceae, Glossopteridaceae, Peltaspermaceae, Corystospermaceae* and *Caytoniaceae.* Of these, the family Lyginopteridaceae *(Lygnodendron, Lagenostoma* and *Heterangium)* is discussed here :

Fig. 12.1. **Lyginopteris.,**
Reconstruction of the plantbody of L. oldhamia.

LYGINOPTERIS

Lyginodendron is the stem genus of the plant *Lyginopteris oldhamia (Calymmatotheca hoeninghausi).* The stem genus is based upon the cortical impressions. This was first described in 1866 by Binney under the name *Dadoxylon oldhamia*. The various parts of the plant *Lyginopteris* have been described under the following names:

Stem *Lyginodendron*

Rachis *Rachiopteris aspera*

Root *Kaloxylon hookeri*

Frond *Sphenopteris hoeninghausi*

Pollen beanag organs *Crossotheca hoeninghausi*

Seed *Lagenostoma lomaxi*

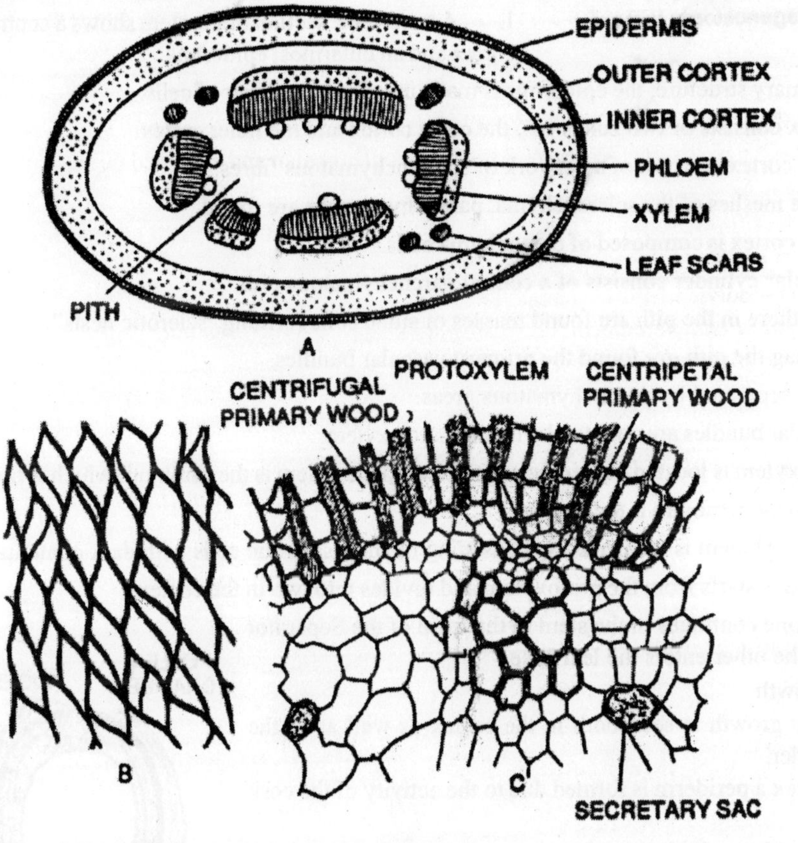

Fig. 12.2. **Lyginopteris**

A-T. S of Stem, B. Outer cortex showing fibrous strands. C. Primary vascular bundle (magnified)

Structure of the Plant

1. The plantbody of *Lyginopteris* had a long and slender stem.
2. The diameter of the stem varied from 2 mm. 4 cms.

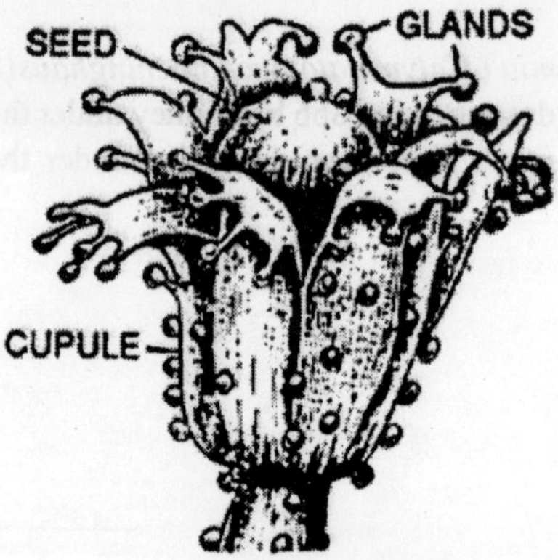

Fig. 12.3. **Lyginopteris**
Seed of lagenostoma lomaxii.

3. The stem was frequently branched and had adventitious roots which acted like prop roots.
4. The leaves were borne spirally and measured nearly 50 cms.
5. The rachis had glands similar to those on the seed.
6. The leaves were bipinnate to tripinnately compound and had pinnae at right angles to the rachis.
7. Pinnae were opposite to each other. Each pinnae in turn bore pinnules or leaflets which were connate and arranged alternately.
8. The stem was radially symmetrical showing that it grew erect, though it had to recline and seek the support of the neighbouring plants due to the heavy weight of the foliage.

Internal structure of the stem

1. A transverse section of the stem shows a central stele, middle cortex and an outermost epidermis.
2. In the primary structure, the epidermis is made up of a single layer of cells.
3. The cortex consists of two zones *viz.*, the outer cortex and the inner cortex.
4. The outer cortex consists of a network of sclerenchymatous fibres.
5. Within the meshes of the sclerenchyma, parenchyma cells are found.
6. The inner cortex is composed of parenchyma cells.
7. The vascular cylinder consists of a central parenchymatous pith.
8. Here and there in the pith are found masses of stone cells forming 'sclerotic nests".
9. Surroundiag the pith *are* found the primary vascular bundles.
10. These are separated by parenchymatous areas.
11. The vascular bundles are collateral with measarch xylem.
12. The protoxylem is located nearer the edge . Next to the xylem is the cambium which is rarely preserved.
13. External to the cambium is the phloem.
14. Outside the phloem is the pericycle consisting of short sclerotic cells with dark contents.
15. The leaf trace starts from the protoxylem and divides into two in the coctex.
16. Of these, one continues in the stem in the form of the Separator strand, while the other enters the leaf base.

Secondary growth

1. Secondary growth is seen both in the cortex as well as in the vascular cylinder.
2. In the cortex a periderm is formed due to the activity of the cork cambium.
3. In the vascular cylinder the fascicular cambium produces the secondary vascular tissues.
4. The secondary xylem is traversed by secondary medullary rays.
5. In addition to the medullary rays, leaf traces also disturb the vascular ring.
6. External to the secondary xylem is the secondary phloem.
7. The phloem cells are both large and small which are arranged alternately.

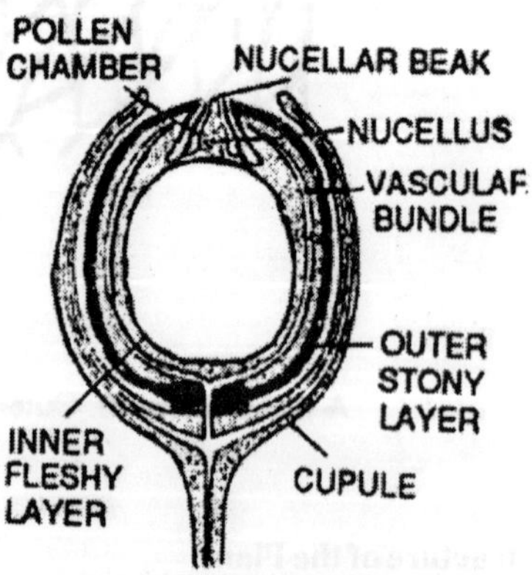

Fig. 12.4. **Lyginopteris**
Long section of the seed of lagenostoma lomaxii.

REPRODUCTION

1. The male and female reproductive structures probably arose on different plants.

2. The microsporangiate organ possessed fertile pinnules with each pinnule having six or more microsporangia.

SEED

1. The seed of *Lyginopteris* has been given the name *Lagenostoma lomaxi.*
2. In 1903, Oliver and Scott first showed the relationship between the stem and the seed.
3. The seed is terminal on the ultimate and naked branches of the frond.
4. Each seed is a small barrel shaped structure measuring approximately 4.25 mm. across and 53 mm. in length.
5. The ovule is orthotropous and 4.25 mm. across and *53* mm. in length and symmetrical.
6. Surrounding the ovule and covering it partially is a lobed husk or cupule.
7. The cupule has gland on the exterior resembling those present on the rachis.
8. A long section of the seed shows the following features.
9. The outermost layer is the integument which is traversed by nine vascular strands.
10. There are two regions in the integumer *viz.,* outer *sclerotesta* and inner *endotesta.*
11. The integument is fused with the nucellus except at the micropylar end where it forms a canopy that surrounds and overhangs the pollen chamber.
12. The canopy is divided into nine locules with a vascular strand extended into each.
13. It is believed that the loculed structure helps in the supply of water to the apex of the seed.
14. The pollen chamber is complex and in the apical cavity between the free part of the integument and the nucellus flask shaped prolongation arises from the nucellus and extends into the micropyle.
15. Within this prolongation arises a column of nucellar tissue for reception of the pollen.
16. This flask shaped structure called *Lagenostome* is responsible for the generic name.
17. Internal to the nucellus is the female gametophyte. A few archegonia have been reported by Lang (1944) in *L.ovoides.*

Systematic postion

CYCADOPSIDA

PTERIDOSPERMALES

LYGINOPTTERIDACEAE

LYGINOPTERIS

HETERANGIUM

Heterangium was first discovered from the coal balls of Bohemia, of middle carboniferous age (Corda 1845). The genus consists of about a dozen species of which the best known is *H. grievii.*

The stem of *H. grievii* was of considerable height and had a diameter of 1.5-2 cms. Most of them were straight, slender and unbranched. Very rarely branching was observed. The stem bore foliage known as *Sphenopteris elegans.*

Internal strucutre of the stem

1. A transverse section of the stem showed a large protostele.
2. The vasculature instead of forming a solid core at the centre had many gaps which were filled by parenchyma.
3. The tracheids had multiseriate bordered pits.

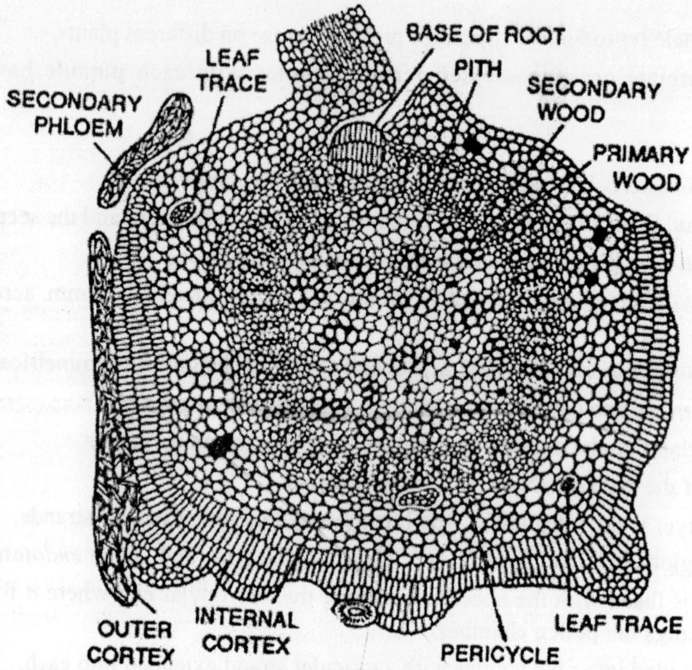

Fig. 12.5. **Heteragium**
T.S. of stem of H. grievii

4. The xylem groups (about twenty) were mesarch in arrangement.
5. The primary xylem was surrounded by a zone of secondary wood which was scarcely more than a few mm. thick.
6. The tracheids of the secondary wood also had multiscriate bordered pits.
7. The leaf traces started as a single strand having a single protoxylem point which immediately split into two.
8. The vasculature of the leaf trace was collateral at the base but concentric a little above.
9. Surounding the vasculature was a cortex whose outer zone had multiple sclerenchyma strands lying parallel to each other. Internal to this was a zone containing transverse plates of stone cells.
10. Next to this zone was the parenchymatous region of the cortex. The outermostt layer was the *epidermis*.

Systematic position

CYCADOPSIDA

PTERIDOSPERMALES

LYGINOPTERIDACEAE

HETERANGIUM

CHAPTER 13

CYCADOPSIDA

CYCADEOIDEA(=BENNETTITALES)

Of all the fossil gymnosperms no other genus perhaps has aroused so much of interest and attention as *Cycadeoidea*. This is because *Cycadeoidea* has reproductive organs that are very much similar to the flowers of angiosperms. This resemblance is the basis of a theory for the origin of angiosperms. (Arber and Parkin, 1907).

The genus *Cycadeoidea* consists of about 30 species being found in various places of the world in the form of petrified trunks. Geologically *Cycadeoidea* is traceable to the period between upper triassic and middle cretaceous. Some of the common species are *C. marylandica, C. dacotensis, C. wielandii etc.*

MORPHOLOGY

1. The plant body had a columnar or subspherical trunk with a diameter ranging from 5-60 cms. (Fig. 2.5).
2. The plant body attained a height of 3-3.5 meters.

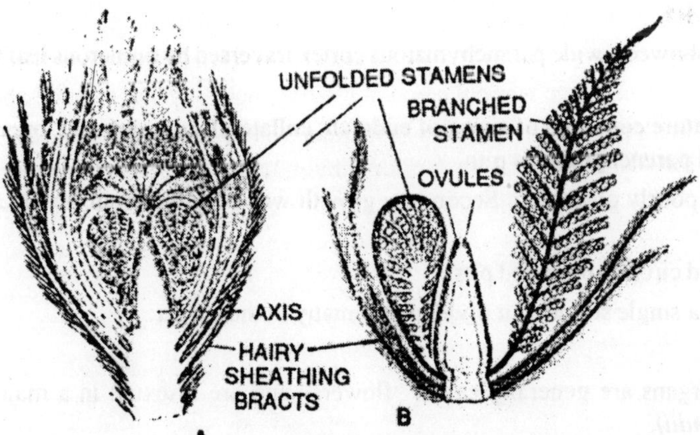

Fig. 13.1. **CYCADEOIDEA**
A-L.S. of strobilius, B-same with unfolded stamen (after Wieland)

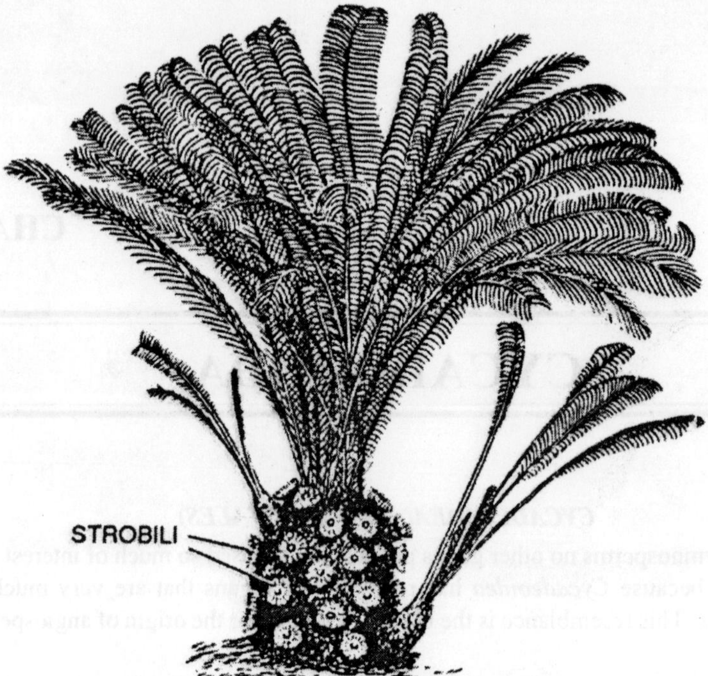

Fig. 13.2. **CYCADEOIDEA**
Reconstruction of the plantbody of C. dacotensis

3. The trunk was studded with persistent leaf bases as in living Cycads. Branching was very infrequent.
4. The stem had a crown of leaves at the apex which were pinnately compound.
5. The leaves were 2-3 meters long.

INTERNAL STRUCTURE

1. The stem internally showed a wide parenchymatous cortex traversed by numerous leaf traces and mucilage canals.
2. The primary vasculature consisted of a ring of endarch, collateral, conjoint and open vascular bundles surrounding a central parenchymatous pith.
3. The epidermis was poorly preserved. Secondary growth was brought about by the activity of vascular cambium.
4. Secondary xylem had circular bordered pits.
5. Leaf traces arose as a single strand, but divided into many in the cortex.

REPRODUCTION

1. The reproductive organs are generally called "flowers" and are bisexual in a majority of the species (unisexual in *C. wielandii).*
2. The flowers arise on short stalks singly at the axil of leaves.

Structure of a "flower"

1. The bisexual "flower" is also called a bisporangiate strobilus.
2. It has a short stalk extending laterally from the main stem.

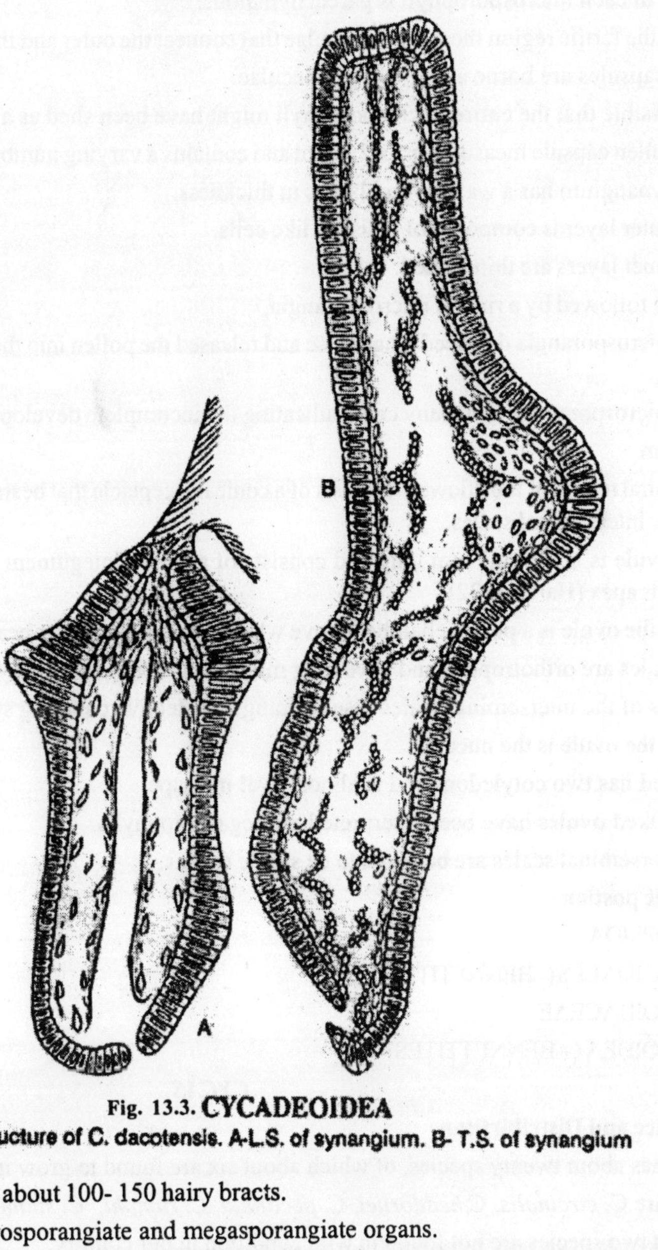

Fig. 13.3. **CYCADEOIDEA**
Synangial structure of C. dacotensis. A-L.S. of synangium. B- T.S. of synangium

3. A little below its tip are borne about 100- 150 hairy bracts.

4. These bracts protect the microsporangiate and megasporangiate organs.

5. At maturity the bracts expand to form cup shaped structures resembling the perianth of angiospernuc flowers.

Microsporophyll :

1. According to the interpretation of Wieland, each microsporophyll bears numerous alternating pinnae.

2. The pinnae bear two rows of kidney shaped, stalked pollen capsules or *synangia.*

3. According to Delevoryas (1963) the microsporophyll never expanded as interpreted by Wieland.

4. As Delevoryas points out, the tip of the microsporophyll was always fused to its base.

5. The tip of each microsporophyll is parenchymatous.

6. Within the fertile region there are trabeculae that connect the outer and the inner walls of microsporophyll.

7.Pollen capsules are borne along these trabeculae.

8. It is possible that the entire microsporophyll might have been shed as a single unit.

9. Each pollen capsule measures 0.5 x 2.5 mm and contains a varying number of microsporangia (20-30).

10. The synangium has a wall several layers in thickness.

11. The outer layer is composed of palisade like cells.

12. The inner layers are thin walled.

13. This is followed by a ring of microsporangia.

14. The microsporangia dehisced lengthwise and released the pollen into the synangium which opened by. an apical slit.

15. The microspores contain many cells indicating the incomplete development of the male gametophyte.

Gynoecium

1. The central region of the "flower" consists of a conical receptacle that bears many stalked ovules interspersed with many interseminal scales.

2. Each ovule is about one mm long and consists of a single integument which is fused with the nucellus except at its apex (Harris 1932).

3. Within the ovule is a pollen chamber above which there is a nucellar beak.

4. The ovules are orthotropous and have long micropylar beaks formed by the extended integument.

5. The tips of the interseminal scales fuse forming a protective covering surrounding the seeds.

6. Within the ovule is the nucellus.

7. The seed has two cotyledons and is almost oval in shape.

8. The stalked ovules have been interpreted as megasporophylls.

9. The interseminal scales are believed to be sterile ovules.

Systematic postion

CYCADOPSIDA

CYCADEOIDALES (=BENNETITITALES)

CYCADEOIDACEAE

CYCADEOIDEA (=BENNETTITES)

CYCAS

Occurrence and Distribirtiton

1. *Cycas* has about twenty species, of which about six are found to grow in our country.

2. These are C. *circinalis, C.heddornei, C. pectinata, C. rumphi, C. siamensis, and C. revoluta.*

3. The last two species are not found in wild condition in our country.

4. In Karnataka state *C. circinalis* grows wild near Nagamangala of Tumkur District.

5. *Cycas* is a very slow growing plant and grows under xerophytic conditions, like the exposed slopes of hills and other sunny places where there is scarcity of water.

Morphology of the plant

1. The plant body consists of an aerial, erect stem which is anchored to the soil with the help of a tap root as stout as the main stem.

2. The plant attains a height of 3-6 metres resembling a small palm tree.

3. Some of the species may reach a height of 20 metres.
4. The stem at its tip bears a crown of graceful leaves.
5. The stem has a diameter of 0.5-1 metre at the base.
6. It has a thick armour of .persistent leaf bases.
7. The stem is generally unbranched but branched specimens are also not very rare.
8. Branching is dichotomous when it occurs.
9. Very rarely three branches may arise at a point.
10. The leaves are borne at the apex of the stem in a spiral manner.
11. They are of two types *viz., scale leaves* and *foliage leaves.*
12. Scales leaves are thick and short and brownish in colour.
13. In the axils of scale leaves, the foliage leaves develop.
14. The foliage leaves are more than a metre long when fully grown and exhibit *cininate vernation* which is unusual to gymnosperms and recalls a fern character.
15. The leaves are pinnately compound and have a stout rachis.

Fig. 13.4. **Cycas**
Habit of C. rumphii (After Brown)

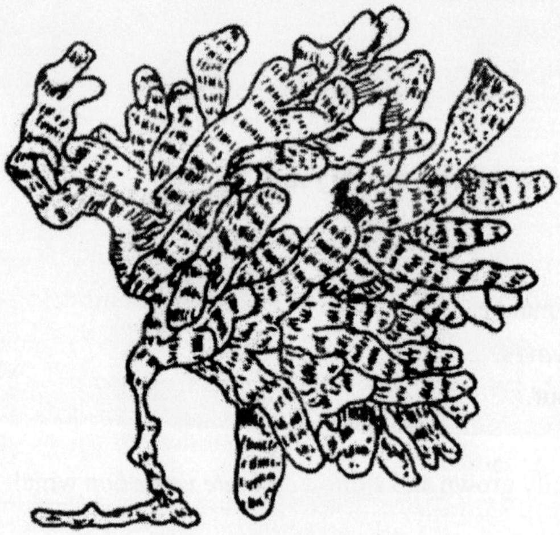

Fig. 13.5. **Cycas**
Coralloid roots

16. Some of the basal pinnae are modified into spines.
17. The pinnae are linear and have a central unbranched vien.
18. The stem is anchored to the soil with the help of a taproot which is as stout as the stem.
19. The tap root bears a number of secondary roots.
20. Some of these secondary roots are modified into *Coralloid roots.*
21. These are irregularly branched clusters of roots that grow apogeotropically and are found a few- inches beneath the soil.
22. The coralloid roots are greyish green in colour and owe their name to their irregular coral like branching.
23. On the surface of the root, small lentice like openings are found probably helping in respiration.
24. Internally the coralloid roots have an endophytic blue-green *(Anabena cycadacearum)* in the cortical region.
25. The relationship between the alga and the roots of *Cycas* is believed to be symbiotic.
26. While alga provides nitrogenous material (it has nitrogen fixing capacity) to the plant it gets protection and shelter in turn.
27. The manner in which a secondary root develops into a coralloid root is as follows.

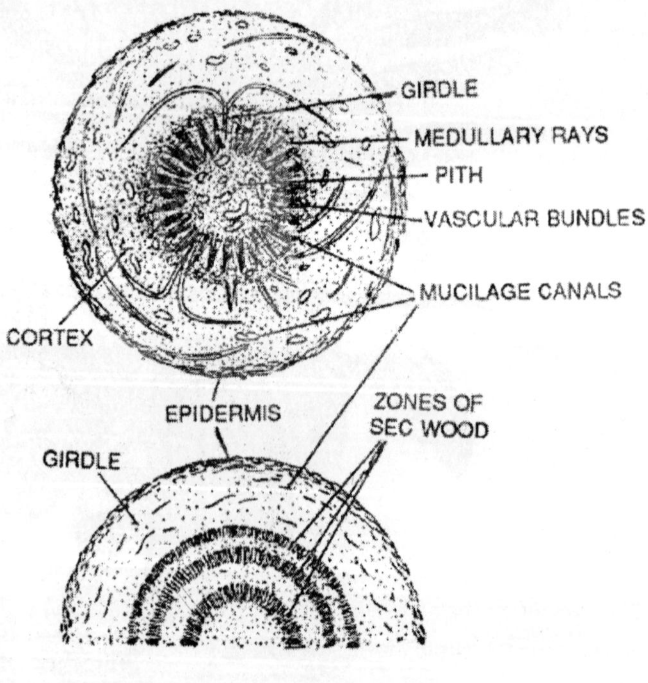

Fig. 13.6. **Cycas**
A- T.S. of young stem B- T.S. of old stem
(Note the rings of secondary Xylem)

28. Early in its development a secondary root gets the infection of the soil inhabiting bacterium.
29. Due to the infection of bacteria which come to lie in the cortical region, irregular branching sets in, producing the familiar coralloid roots.
30. The bacteria present in the cortex die very soon and the empty space left by them is filled by the blue green algae.

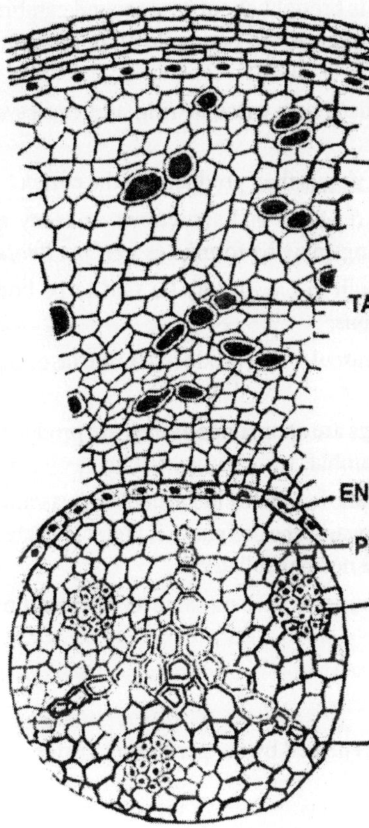

Fig. 13.7. **Cycas**
T.S. of normal root
(A portion enlarged)

INTERNAL STRUCURE

Stem

1. A transverse section of a young stem shows the following features viz., epidermis, cortex and stele.

2. Epidermis is single layered and is discountinous due to the presence of leaf bases.

3. It is short-lived and very soon will be replaced by the periderm.

4. Cortex is very wide, surrounding a narrow vascular cylinder (manoxylic).

5. The entire cortical region is filled by thin walled parenchyma cells rich in starch.

6. The cortical region also consists of a number of mucilage ducts and leaf traces.

7. The leaf traces are of two types *viz., direct* or *radial traces* and *girdle traces.*

8. Direct traces arise from the primary vascular bundles, in duplicate and enter the leaf base.

9. The girdling traces which also arise in duplicate pass obliquely into the cortex, grow in opposite direction meet and then enter the leaf.

10. This forms a sort of girdle.

11. The girdle traces are linked by cross connections with each other, with other leaf traces and also with the radial traces forming a complicated network. Before entering into the leaf base the traces divide and form a large number of bundles.

12. The vasculature in the primary structure consists of a number of collateral, open vascular bundles with endarch xylem. 13. The vascular bundles surround a central parenchymatous pith and are separated by medullary rays.

14. Mucilage ducts are also found in the pith.

Secondary growth

1. This takes place both in the cortex as well as in the vasculature.

2. In the outer layers of the cortex, arise a *cork cambium* (phellogen) which produces the periderm.

3. Towards its inner face the cork cambium produces *phellem* or the cork tissue.

4. The three together constitute the periderm.

5. Secondary growth taking place in the vasculature produces many rings of vascular tissue.

6. But these are not annual rings as their development is irregular.

7. At the initiation of the secondary growth, the strips of fascicular cambia present in the vascular bundles

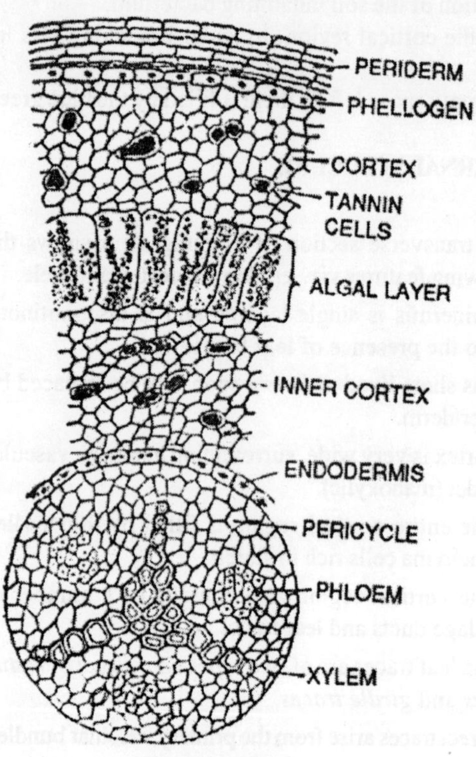

Fig. 13.8. **Cycas**
T.S. of coralloid root
(A portion enlarged)

radially extend and form a completely closed ring.

8. This produces secondary phloem to the exterior and secondary xylem to the interior.

9. Soon after the formation of the first vascular ring, the activity of the cambium ceases.

10. Further growth is brought about by a second cambial ring which arises either in the pericycle or in the inner layers of the cortex.

11. This also produces one vascular ring and ceases to function.

12. A third cambial ring is then produced in the cortex.

13. The process is repeated and sometimes as many as fourteen vascular rings may be formed as in *C.pectinata*.

14. Secondary growth in *Cycas* may be said to be both normal and anomalous.

15. The inner most vascular ring produced by the fasoeular cambium is nornal.

16. All the other rings are anomalous as they are produced by extrafascicular cambia.

17. Another important feature in the secondary vascular bundles produced towards the periphery is their concentric nature instead of the normal collateral.

18. The concentric bundles have mesarch xylem recalling the fern features.

Root

1. Anatomically the root resembles dicotyledons.

2. A transverse section shows an outermost epidermis (which is soon replaced by the periderm), cortex and stele.

3. The epidermis or epiblema is single layered

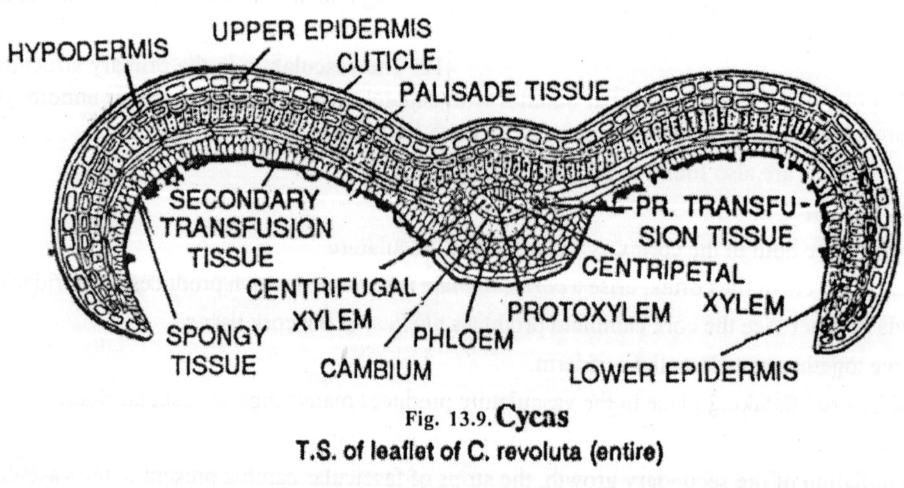

Fig. 13.9. **Cycas**
T.S. of leaflet of C. revoluta (entire)

4. The cortex consists of many layers of parenchyma cells interspersed with nucilage cells.
5. Tannin cells also found in the cortex.
6. The stele is surrounded by single layered endodermis followed by a layer of pericycle.
7. The vascular bundles are radial with diarch or triarch and exarch xylem.
8. There is no pith.
9. Conjunctive parechyma is present between phloem and xylem.
10. The xylem consists of tracheids only, with spiral, scalariform and pitted thickenings.
11. Phloem consists of sieve tubes and phloem parenchymma.
12. In older roots phloem may have bast fibres.
13. Secondary growth also takes place in the root much the same way as in a dicot root.

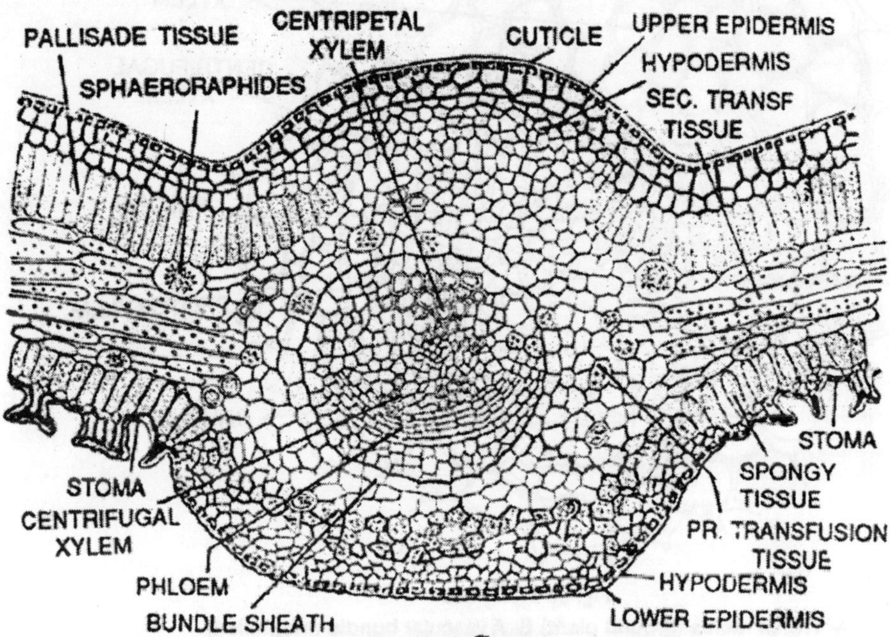

Fig. 13.10. **Cycas**

T.S. of leaflet of C. revoluta (a portion magnified)

14. In some cases, two peridsrm layers are found separated by the cortex.
15. This is due to the formation of two separate phellogens.
16. Lenticels are found in the periderm.

Coralloid root

1. It is similar to the normal roots in all respects except for the presence of an algal zone in the cortex.
2. The algal zone is known to consist of several blue-green algae *(Anabena cycadacearum, Nostoc punctiforme, Oscillatoria sp.*, etc.) and bacteria like *Azatobacter, Pseudomonas* etc.

LEAF

1. The pinnae arise on a stout primary rachis.

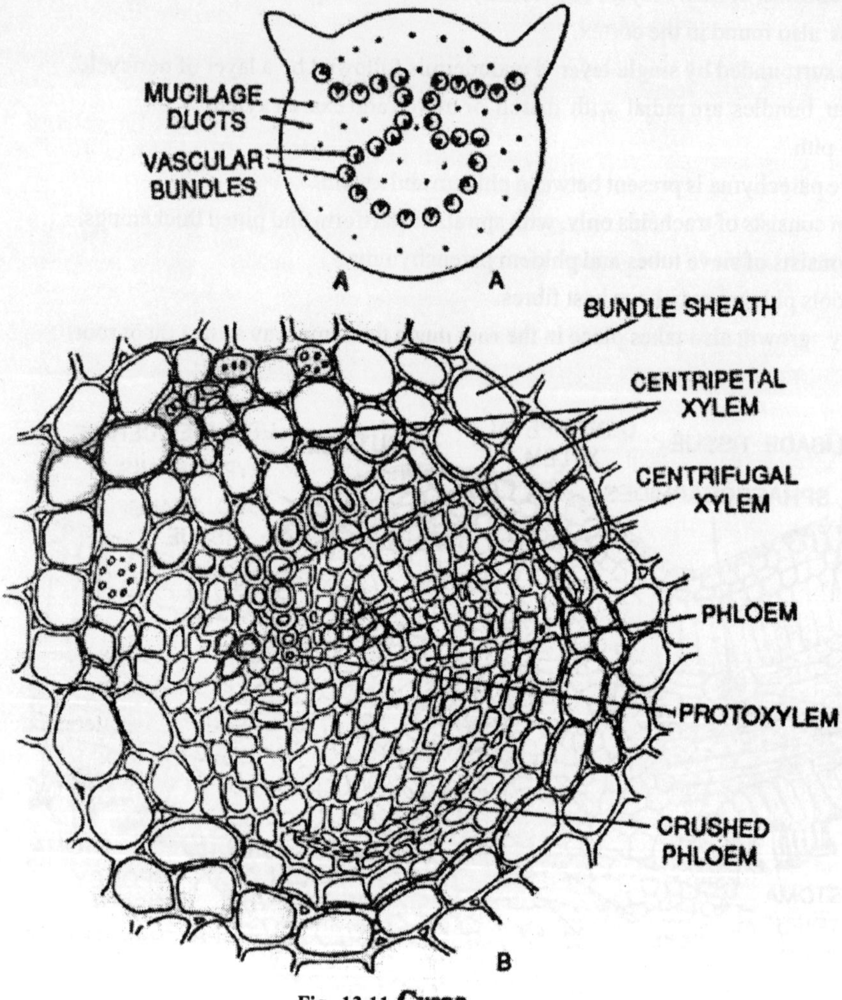

Fig. 13.11. **Cycas**

A- T.S. of rachis (ground plant) B- A vascular bundle (magnified)

2. The pinnae anatomically show bifacial arrangement.
3. It has mainly the following regions.

Cuticle: It is deposited above the upper and lower epidermal layers.

Upper- epidermis

It is single layered and consists of thick walled parenchyma cells.

Hypodermis

Internal to the epidermis is the one or two layered sclerenchymatous hypodermis.

Mesophyll

1. It is differentiated into *palisade* and *spongy parenchyma.*
2. The cells of the palisade parenchyma are columnar and are closer packed being present below the upper epidermis.

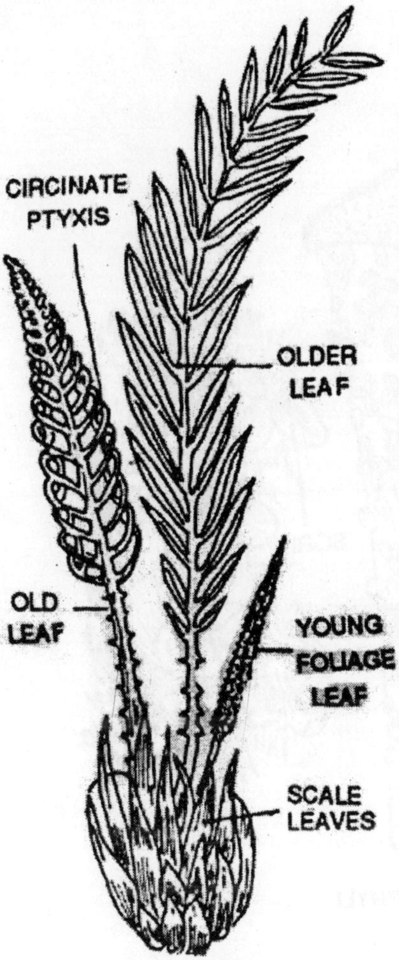

Fig. 13.12. Cycas
A germinating bulbil

3. The cells of the spongy parenchyma are loosely arranged and are present above the lower epidermis.

4. The palisade parenchyma has plenty of chloroplasts and is the chief photosynthetic tissue of the plant.

5. A few chloroplasts are present in the cells of the spongy parenchyma also.

Vein

1. A vein cut in the transverse section shows an outermost bundle sheath surrounding the vasculature which shows the *X/P* arrangement.

2. The xylem consists of both centripetal (exarch) and centrifugal (endarch) elements and is mesarch.

3. The pholem is present below the xylem and consists of sieve tubes and phloem parenchyma.

Transfusion tissue

1. There are two types of transfusion tissues- primary and secondary.

2. The primary transfusion tissue consists of masses of tracheids having bordered pits on their walls.

3. It is present on the sides of the vascular bundle and embedded in the parenchyma that surrounds the vein.

4. According to Woresdell (1887), it is a lateral extention of the xylem.

5. According to Carter (1911) it is derived from leaf parenchyma.

6. In between the spongy and palisade parenchyma radially elongated cells are present constituting the secondary transfusion tissue.

7. These cells also have pits on their radial walls and are supposed to help in lateral conduction in the absence of lateral veins.

8. Some of the parenchyma cells surrounding the vasculatore have *spheroraphides*.

Lower epidermis

1. It is also single layered, but is discontinuous due to the preisence of small pits in which are found the the sunken stomata of *the amphicyclic* type.

2. The guard cells are surrounded by a ring of subsidiary cells.

Xerophytic feature seen in the leaflet

1. The leaflet shows many xerophytic adaptations.

2. These are : (a) thick cuticle, (b) thick walled epidermal cells, (c) presence of well developed mechanical tissue, (d) sunken stomata restricted to the ventral surface and (e) shiny and leathery nature of the leaflets.

Rachis

1. A transverse seclion show's an epidermis covered Itide.

2. Next to the epidermis is a band of chlorenchyma. Next of this, is a few layered sclerenchyma.

3. The entire inner region of the rachis is fillled by a mass of parenchymatous ground tissue in which are

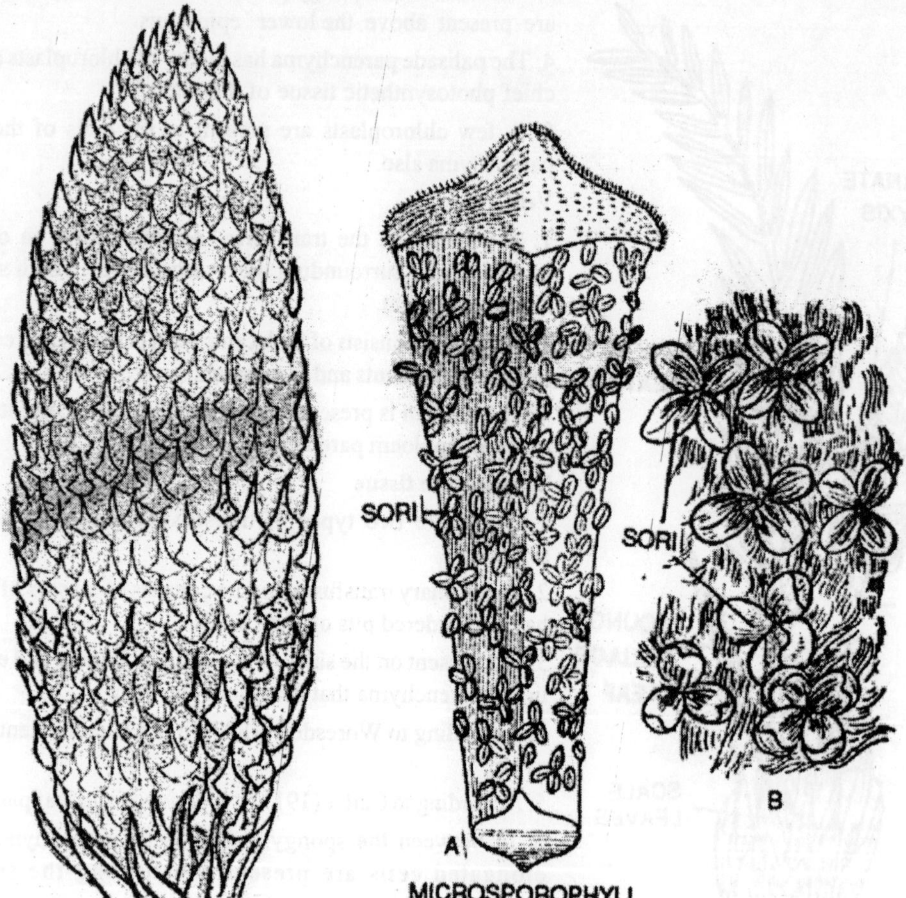

Fig. 13.13. Cycas
Male reproductive structure
Left - Entire male cone
Middle - microsporophyll
Right - Sori enlarged

embedded a number of vascular bundles arranged in an inverted omega fashion.

4. The vascular bundles are open and collateral.

5. The xylem lies on the inner side towards the centre of the petiole and consists of both centripetal and centrifugal elements (Diploxylic).

6. Below the xylem, phloem is found.

REPRODUCTION

1. The sporophytic plantbody reproduces by vegetative propagarion as well as by spore production.

2. Vegetative propogation is brought about by the formation of bulbills which are formed mostly in the basal part of the stem.

3. The bulbil consists of a condensed axis surrounded bv a few fleshy, scaly leaves.

4. On falling upon a suitable substratum a bulbil develops into an independent plant.

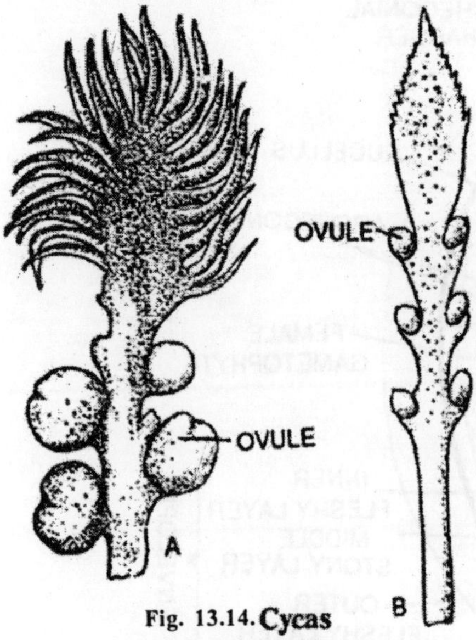

Fig. 13.14. **Cycas**
Megasporophylls
A- C. revoluta, B- C. circinalis

SPORE PRODUCTION

1. *Cycas* is strictly dioecious.
2. Micro and megasporophylls are borne on separate plants.
3. The growth pattern in the male and female plants differs.
4. The male cone is produced by the apical meristem and as a result growth comes to a stop.
5. Further growth is brought about by an accessory meristem (sympodial).
6. In the female plants, only the derivatives of the apical meristem develop into megasporophylls; hence the growth is uninterrupted (monopodial).

Male reproductive structures

1. The male cone is a compact structure and measures in length from 40-50 cms.
2. It has a central cone axis on which are spirally arranged the microsporophylls.
3. Each microsporophyll, is a flattened, wedge shaped structure consisting of a narrow lower portion which broadens above.
4. The upper portion is drawn into a projection called the *apophysis.*
5. The microsporophylls on their abaxial surface bear a number *of. pollen sacs* or *microsporangia.*
6. The microsporangia are present in two large goups separated by a median sterile line.
7. In each group there are a number of aggregations of microsporangia called *sori.*

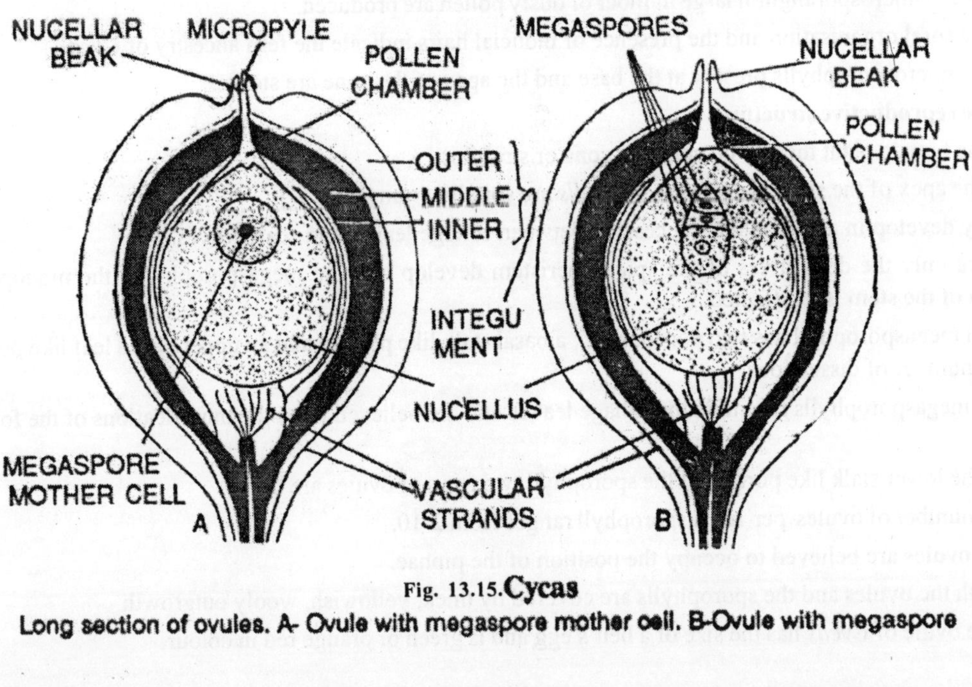

Fig. 13.15. **Cycas**

Long section of ovules. A- Ovule with megaspore mother cell. B-Ovule with megaspore

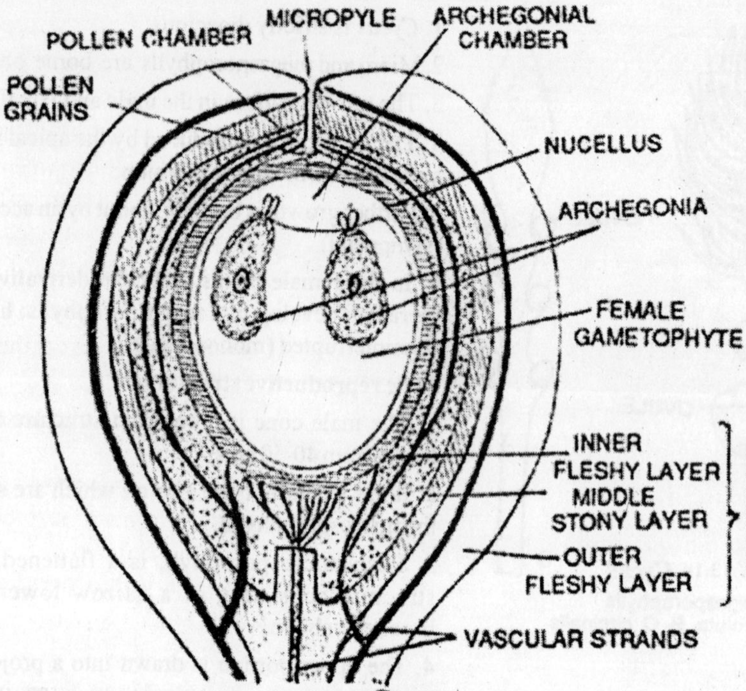

Fig. 13.16. **Cycas**
L.S. of ovule of C. revoluta (before fertilisation)

8. Each sorus has an elevated central portion around which are found *3-6* micropsorangia.
9. Between the microsporangia, inducial hairs are present.
10. In each microsporangnim large number of dusty pollen are produced.
11. The soral organisation and the presence of inducial hairs indicate the fern ancestry of *Cycas*.
12. The microsporophylls present at the base and the apex of the cone are sterile.

Female reproductive structures

1. In the female plant there is no compact cone or strobilus.
2. At the apex of the stem the *megasporophylls* are spirally arranged.
3. They develop in acropetalous succession between foliage leaves and scale leaves.
4. Since only the derivatives of the apical meristem develop into the megasporophylla, the monopodial growth of the stem is unaffected.
5. Each megasporophyll has two portions *viz.*, a basal stalk like portion and an upper broad leaf like portion with a number of dissections.
6. The megasporophylls resemble the foliage leaves and are believed to be the modifications of the foliage leaves.
7. On the lower stalk like portion of the sporophyll two rows of ovules are borne.
8. The number of ovules per megasporophyll ranges from 2-10.
9. The ovules are believed to occupy the position of the pinnae.
10. Both the ovules and the sporophylls are covered by thick, yellowish, wooly outgrowth.
11. The ovule of *Cycas* has the size of a hen's egg and is green or orange red in colour.

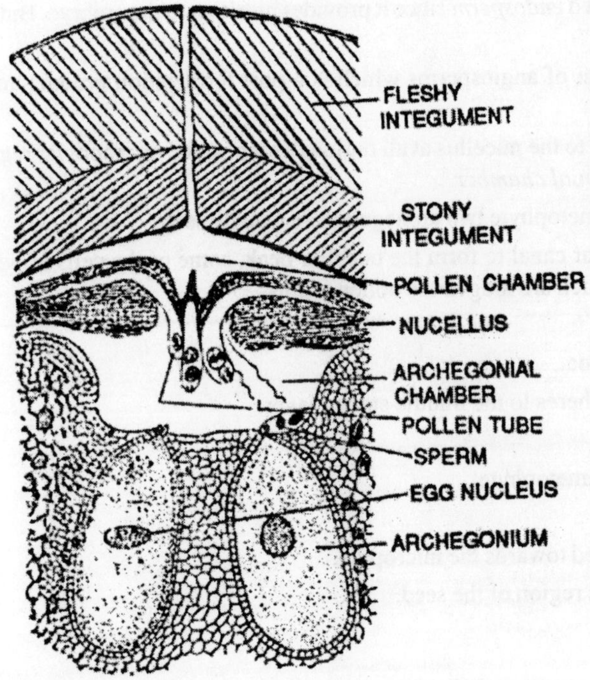

Fig.13.17. Cycas Diagramatic L.S. of micropylar portion of ovule at the time of fertilisation (after Chamberlain)

12. It is the *biggest ovule* in the plant kingdom.

13. The different species of *Cycas* show variations in the nature of the megasporophyll and in the number of ovules per sporophyll.

14. The megasporophyll has a broad upper portion in *C. revoluta* whereas in *C.circinalU* the upper portion is also narrow.

Structure of a mature ovule

1. In a long sectional view the ovule shows the following features.

2. The integument is very thick and is made up of three layers.

3. The integument leaves a small opening called the micropyle.

4. Since the integument is very thick the micropyle forms a narrow tube called the micropylar canal.

5. The ovule is supplied by three bundles *viz.*, a central bundle, an outer bundle and an inner bundle.

6. The vasculature enters into the ovule and out of the three bundles two of them reach the tip of the integument.

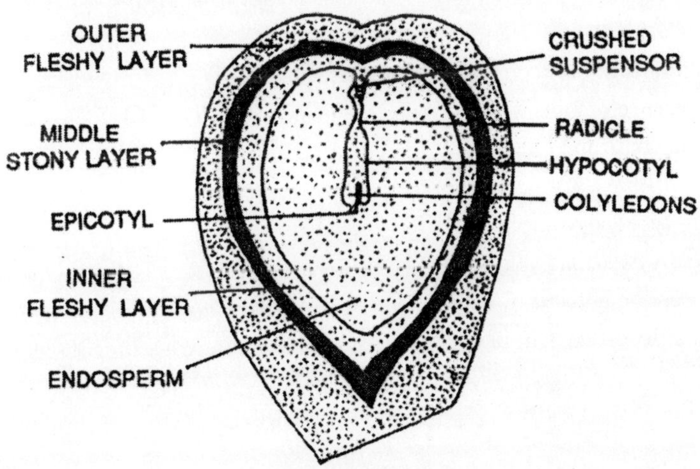

Fig. 13.18. Cycas
L. S. of seed

7. Internal to the integument is found the nucellus in the form of a few layers of cells, the bulk of which having been consumed during the development of the female gametophyte.

8. Internal to the nucellus is the large cellular female gametophyte occupying most of the interior of the ovule.

9. The cells of the female gametophyte are rich in starch content and other nutrient materials.

10. The female gametophyte is very often called *endosperm* since it provides nutrition to the embryo. But this term does not seem to be very appropriate.

11. It is totally, different from the endosperm of angiosperms which is a post fertilization product and is generally triploid.

12. The female gametophyte will be adhering to the nucellus at all regions except at the microphylar region, where a small space is left called the *archegonial chamber*.

13. In the micropylar region of the famale gametophyte two archegonia are usually seen.

14. The nucellus protrudes into the micrppylar canal to form the nucellar beak, some of the cells of which disintegrate to form the pollen chamber in which are lodged the pollen grains.

Structure of the seed :

1. The integument hardens to form the seed coat.
2. The outer fleshy layer gets dried up and adheres to the middle stoney layer.
3. The inner fleshy layer soon shrivels up.
4. Within the seed coat is found the female gametophyte.
5. Embedded in this is the embryo.
6. The radicile portion of the embryo is directed towards the micropyle.
7. The plumule is directed towards the chalzal region of the seed.

Systematic position

CYCADOPSIDA

CYCADALES

CYCADACEAE

CYCAS

CHAPTER 14

CONIFEROPSIDA

Of all the gymnosperms perhaps the most dominant and the most beautiful are the coniferous plants. The popular name *"Conifers"* is due to the cone like appearance of the plantbody. Tle branches are long towards the base, succeedingly they become shorter giving a cone like appearance to the plant body. The coniferous forests are quite extensive, particularly in the temperate regions of the world. Even in tropical countries the vegetation at higher altitudes always comprises coniferous plants. It has been estimated that conifers occupy about 8% of the land on Earth.

Geologically, conifers date back to the Carboniferous period. But some, like Cordaitales were found as early as the Devonian period.

Coniferopsida differ from Cycadopsida in having.(a) simple leaves (b) narrow pith and cortex, (c) active fascicular cambium producing annual rings, (d) pycnoxylic wood and (e) bilaterally symmetrical ovules and seeds.

Sporne, (1965) has classified coniferopsida into four orders *viz., Cordaitales, Ginkgoales, Coniferales,* and *Taxales.* Of these, Cordaitales and Confierales are discussed here.

CORDAITALES

1. This fossil order constitutes the most ancient and primitive order of the class.
2. Cordaitales are believed to have taken their origin during the Devonian period.
3. As revealed by the abundance of fossils, the order reached the peak of development during the Carboniferous and Permain periods, forming a dominant arborescent vegetation.
4. The plants were arborescent and had a monopodial branching.
5. Secondary wood was abundant and pycnoxylic.
6. Strobili were monosporangiate.

CORDAITES

The genus *Cordaites* is known from the assemblage of dismembered parts which are known under various names.

Stem and wood - *Cordaites, Dadoxylon, Mesoxylon, Parapitys* etc.

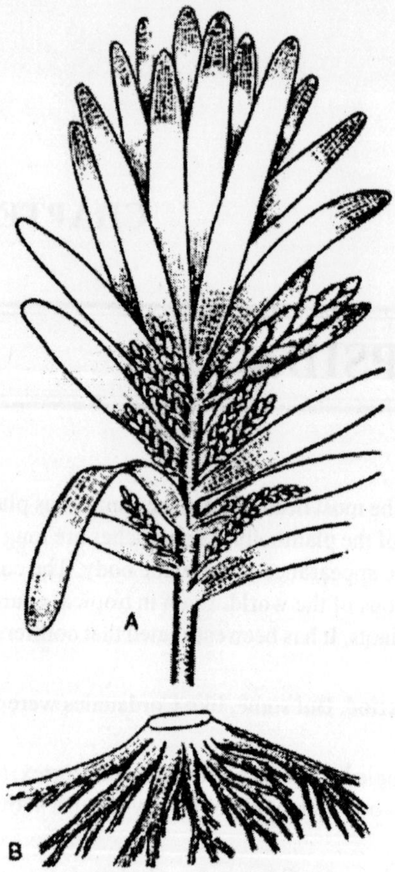

Fig. 14.1. **Cordaites**
Reconstruction of the plantbody.
A. A branch bearing foliage and inflorescence, B Root system

Leaves - *Cordaites* - Defoliated branches - *Cordaicladus*.

Roots -*Amyelon*.

Strobilus - *Cordainthus*.

Seeds - *Cordiaicarpus*. (= *Cordiocarpus*).

The plant body of *Cordaites* was a tall tree reaching a height of 30 metres and grew in dense forests. The stems were slender and attained a diameter of 65-70 cms. The branching was lateral and arose only at the top part of the plant.

Leaves (compressions) were found on the branches and were spiral in arrangement. They were petiolate, subulate or spatulate or grass like and had an entire margin. Venation in the leaf was parallel. Based on the nature of the leaves the genus is divided into three sub-genera *viz.*, *Eucordaites* (Leaves broad with a blunt apex *Dorycordaites* (leaves with acute apex) and *Poacordaites* (Leaves were grass like and small)..

The plant was anchored to the soil with the help of repeatedly branche root system.

INTERNAL STRUCTURE

Stem

1. As has alteady been said, Cordaitalean stems are known under various names, In *Cordaites* the stem has an extensive cortex which had parenchyma with interspersed scelernchyma.

2. Secretory sacs were also present in the pith. Pith had large air chambers separated by diaphragms.

3. Pith casts have given the name *Artisia*.

4. The vasculature in its primary structure consisted of a large number of endarch or mesarch protoxylem elements.

5. Secondary growth in the stem resulted in the formation of compact (pyconoxylic) wood.

6. The tracheids in the secondary wood were long and narrow and had uniseriate bordered pits.

7. Interspersed with the wood elements, were the parenchymatous rays which were simple and one cell wide.

8. Some of the cells of the ray had resinous material. However there were two resin canals as in *Pinus*.

9. The leaf traces resembled those of cycads in the *pseudomesarch* and *diploxylic* nature.

10. Each leaf was provided with a pair of traces.

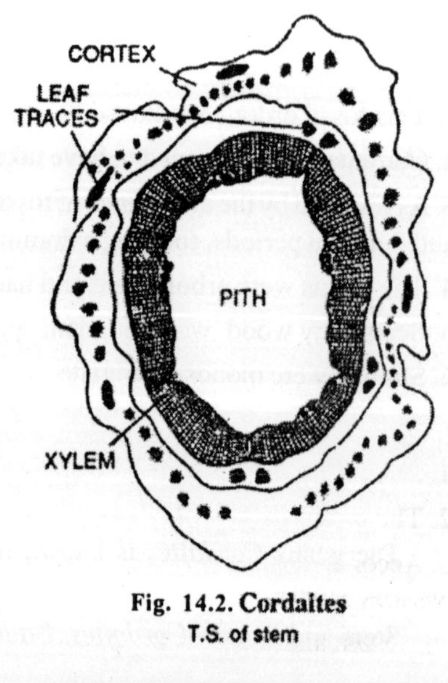

Fig. 14.2. Cordaites
T.S. of stem

Leaf

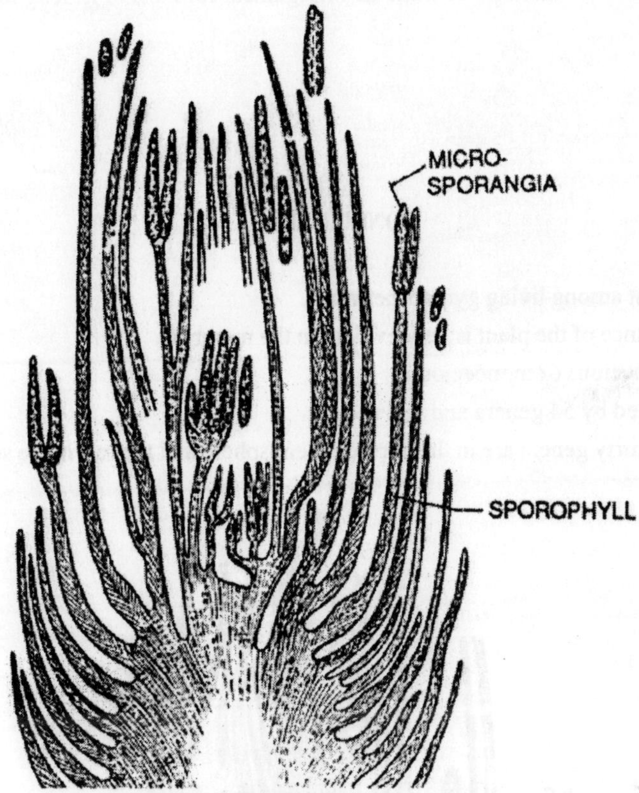

Fig. 14.3. **Cordaites**
L.S. of male reproductive structures (Cordaianthus penjonii) (after Renault)

1. Anatomically the leaves exhibited differentiated mesophyll in most cases.
2. The epidermis had stomata distributed in longitudinal bands. Guard cells were surrounded by 4-6 subsidiary cells.

Root

1. Internally the root showed a diarch to tetrach xylem.
2. Stele was protostelic.

REPRODUCTION

1. Cordaitean plants were *dioecious.*
2. The strobili, called *Cordaianthus* were unisexual and 1 cm long.
3. The male strobili had a central axis which had a few microsporophylls and numerous sterile bracts. Each microsprophyll bore 1-6 pollen sacs.
4. The female strobili like the male ones had a stout central axis on which were borne spirally arranged bracts.
5. The length of the strobili varied from 25-30 cms.
6. In the axis of the bracts arose short, fertile, secondary shoots which had spirally arranged fertile and sterile *bracteoles* (megasporophylls).
7. Each megasporophyll had 2 or more ovules at its apex.

8. In the ovules, the nucellus remained free from the intergument for most of its length.

Systematic position

CONIFEROPSIDA

CORDIATALES

CORDIATACEAE

CORDAITES

CONIFERALES

1. The order *is* the largest among living gymnosperms.
2. The cone like appearance of the plant is very evident in the members.
3. The plants may be dioecious or monoecious.
4. The order is represented by 54 genera and 482 species.
5. Among these, about thirty genera are in the northern hemisphere and the rest in the southern hemisphere.

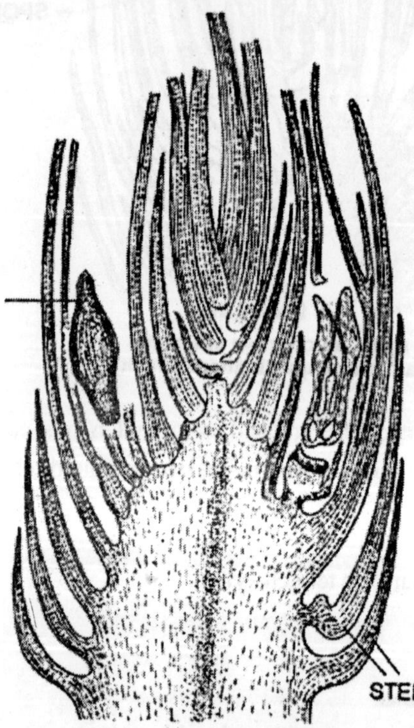

Fig. 14.4. **Cordaites**
L.S. of female reproductive structures, (Cordaianthus williamsonii)
(after Renault)

6. The plants flourish very well in a temperate climate. In habit, the plants range from shrubs to gigantic trees.
7. Some of the tallest trees of the plant kingdom *(Sequoia, Sequoiadendron)* belong to this order.
8. Some of these are of great antiquity being more than 4000 years old.
9. Some of the pinp trees which are living even today were already old when the Roman empire was young.

PINUS

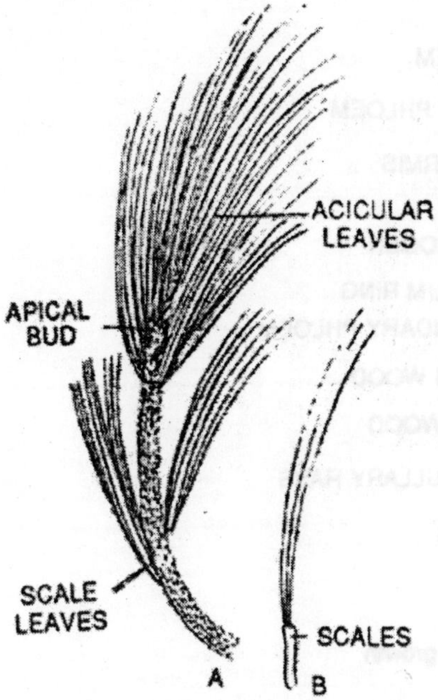

Fig. 14.5. **Pinus**
A. Shoot of unlimited growth.
B. Drawf shoot

Occurrence and distribution

1. The genus consists of 90-95 species spread over the northern hemisphere in temperate regions.

2. About 1/3 of the species are endemic to the old world and the rest in the new world.

3. The plants grow on the slopes of hills and form extensive evergreen forests.

4. Some of the species of *Pinus* are of great economic value.

5. In India, the genus is represented by six species *viz., P. wallichiana, P. armandi, P. gerardiana, P. insularis, P. roxburghii* and *P. merkusii.*

6. The classification of the Indian species is based on the number and form of leaves (Raizada and Sahni, 1960).

7. The Indian species of *Pinus* grow in the northeast and northwest Himalayas.

8. In the hills of south India *Pinus* is cultivated.

Morphology

1. The plant body *of Pinus* is a tall tree varying in height from 70-200 *feet.*

2. The girth of the stem maybe as much as 12 *feet.*

3. Some of the species of *Pinus (P. aristata)* growing in California are as old as 4000 years.

4. The stem is erect, woody and covered with rough scaly bark, which occasionally peels off.

5. The tip of the stem has a large apical bud which brings about the growth in length

6. The branching of the stem is monopodial.

7. It is very regular and symmetrical and arises laterally from the axis of scale leaves.

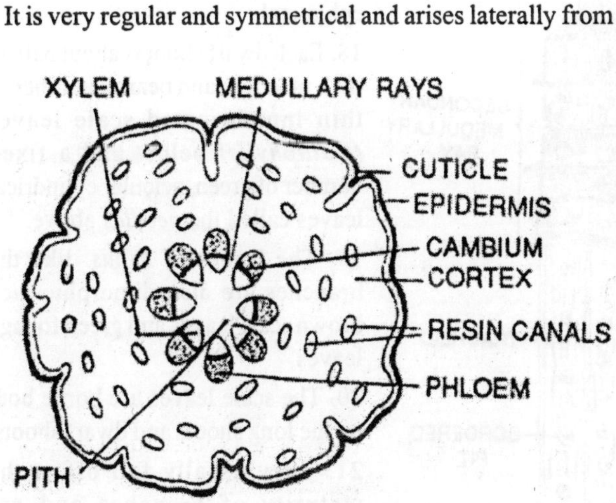

Fig. 14.6. **Pinus**
T.S. of young stem

8. The branches are longer on the lower part of the stem and become shorter towards the apex giving the familiar cone like appearance.

9. In some old stems the symmetry may be lost due to the death of some branches and irregular growth of the others.

10. The branches are dimorphic *viz.,* branches of *unlimited growth* or *long shoots* and branches of *limited growth* or *dwarf shoots.*

11. The long shoots arise from lateral buds on the main stem in the axils of scale leaves.

12. As the name indicates, they grow indefinitely by means of an apical bud (Fig. 2.34 A).

13. It is these branches which impart the characteristic appearance to the plant.

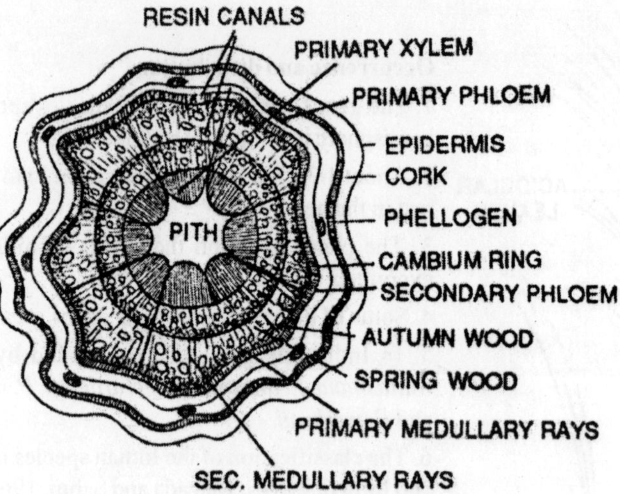

Fig. 14.7. **Pinus**
T.S. of old stem showing secondary growth

14. In some cases when the apical meristem of the main stem is injured, one of the long shoots at the apex becomes dominant and grows like a main stem.

15. The long shoots are covered with brown scale leaves in the axils of which arise the dwarf shoots.

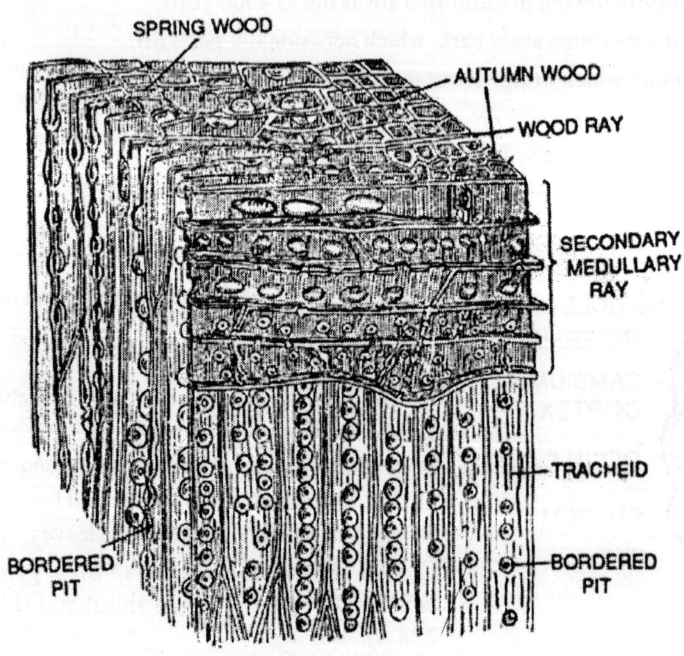

Fig. 14.7. **Pinus**
Three dimensional model of a block of wood

16. The drawf shoots never arise on the main stem.

17. They are numerous in number and as the name indicates, are of limited growth, because the apical bud is ephimeral.

18. Each dwarf shoot is about half an inch in length and bears a number of thin innmbricated scale leaves *(Cataphylls)* below and a fixed number of green, acicular cylindrical leaves called the *needles* above.

19. The leaves of *Pinus,* like the branches are also dimorphic *viz.,* brown scale leaves and green foliage leaves.

20. The scale leaves are borne both on the long shoots and dwarf shoots.

21. They usually fall off at the maturity of branches and are supposed to help in the conservation of water around the young branches.

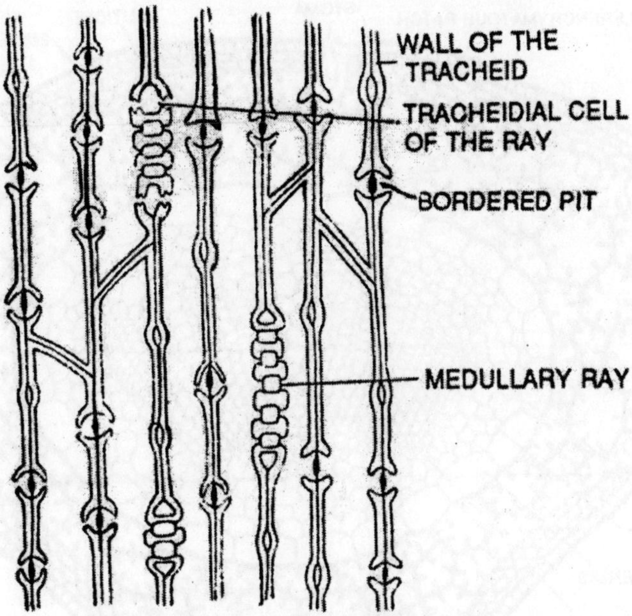

Fig. 14.8. **Pinus**
Longitudional tangential section of wood.

22. The foliage leaves *i.e* needles are produced on the dwarf shoots only, persist for a number of years and fall off only when a *foliar spur* (dwarf shoot with needles) is shed.

23. The number of needles varies in different species of *Pinus*.

24. It may be three *(P. longifolia, P.gerardiana* etc.), four (*P. quardrifotia)* or five *(P. excelsa)*. Even two leaved *(P.penaster)* and single leaved (*P. monophylla)* foliar spurs are also found.

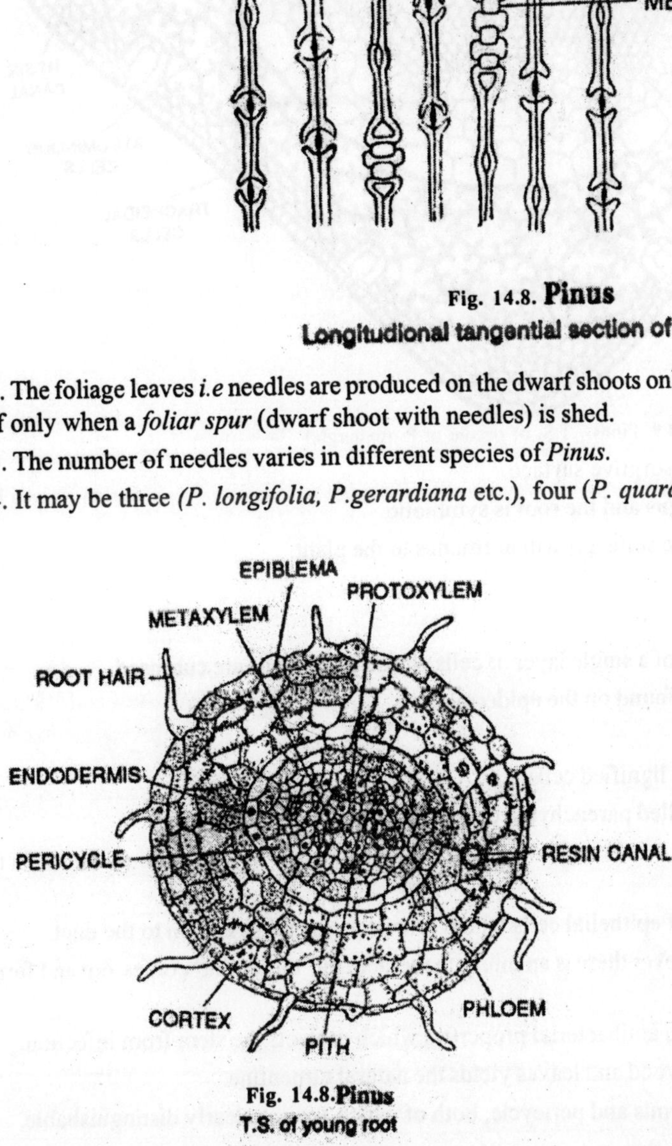

Fig. 14.8.**Pinus**
T.S. of young root

25. The phyllotaxy depends on the number of needles. It may be opposite or whorled.

26. The plant is anchored to the soil with the help of a promiment tap root.

27. The tap root produces a number of well developed lateral roots.

28. The root system is extensive, but does not grow very deep in the soil.

29. The root hairs are not well developed and sometimes even absent on the ultimate branches.

30. The roots have the mycorrhizal association.

31. The fungus is ectotrophic and

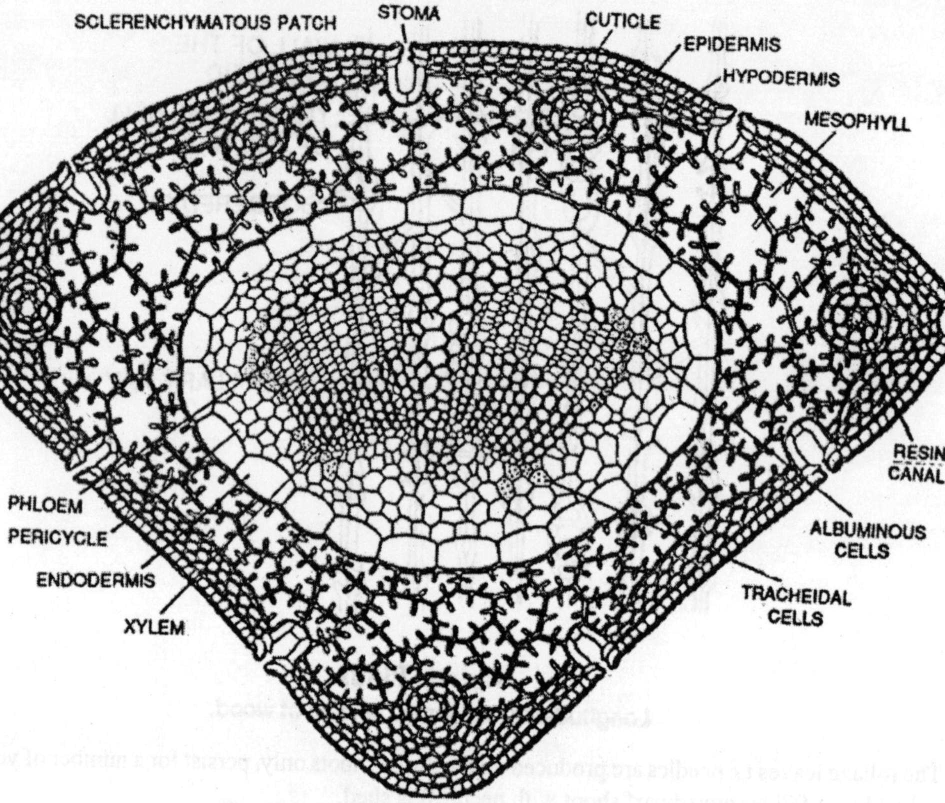

Fig. 14.9. Pinus. T.S. of needle of P. roxburghil

provides the root with an effective absorptive surface.

32. The relationship between the fungus and the root is symbiotic.

33. It is believed that the fungi provide some growth hormones to the plant.

INTERNAL STRUCTURE

Stem

1. Epidermis is irregular and made up of a single layer of cells whose outer walls are cutinised.
2. Occasionally scattered stomata are found on the epidermis.
3. Cortex has two regions.
4. The outer region *i.e* hypodermis has lignified cells.
5. The inner cortex consists of thin walled parenchyma cells with some chlorplasts.
6. Distributed in the cortex are found a number *of resin ducts,* which extend as longitudinal canals in the stem.
7. Each resin duct is lined by a layer of epithelial cells which secrete resin and pass it on to the duct.
8. The resin has *turpentine* and whenever there is an injury to the stem the turpentine comes out and forms a protective covering for the wound.
9. It is known that the turpentine has antibacterial properties which protects the stem from infection.
10. Economically, distillation of pine wood and leaves yields the natural turpentine.
11. The *stele is* surrounded by endodermis and pericycle, both of which are not clearly distinguishable.

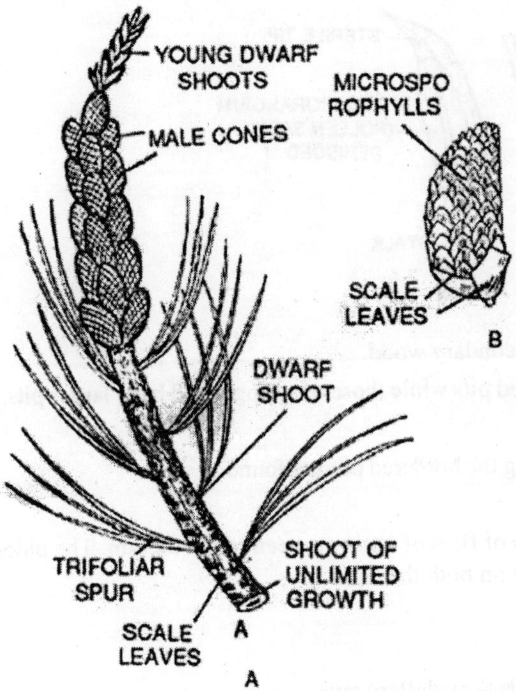

Fig. 14.10.**Pinus**
A. A shoot bearing male cones.
B. A male cone enlarged

12. The vascular tissue in its primary structure consists of 6 -8 collateral, conjoint, open, vascular bundles surrounding a central pith.

13 The xylem is endarch.

14. There are no vessles or wood fibres in the xylem.

15. The protoxylem consists of spiral and annular tracheids with a fewbordered pits.

16. The metaxylem trachieds have numerous bordered pits on-their radial walls.

17. A few wood rays are found in the xylem.

18. Phloem consists of sieve tubes, phloem parenchyma and abuminous cells.

19. The last mentioned cells are in lieu of companion of cells.

20. Between the vascular bundles are present very narrow primary medullary rays.

21. In its essentials, the anatomy of the stem resembles that of dicotyledons.

22. In a transverse section of the stem the primary structure reveals an epidermis, cortex and stele.

Secondary Growth

1. This is brought about by the activity of the fascicular cambium.

2. The cambium which is present in the form of strips in the primary vascular bundles extends radially and forms a closed ring of cambium, the cambial cells divide repeatedly in a tangential direction.

3. At each division two cells are formed. One remains cambial, while the other develops into either xylem or phloem.

4. Secondary xylem is formed towards the interior of the cambial cell while secondary phloem is formed towards the exterior.

5. As a result of secondary growth a completely closed ring of vasculature is formed.

6. Unlike in *Cycas* the cambium here remains as a permanent meristem.

7. Due to the activity of the cambium annual rings are produced.

8. Each annual ring has tracheids of two different sizes *viz.,* spring wood and autumn wood.

9. The former has abroad lumen while the latter has a narrow lumen in their tracheids.

10. Unlike in *Cycas,* the annual rings are regular and they give an approximate idea of the *age* of the plant.

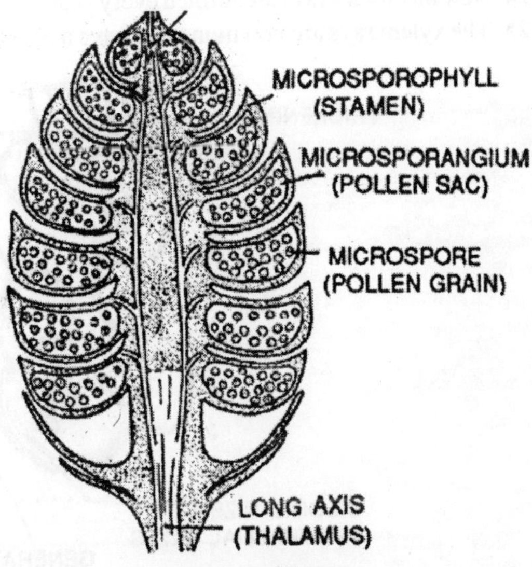

Fig. 14.11.**Pinus**
Long section of a male cone

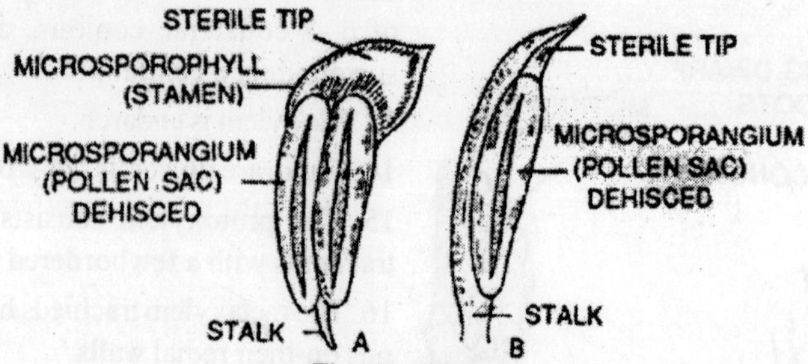

Fig. 14.12. **Pinus**

11. Risin ducts are present both in the primary and secondary wood.
12. The tracheids of autumn wood have small bordered pits while those of spring wood have larger pits.
13. The pits have a distinct torus.
14. Special thickenings of primary lamella surrounding the bordered pits are found.
15. These are called Bars or rims of sanio.
16. When the pits are too close to each other the pairs of Bars of sanio are seen round the rim. The phloem consists of radial rows of long sieve cells which taper on both the ends.
17. The sieve plates occur mostly on the radial walls.
18. The sieve pores may become covered with callus.
19. The primary medullary rays are replaced by secondary medullary rays.
20. Like the secondary vascular tissues they are also formed by cambium.
21. They travel through the secondary xylem and phloem at certain places.
22. The portion of the ray that passes through the secondary xylem is called the xylem ray and the one that passes through the secondary phloem is called the phloem ray.
23. The rays are one celled broad and 2 to 12 cells in height. Occasionally the rays may become multi seriate 'when they come in contact with a resin canal.
24. New medullary rays are formed every year.
25. The xylem rays are non living and have a lignified wall which may protrude into the lumen of the cells.

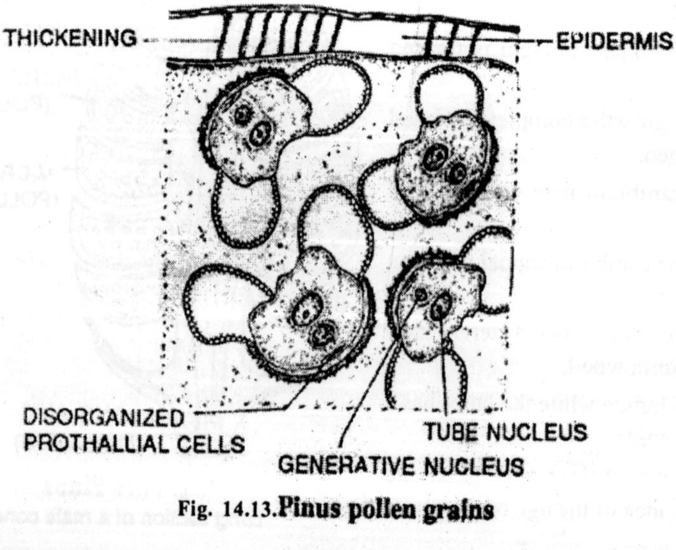

Fig. 14.13. **Pinus pollen grains**

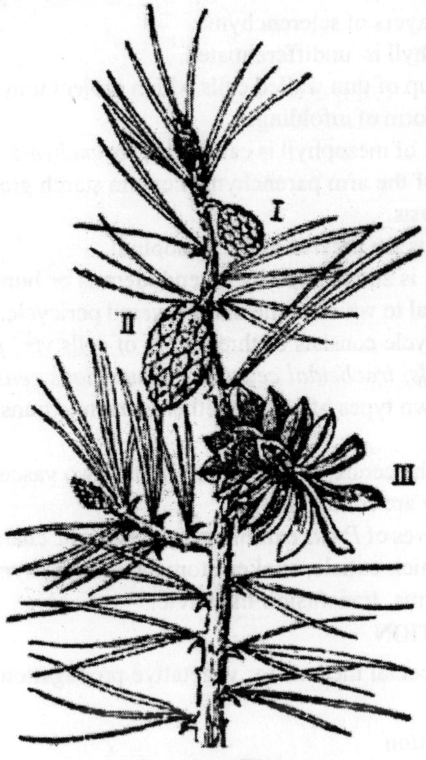

Fig. 14.14. **Pinus**
A shoot bearing female cones

26. The walls have bordered pits. In between the cells of the ray are situated a few living ray parenchyma cells.
27. The phloem rays have albuminous cells in place of ray tracheids.
28. They have a prominent nucleus and dense cytoplasm.
29. The albuminous cells function like the campanion cells of Angiosperms.
30. Between albuminous cells, occur cells of ray parenchyma.
31. The structure of the wood (pycnoxylic) and the secondary medullary rays maybe studied by taking sections of the stem in different planes.
32. These are, (1) Cross section, (2) Radial longitudinal section (RLS) and (3) Tangential longitudinal section (TLS).
33. In a *cross section*, both spring and autumn wood are seen cut transversely. Tracheids of autumn wood are squarish, while those of spring wood are polygonal in outline.
34. Wood rays are cut lengthwise and the breadth of the rays can be seen in this section.
35. In a **R.L.S.**, tracheids are cut lengthwise. Bordered pits are seen showing the surface view.
36. Secondary medullaiy ray is cut at its side, lengthwise.
37. In this, the height (thickness) of the rays can be seen.
38. Pits are seen in the medullary rays also. Both xylem rays and phloem rays may be seen.
39. In a **R.L.S.** the tracheids are cut lengthwise at their side.
40. Sectional views of the bordered pits, with torus is clearly noticeable.
41. Secondary growth in the cortex takes place by the activity of a cork cambium.
42. This produces the periderm tissue.

Root

1. A transverse section of the root shows that it is identical with the dicot root.
2. Epiblema is single layered, from which arise the root hairs.
3. This layer surrounds the parenchymatous cortex.
4. Here and there resin canals are seen. Bounding the stele is the singlc layered endodermis.
5. It is usually impregnated with tannin.
6. Internal to the endodermis is the many layered pericycle.
7. Vascular bundles are radial with exarch and diarch or triarc xylem.
8. The pith is very narrow.
9. In the mycorrhizal roots, the hyphae may be seen in the intercellular spaces of the cortex.
10. Secondary growth in the root takes place by the activity of interfascicular cambium in the stele, and by cork cambium in the cortex.

Leaf

1. The outline of the section depends of the number of needles in the foliar spur.
2. In single leaved spurs *(P. monophylla)* it is circular, in two leaved spurs-semicircular *(P. sylvestris)* and in three leaved spurs it is triangular *(P. longifolia)*.
3. The outermost layer is the epidermis whose cells have thick and heavily cutinised walls.
4. The continuity of the epidermis is interrupted by sunken stomata which are present on all sides of the leaves.

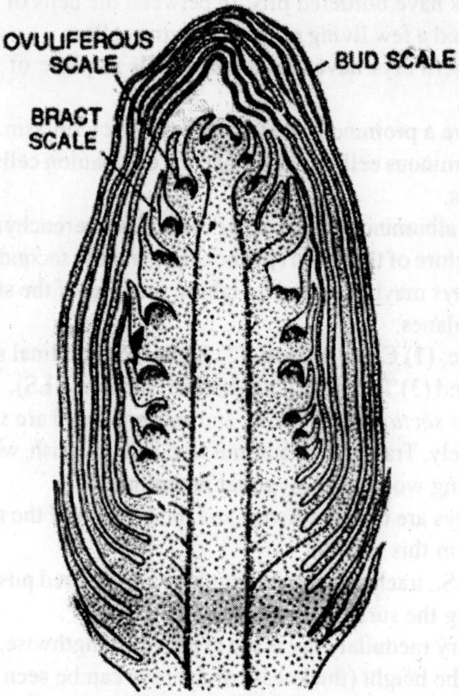

Fig. 14.15. Pinus

L.S. of a young female cone of P. wallichiana

5. Internal to the epidermis is the hypodermis consisting of two to three layers of sclerenchyma.
6. The mesophyll is- undifferentiated.
7. It is made up of thin walled cells which project into the cavity in the form of infoldings.
8. Such a type of mesophyll is called arm *parenchyma.*
9. The cells of the arm parenchyma contain starch grains and chloroplasts.
10. Resin ducts are present in the mesophyll.
11. The vein is surrounded by an endodermis or bundle sheath, internal to which is the many layered pericycle.
12. The pericycle consists of three types of cells *viz., parenchyma cells, tracheidal cells, and albuminous cells.*
13. The last two types of cells together constitute transfusion tissue.
14. Lying in the centre of the vien are found two vascular bundles. They are collateral.
15. The leaves of *Pinus* exhibit many *xerophytic* characters such as thick cuticle, sunken stomata, sclerenchymatous hypodermis, transfusion tissue etc.

REPRODUCTION

There is no special method for vegetative propagation in *Pinus.*

Spore Production

1. The plants *otPinus* are monoecious, but the cones are monosporangiate.
2. The two cones are produced on different branches.

Male cone

1. The male cones are produced on the plant earlier than the females ones.
2. They are usually produced in the month of March or April in hills and January in plains.
3. The male cones are produced in clusters in the axils of scale leaves on the branches of the unlimited growth.
4. On the long shoots, the cones are found at the tip of the youngest shoots.
5. Since the male cones occupy the place of dwarf shoots they are supposed to be the modifications of the latter.
6. A young male cone is ovate in outline measuring about 3-4 cms in length and 0.65 cms in diameter.
7. The cone consists of a central axis on which are spirally arranged microsporophylls which are 60-135 in number.
8. Each microsporophylls is somewhat tri- angular in outline and has a short stalk.
9. The terminal portion of the sporophyll is sterile, pointed

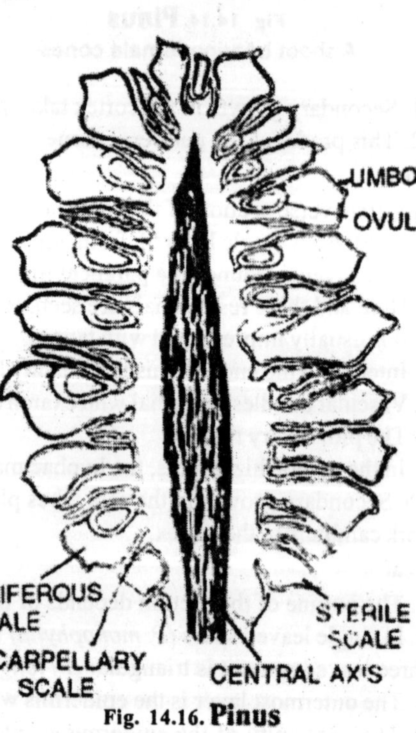

Fig. 14.16. Pinus

L.S. of mature female cone of P. roxburghii

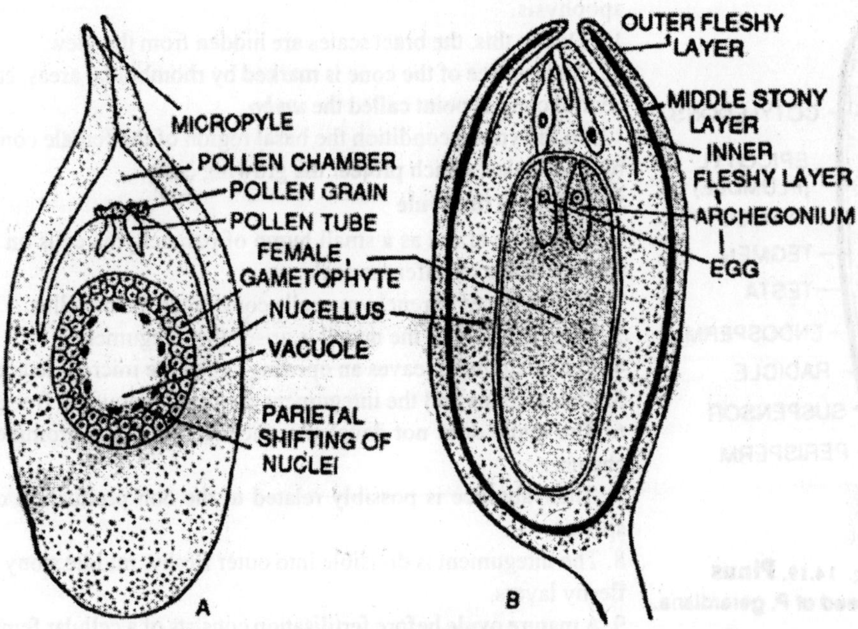

Fig. 14.17. **Pinus**
L.S. of ovules, A. ovule with young female gametophyte, B. ovule with female gametophyte ready for fertilisation

and slightly bent upwards.

10. The basal portion of the sporophyll on its abaxial surface bears two microsporangia.

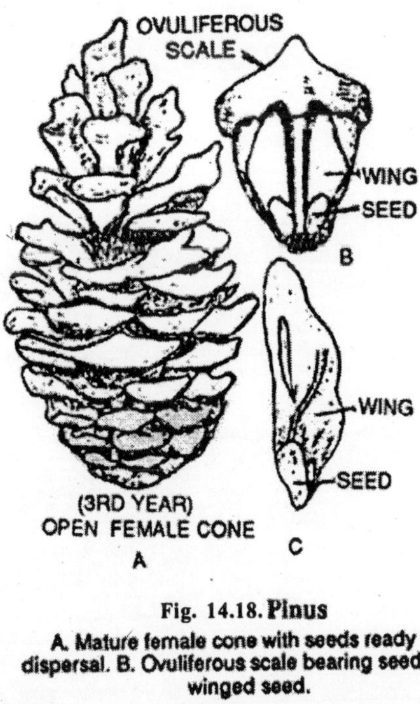

Fig. 14.18. **Pinus**
A. Mature female cone with seeds ready for dispersal. B. Ovuliferous scale bearing seeds. C A winged seed.

Female cone

1. This is also known as the- *ovulate cone.*
2. The female cones are produced on the uppoer part of the stem in the axils of scale leaves.
3. They take the position of a shoot of unlimited growth and are apparently the modifications of the latter.
4. On the newly formed long shoots which do not bear male cones, the female cones occur below the terminal bud, each representing a newly formed long shoot.
5. The young female cone is a small reddish green structure consisting of a central axis on which are arranged spirally the megasporophylls.
6. Each megasporophyll has two appendages *viz., the bract scale* (carpellary scale) and the *ovuliferous scale* (seed scale).
7. The bract scale is leathery and is directly attached to the cone axis.
8. The ovuliferous scale is larger than the bract scale and stouter and harder.
9. It develops from the upper surface of the bract scale and bears two ovules as swellings.
10. The ovuliferous scale is woody, wedge shaped and has a broad sterile end directed outwards, which is called the

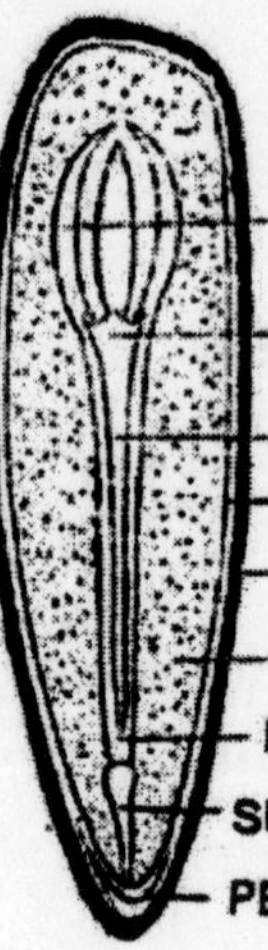

Fig. 14.19. Pinus
L.S. of seed of P. gerardiana.

apophysis.

11. Due to this, the bract scales are hidden from the view.

12. The surface of the cone is marked by rhomboidal areas, each with a central concial point called the *umbo*.

13. In the young condition the basal region of the female cone covered by bud scales, which project the growing cone.

Structure of the ovule

1. The ovule arises as a small hump of parenchyma cells on the upper surface of the ovuliferous scale.
2. This hump of parenchyma cells constitutes the nucellus.
3. From the base of the nucellus arises the integument.
4. The integument leaves an opening called the micropyle at one end.
5. The nucellus and the integument are fused below but free above.
6. The ovule does not develop a nucellar beak or a prominent pollen chamber.
7. Their absence is possibly related to the non production of ciliated sperms,
8. The integument is divisible into outer fleshy, middle stony and inner fleshy layers.
9. A mature ovule before fertilisation consists of a cellular female gametophyte surrounded by a thin layer of nucellus. 2-6 archegonia are found near the micropylar region of the female gametophyte.
10. A small space is left between the nucellus and the female gametophyte to form the archegonial chamber.
11. The pollen chamber is not properly defined.
12. Lodged in it are found a few pollen grains.
13. There is no cellular beak as in *Cycas*.
14. The integument is three layered.

Systematic position

CONIFEROPSIDA

CONIFERALES

PINACEAE

PINUS

Chapter 15

GNETOPSIDA

1. The class **Gnetopsida** comprises some of the most advanced members of gymnosperms.
2. The class consists of an order Gnetales.
3. But some botanists identify three orders namely *Ephedrales, Welwitschiales* and *Gnetales.*

GNETALES

1. The members of the order are woody plants: they may be trees, shrubs or lianes.
2. Many of the plants resemble dicotyledons in their general morphology.
3. Plants may be monoecious or dioecious and reproduce by means of "flowers".
4. Members have vessels in the secondary wood.
5. The order consists of three families *viz.* , *Gnetaceae, Welwitschiaceae* and *Ephedraceae.*

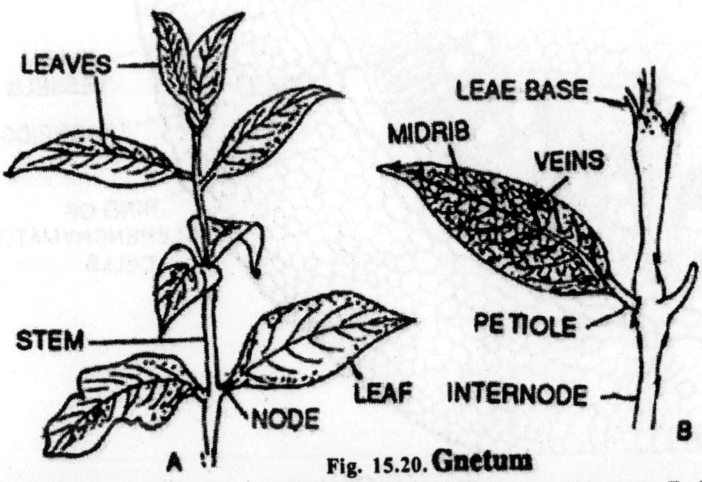

Fig. 15.20. **Gnetum**

Habit of G. gnemon. A. A twig showing arrangement of leaves. B. A leaf showing venation.

GNETUM

The family *Gnetaceae* consists of a single genus *Gnetum,* with about *40* Species, which are distributed all over the tropics, mostly in Asia, America and Africa. In our country *Gnetum* is represented by six species (*G. montana, G. gnemon, G. latifolium, G. contractum* and *G.oblongum*). In India the distribution of *Gnetum* is mostly in the eastern Himalayas and southern peninsula between altitudes of 1000-2000 meters. In Karnataka, *G. ula* is common in western ghats.

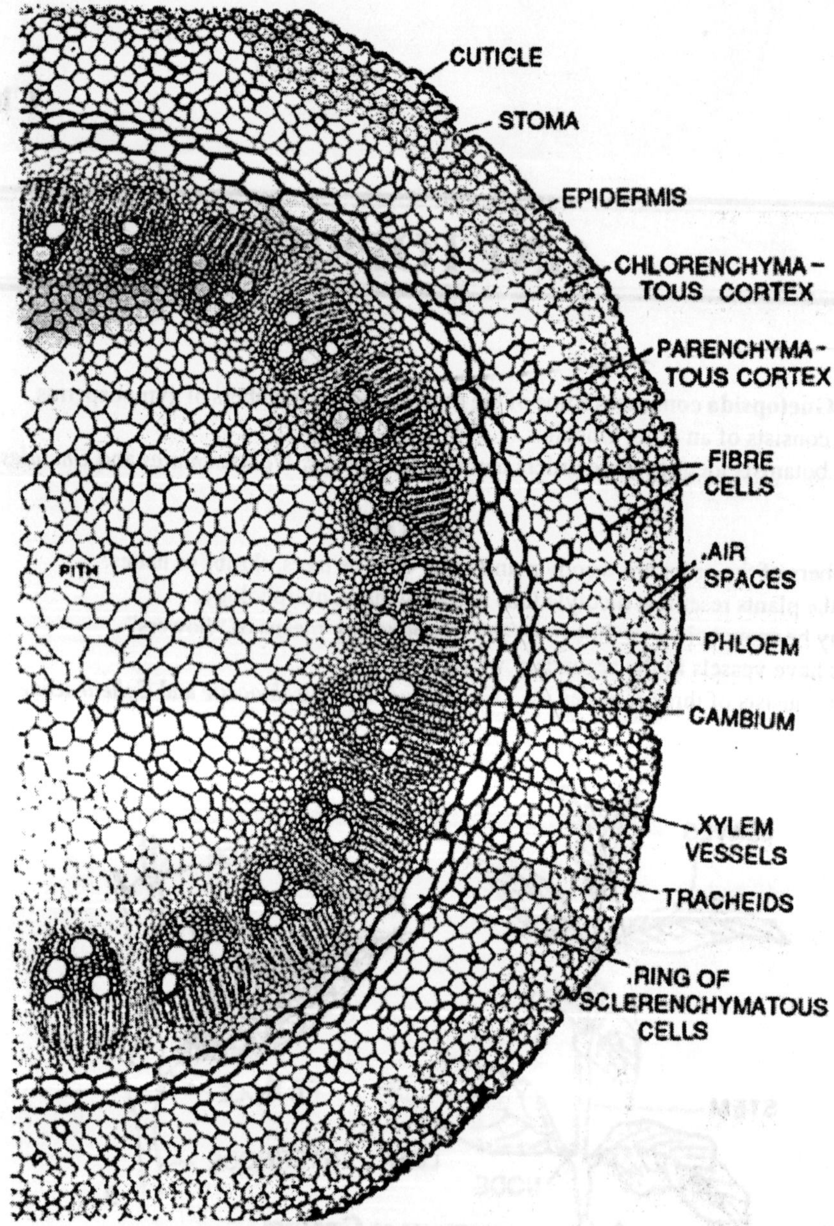

Fig. 15.21. **Gnetum**
T.S. of young stem of G. ula. (a portion enlarged)

Morphology

1. In its general habit, *Gnetum* remarkably resembles a dicotyledolums plant.

2. The nature and habit of the plant is varied.

3. A great majority of the spedes are climbers with twining stems (G. *neglectum, G. latifolium* etc.), while others may be shrubs or trees.

4. Tree habit is found in G. *gnemon,* while G. *ula* is a liane (woody climber).

5. In the climbing species the lower part of the stem is usually devoid of leaves.

6. The stem bears two types of branches, *namely* branches of *unlimited growth* and branches of *limited growth.*

7. The difference between the two, is not as pronounced as in *Pinus.*

8. The stem is differentiated into nodes and internodes. Internodes are present even on the branches of limited growth.

9. The stem in several species is provided with articulations. It is at these joints that the branches are shed

10. The leaves of *Gnetum* are exactly similar to those of dicotyledons.

11. In climbing species, the foliage leaves are borne only on the branches of limited growth.

12. At each nodal region two opposite leaves are present.

13. About 9-10 leaves are arranged decussately on each branch.

14. Leaves are petiolate and have a large, oval and entire lamina with reticulate venation.

15. The leaves are leathery in texture.

16. The stem is anchored to the soil with the help of the tap root system.

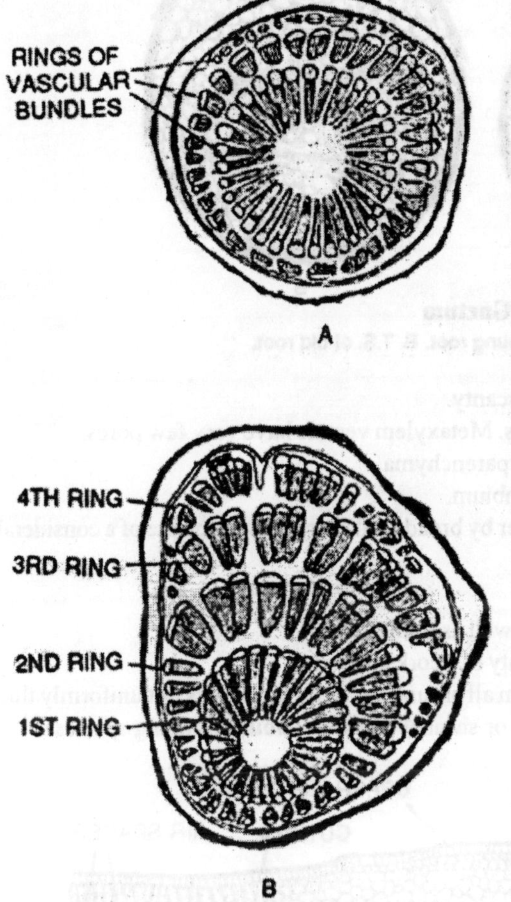

Fig. 15.22. **Gnetum**
Secondary growth. A. T.S. of stem showing concentric rings of vasculature. B. T.S. of stem showing eccentric rings of vasculature.

INTERNAL STRUCTURE

Stem

1. Transverse section of a young stem in its primary structure reveals the following features.

2. The outermost layer is the epidermis.

3. It is made up of rectangular cells which bear papillate outgrowths.

4. There is a deposition of the cuticle above the epidermis.

5. The continuity of the epidermis is interrupted by the presence of stomata.

6. Internal to the epidermis is the cortex, consisting of three distinct regions.

7. The outer cortex is chlorenchymatous and made up of 4-6 layers of oval chlorenchyma cells.

8. The middle cortex is parenchymatous, having compactly arranged, thin walled, polygonal cells.

9. The innermost cortex is made up of two to four layers of sclerenchymatous cells.

10. Next to the cortex, internally are present inconspicuous endodermis and pericycle.

11. The stele consists of a number of open, collateral vascular bundles arranged in the form of a ring surrounding a central pith. They xylem is endarch and consists of mostly tracheids and a few vessels.

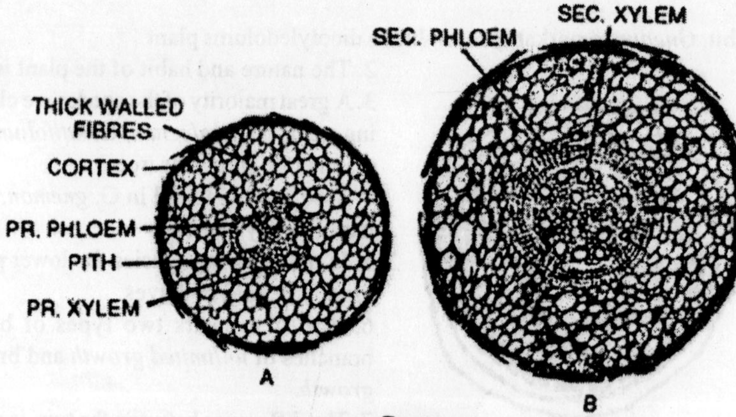

Fig. 15.23. Gnetum
Anatomy of root. A. T.S. of young root. B. T.S. of old root.

12. Xylem fibres are absent and xylem parenchyma is scanty.
13. The protoxylem vessels have circular bordered pits. Metaxylem vessels have very few pores.
14. Primary phloem consists of sieve cells and phloem parenchyma.
15. Between xylem and phloem is present a strip of cambium.
16. The vascular bundles are separated from one another by broad medullary rays which are of a considerable height.

Secondary growth

1. Secondary growth takes place both in the cortex as well as in the stele.
2. In the cortex periderm formation occurs by the activity of a cork ambium.
3. Since the activity of the cork combium is not uniform all around the stem, the cork is not uniformly thick.
4. Secondary growth in the stele is both *normal* (trees or shrubs) and *anomalous* (climbing species).

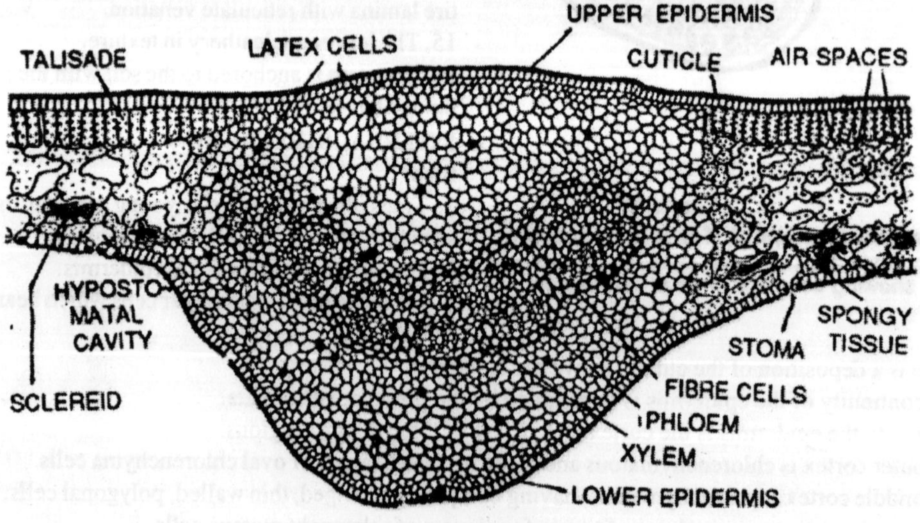

Fig. 15.24. Gnetum
T.S. of leaf

5. Normal secondary growth is brought about by the activity of a fascicular cambium.
6. Secondary medullary rays are present.
7. Secondary xylem consists of three types of elements *viz., tracheids, xylem parenchyma,* and *vessels.*
8. Tracheids have circular bordered pits which are uniseriate in arrangement.
9. Xylem parenchyma cells are living thin walled cells.
10. Vessels are elongated cells with tapering ends and possess bordered pits.
11. The end walls are perforated.
12. The number of perforations varies with species.
13. In *G. ula* there is a single large perforation at the end walls.
14. The vessles develop from tracheids whose end walls have a number of bordered pits.
15. By the elimination of the end walls a vessel with a single large perforation is produced.
16. Secondary phloem also consists of three types of elements *viz.*, sieve cells, phloem parenchyma and albuminous cells.
17. Anomalous secondary growth is seen in climbing species like *G. ula.* Here the fascicular cambium ceases to function soon after the first vascular ring is formed.
18. Successive cambial rings arise at different positions in the cortex and produce the vascular rings.
19. Sometimes the cambia do not form a closed ring but remain as patches, due to the presence of secondary medullary rays with the result long vascular bundles with secondary xylem and phloem are formed.
20. These bundles in a transverse section resemble the canine teeth.
21. In some cases the cambium may form a closed ring but its activity is not uniform all round.
22. As a result, *eccentric* rings of vasculature is formed.

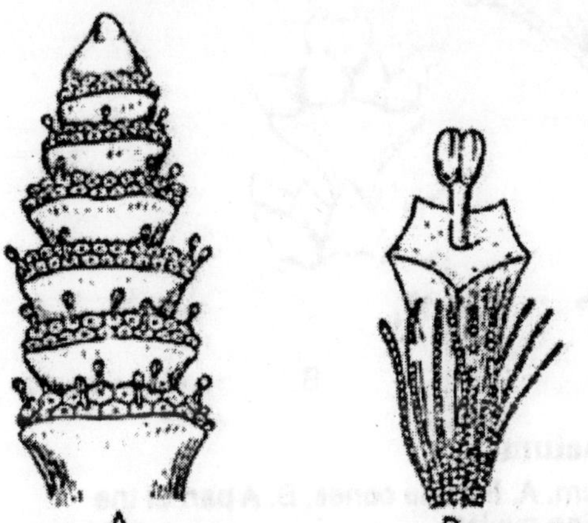

Fig. 15.25. **Gnetum**
Male reproductive structures, A. A portion of male cone, B. A single male flower.

Root

1. A transection of a young root shows a single layered epidermis surrounding a many layered parenchymatous cortex.
2. The cells are filled with starch.
3. The stele is surrounded by a pericycle and endodermis.
4. Vascular bundles are radial, two in number with exarch and diarch xylem.
5. Primary xylem consists of both tracheids and vessels.
6. There is a parenchymatous pith n the centre.
7. Secondary growth takes place by the activity of a fascicular cambium.
8. It is normal and forms a completely closed ring of vasculature.
9. The secondary xylem is similar to that of stem in its essentials.
10. Secondary medullary rays are multiseriate.

Leaf

1. A transverse section of the leaf shows a bifacial (dorsiventral) arrangement.
2. The upper epidermis is single layered and has cells with undulating walls; the cells of the lower epidermis are similar in structure.
3. There is a deposition of cuticle above the epidermal layers but it is comparatively thin towards the lower epidermis.
4. Leaves are hypostomatic.
5. The type of stomata may be *syndetocheilic* (a single mother cell produces both the guard cells and the two

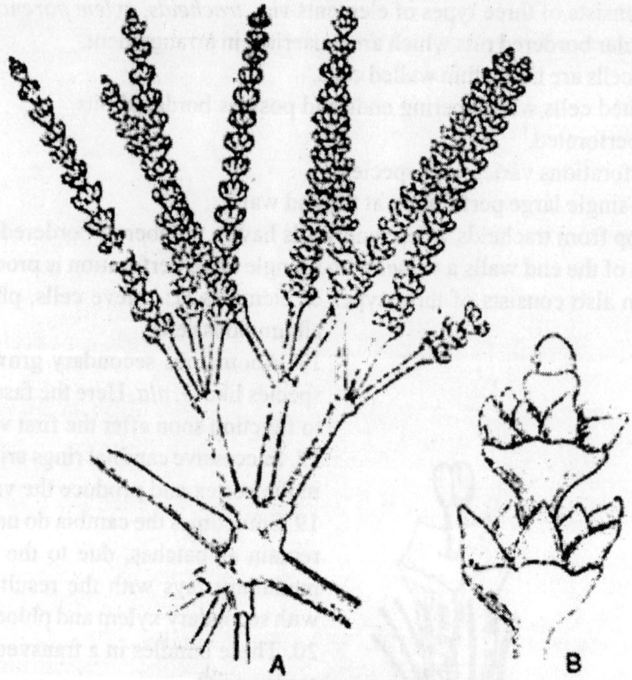

Fig. 15.26. **Gnetum reproductive structures of G. latifolium, A. Female cones, B. A part of the female cone showing ovules.**

subsidiary cells) as in *G.gnemon* or *haplocheilic* (stomatal initial produces only guard cells) as in *G. ula*.

6. The mesophyll consists of distinct palisade and spongy parenchyma.
7. However surrounding the midrib the cells are compact and the mesophyll is undifferentiated.
8. The vascular region in the midrib consists of an arc of vascular bundles.
9. In young leaves vascular bundles are not distinct.
10. The vascular tissues show x/p arrangement.
11. The bundles are conjoint, collateral and endarch.
12. In *G. ula,* each vascular bundle has a patch of stone cells below the phloem.
13. Around the midrib a few crystal cells occur.

REPRODUCTION

1. All the species of *Gnetum* are dioecious.
2. The reproductive organs are grouped into strobili or inflorescences.
3. The strobilus (cone) consists of a stout axis which bears at its base two opposite, connate bracts, in the axils of these bracts arise other inflorescences.

Male cone

1. The cone consists of an elongated axis and is distinguishable into nodes internodes.
2. At the nodal region are borne connate bracts forming a cup like structure called the *cupule* or the *collar*.
3. Above each collar, are found the male flowers arranged in three to six rings.
4. In each ring there will be a number of male flowers.
5. In some cases just above the male flowers there is a ring of imperfect female flowers.

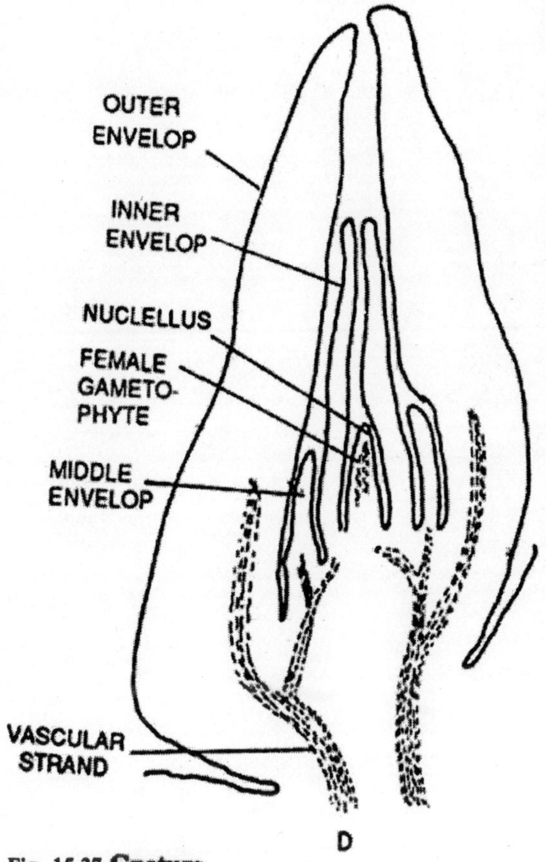

Fig. 15.27. **Gnetum**
the development of ovule in G. ula.

6. The male flower which is enclosed in a sheath like perianth consists of a stalk bearing two unilocular anthers or microsporangia.
7. Surrounding the perianth are found a number of uniseriate, multicellular hairs.
8. Perianth consists of sclerids and laticiferous tubes.
9. A young male cone is completely enclosed within the two basal bracts.
10. Later, the cone axis emerges piercing through the bracts.

Female cone

1. The female cone is also called the ovulate cone.
2. It is constructed on the same general plan as the male cone.
3. It consists of an axis in which are arranged a series of cupules formed by the connate bracts.
4. In the young condition it is difficult to distinguish between the male and female cones.

Within the cupule are found 4-10 female flowers (ovule) arranged in the form of a ring. Each ovule consists of a central mass of nucellus being surrounded by three envelops.

Structure of the ovule at the time of fertilisation

1. The ovule is surrounded by three envelopes. The morphological nature of these envelopes is uncertain.
2. Of the three envelopes, the innermost elongates to form a micropylar tube or style.
3. There is no poollen chamber or archegonial chamber.
4. The envelope encloses the nucellus embedded in which is found the embryo sac.
5. Two or three eggs are found near the micropylar region.
6. The lower portion is free nuclear. But in some cases the chalazal portion may become partly cellular.
7. Below the embryo sac, the nucellus forms a fan shaped tissue called the *pavement tissue*.

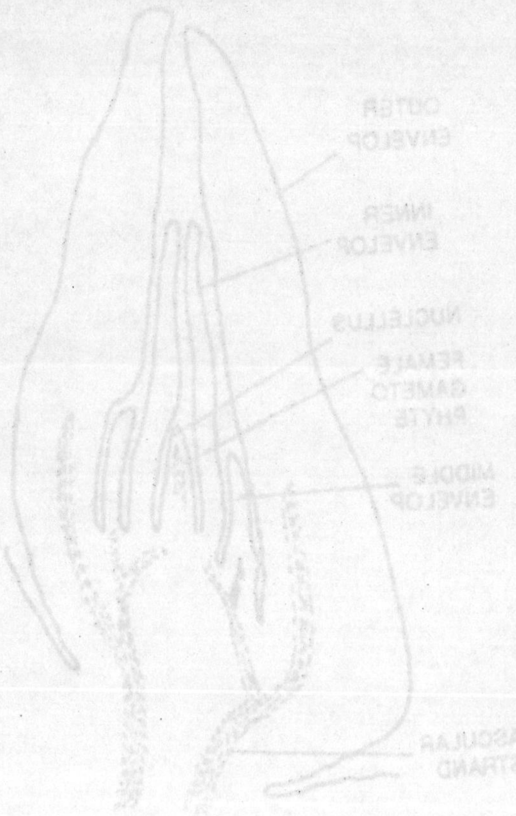

PART THREE
PLANT ANATOMY

Chapter 16

STUDY OF PLANT CELL

Purpose : To study the structure of a typical plant cell under microscope by using thin layer of onion peel.

Materials : Onion bulbs, slides, coverslip, microscope and stain (saffranin).

Procedure : Carefully peel off a thin layer from the bulb of the onion. Initially it may be difficult to get a thin layer but after the practice it will be possible to get a thin layer which is suitable for microscope examination. For this purpose carefully separate one fleshy leaf of the onion bulb. Bend it in the middle so that it incompletely breaks. When you separate the two pieces a thin layer will come out. Immediately after removing place the onion peel in a petridish containing water.

Cut a small piece of the onion peel suitable for observation and place it in a drop of water in the centre of a clean glass slide. Add a drop of stain, put a cover glass and observe under microscope.

Observation and conclusions. The study shows the following structural details of the plant cell.

1. A number of rectangular cells can be seen under the field of the microscope.
2. The cells are many times longer than broad and are joined end to end and side to side to form a layer.
3. The cell wall which is the outermost layer can be clearly demarkated.
4. The cell membrane is not very clearly visible.

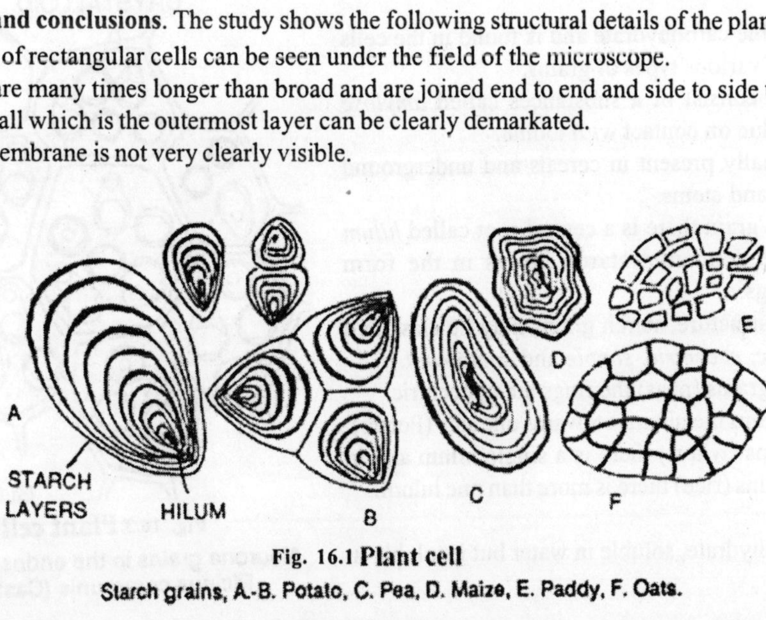

Fig. 16.1 **Plant cell**

Starch grains, A-B. Potato, C. Pea, D. Maize, E. Paddy, F. Oats.

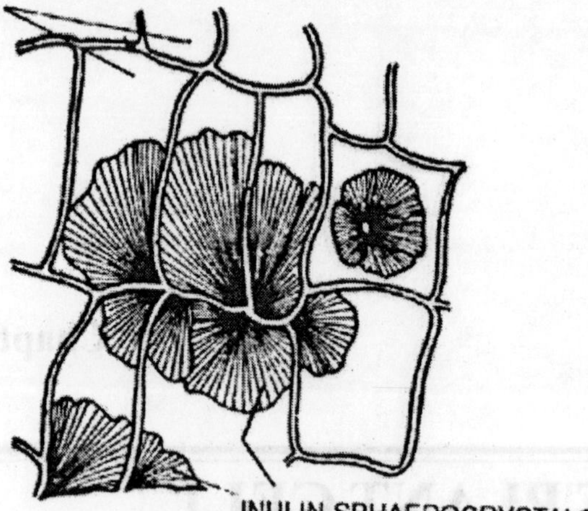

Fig. 16.2 **Plant cell**

Sphaerocrystals of inulin in the roots of Dahlia.

5.Occupying the centre of the cell is a large vacuole.

6.Cytoplasm is peripheral and occupies the space between the cell membrane and the vacuole.

7.Lying in the cytoplasm is a large vacuole which is red coloured (because it has taken saffranina stain).

8.Also distributed in the cytoplasm are seen a number of leucoplasts.

ERGASTIC SUBSTANCES (Non living inclusions):

Many kinds of organic and inorganic compounds occur as non living inclusions in cytoplasm. These are produced during the metabolic activities of the cell and are also known as ergastic substances. Principally, there are three types of ergastic substances, *viz., Reserve food products, Secretory products* and *Waste products.*

1. RESERVE FOOD PRODUCTS: These are synthesised in cytoplasm and are stored as reserve food material. These are classified into three categories *viz., Carbohydrates* (non nitrogenous products), *nitrogenous products* and *fats* and *fatty oils.*

A. Carbohydrates:

1. These are the principal types of food substances containing carbon, hydrogen and oxygen generally in the ration 1: 2: 1.
2. Carbohydrates may or may not be soluble in water.
3. Some of the important carbohydrates are: Starch, Inulin, Hemicellulose, Cellulose and Sugars.

Starch:

1. It is an insoluble carbohydrate and is found in the cells in the form of various types of grains.
2. Starch grains consist of a substances called *amylose* which turns blue on contact with iodine.
3. These are usually present in cereals and underground storage roots and stems.
4. In each starch grain there is a central spot called *hilum* on which are deposited starch layers in the form concentric rings.
5. Based on the structure, starch grains may be classified into *concentric, eccentric, simple* and *compound.*
6. In concentric grains (peas) the rings are concentric.
7. In eccentric grains the hilum is towards one side (Potato), in simple grains (wheat) there is a single hilum and in compound grains (rice) there is more than one hilum.

Inulin:

1. This is a carbohydrate, soluble in water but insoluble in alcohol.

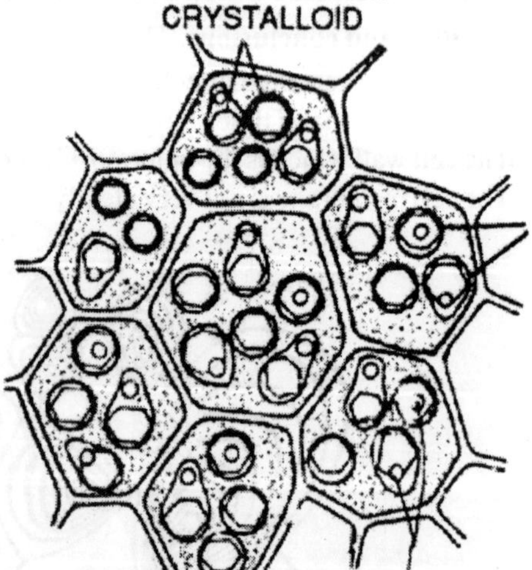

Fig. 16.3 **Plant cell**

Aleurone grains in the endosperm of Ricinus communis (Castor)

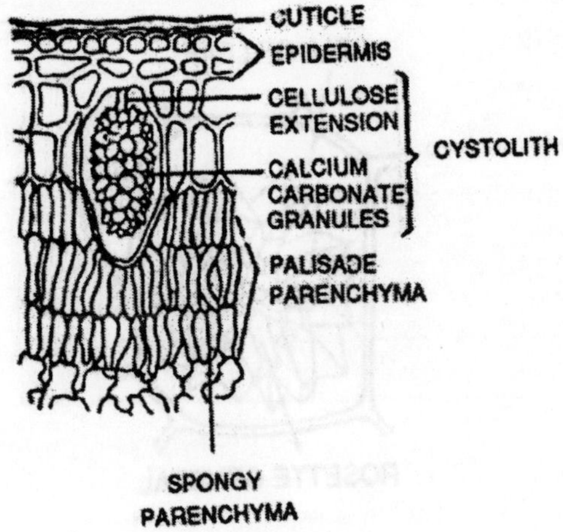

Fig. 16.4. Plant cell
Calcium carbonate cystolith in the leaf of Ficus elastica.

2. It is generally found in the tuberous roots of the members of the family *Asteraceae* such as *Dahlia, Helianthus, Cichorium* etc.
3. In these members inulin forms fan shaped cyrstals.
4. The inulin of *Dahlia* tubers can be precipitated by keeping them in alcohol.

B. Nitrogenous Products:

1. These are reserve food materials containing nitrogen.
2. There are two types of nitrogenous food materials viz., proteins and amino compounds.
3. *Proteins* are very complex macromolecules having carbon, hydrogen, oxygen and nitrogen.
4. Sometimes, in addition to these, sulphur and phosphorous are also present.
5. Proteins are stored in seeds or grains in the form of *aleurone crystals*.
6. These are present in the endosperm of castor, pericarp of maize grain etc.
7. Each aleurone grain (crystal) is a solid, oval or rounded body enclosing a small crystal called *crystalloid* and a small globular body called *globoid*.
8. The crystalloid is proteinaceous and occupies the major portion of the grain, while the globoid consists of phosphates of calcium and magnesium and occupies the rest of the grain.

Amino Compounds occur in the form of amino - acids and amines in the cell sap. These are more frequently present in the growing parts rather than in storage tissues.

C. Mineral Crystals:

1. Various types of mineral crystals are found in plant cells.
2. They may be present either in cell walls or in a cavity in the cytoplasm.
3. Majority of the crystals contain calcium carbonate or calcium oxalate.

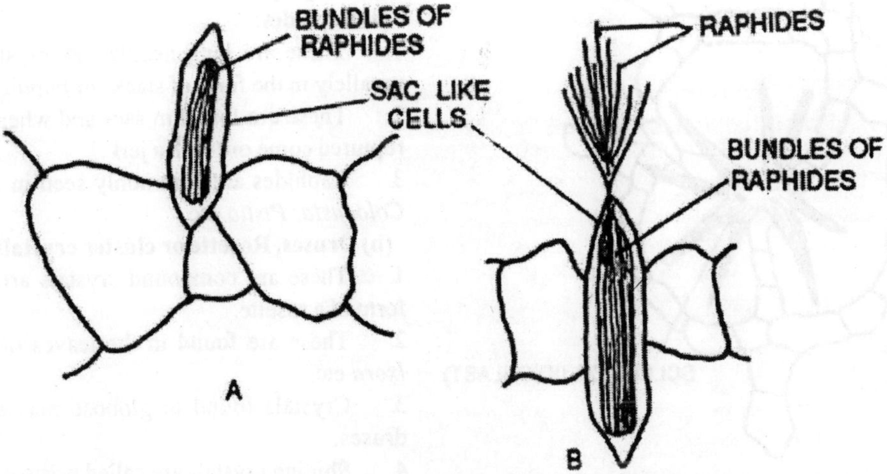

Fig. 16.5 **Plant cell**
Raphides - A. raphides in colocasia, B. ejection of a raphide.

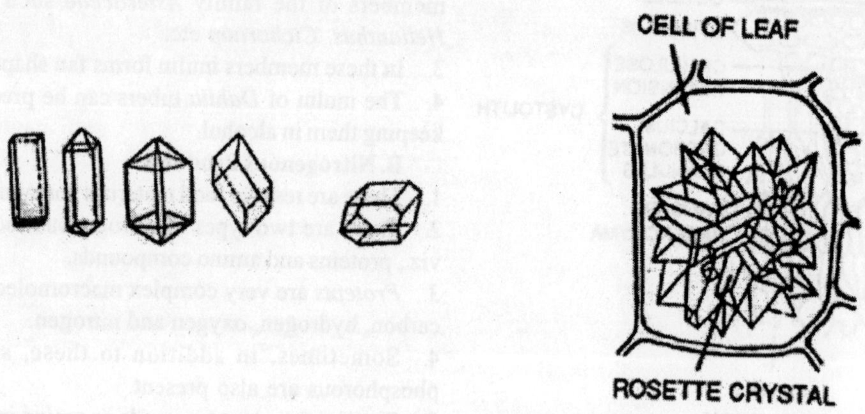

Fig. 16.6 **Plant cell**
Calcium oxalate crystals. Prismatic crystals (left) and Rosette crystal (right)

4. The following are some of the types of mineral crystals.

1. Cystoliths:

1. These are crystals of calcium carbonate, generally found in the leaves of various flowering plants.
2. The members of the family Moraceae, Urticaceae etc., have cystoliths.
3. Each cystolith has a peg like projection of the cell wall into the cell cavity.
4. Cystoliths consist of many small crystals grouped together like a bunch of grapes.

2. Calcium oxalate crystals:

1. These are found in the vacuoles and are of the following types.

(a) Raphides:

1. These are long needle like crystals arranged parallely in the form of stacks or bundles.
2. These are found in sacs and when the sacs are ruptured come out with a jerk.
3. Raphides are commonly seen in the leaves of *Colocasia*, *Pistia* etc.

(b) Druses, Rosette or cluster crystals:

1. These are compound crystals arranged in the form of a rosette.
2. These are found in the leaves of *Eucalyptus*, *Ixora* etc.
3. Crystals found in globose masses are called druses.
4. Shining crystals are called prismatic crystals.

(c) Idioblasts:

1. These are stellate crystals and are usually found in water plants (*Limnanthemum*, *Nymphea* etc.)

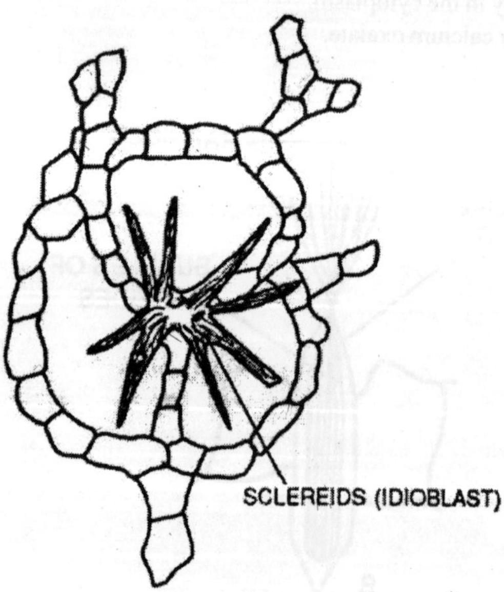

Fig. 16.7 **Plant cell**
Idioblast in a leaf cell

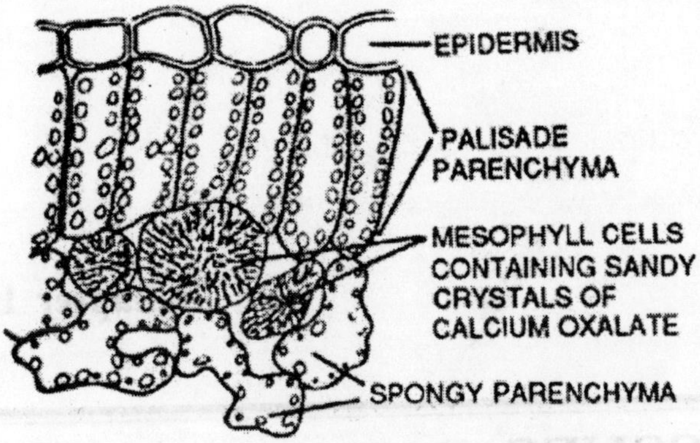

Fig. 16.8 **Plant cell**
Sand crystals of Atropa belladona.

offering mechanical support.

(d) Sand crystals:

1. These are found in masses packed into a cell.
2. Such crystals are seen in the roots and leaves of *Atropa belladona*.

(e) Latex:

1. In some plants, a milky or watery juice comes out of latex vessels or cells.
2. Latex contain starch grains proteins, oils, tannins, gums, resins, alkaloids, salts, enzymes and many poisonous substances.
3. Laticiferous cells are seen in *Carica, Euphorbia, Ficus, Hevea, Calatropis* etc.
4. In some cases the dried and processed latex is the source of natural rubber.

(f) Essential oils:

1. These are volatile oils and are secreted by oil glands.
2. Oil glands are found in almost all parts of the plants, specially in leaves and in the skin of fruits (*Citrus* fruits).
3. Essential oils are also found in tl

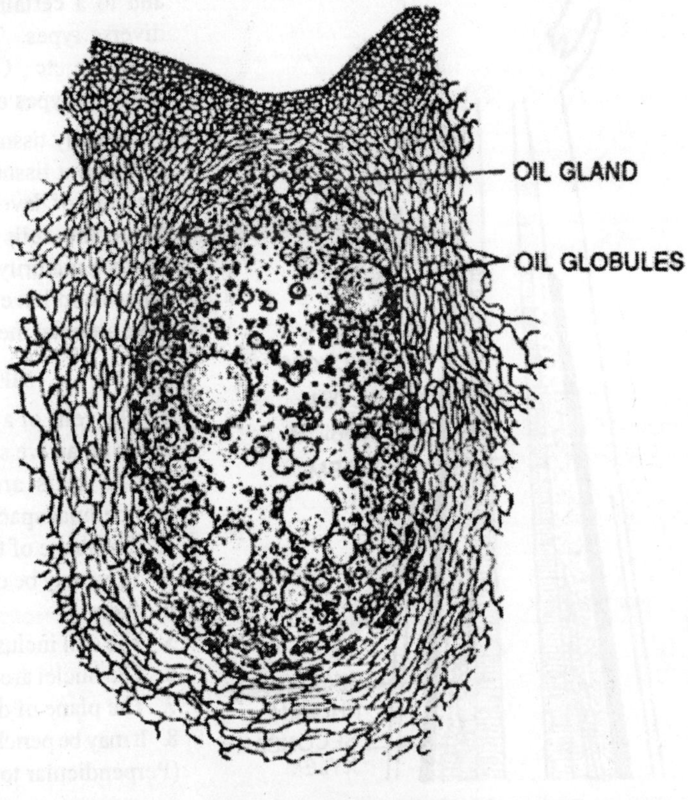

Fig. 16.9 **Plant cell**
Long section of an oil gland from Cirtus aurantiaca.

Chapter 17

TISSUES

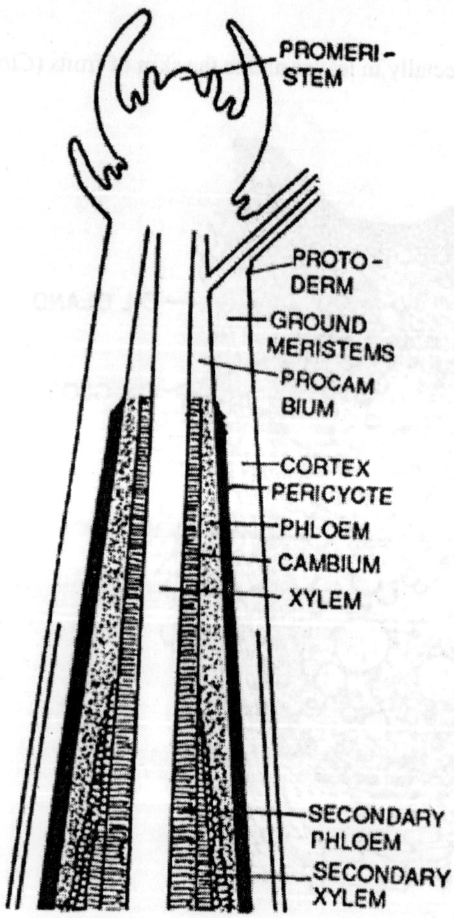

Fig. 17.1 **Meristems**
L.S. of shoot apex showing meristems based on their function

Cells are grouped together to form tissues. Generally cells which are alike in origin, form and functions constitute a tissue. In higher plants like, Gymnosperms and to a certain extent in Pteridophytes cells are of diverse types. They vary in their origin, shape, size, structure etc., Consequently the assemblage of these different types of cells form various types of tissues.

Basically tissues are classified into meristematic and permanent tissues. The basis for this classification is the stage of development of the cells. In a meristematic tissue the cells are actively dividing and have not attained maturity. The cells of a permanent tissue onthe other hand have lost the capacity of cell division and have attained their mature form.

MERISTEMATIC TISSUES

1. The cells of a meristematic tissue are immature and are in an active state of division.
2. The cells are usually closely packed with no intercellular spaces.
3. The shape of the meristematic cells varies.
4. They may be oval, polygonal rounded or rectangular in shape.
5. The cell inclusions will be in their precursor stage.
6. The nuclei are very prominent.
7. The plane of division in the cells varies.
8. It may be periclinal (parallel to the surface) or anticlinal (Perpendicular to the surface).

Classification of meristems:

1. According to the location, there are three types of meristems, viz., apical, intercalary and lateral.

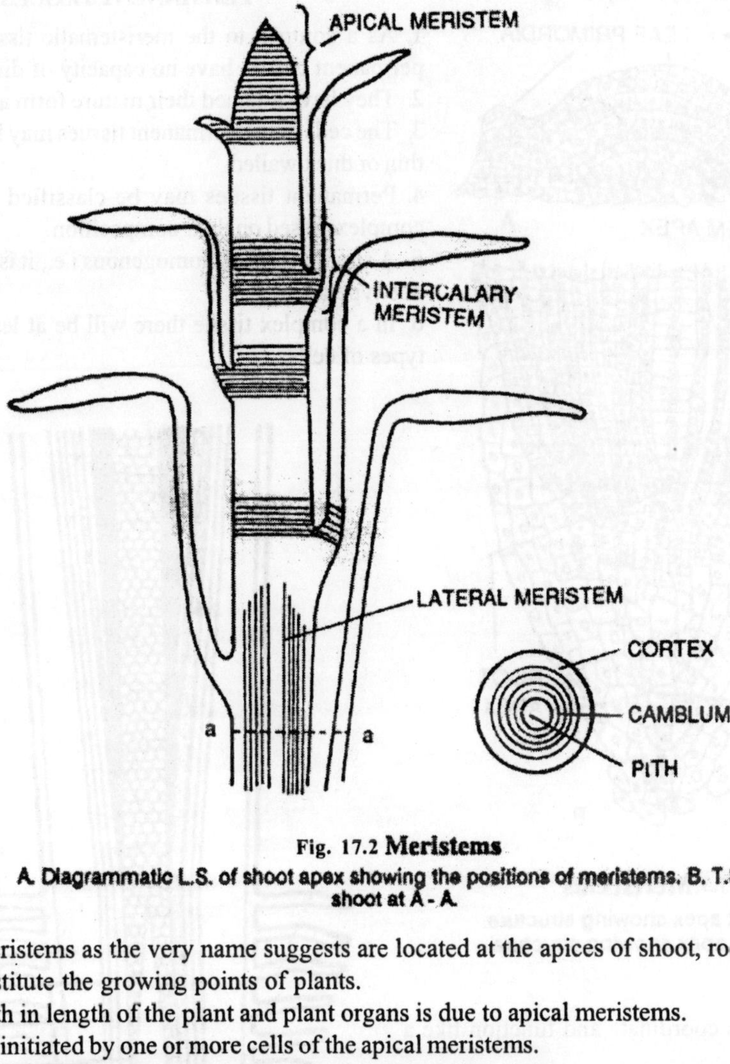

Fig. 17.2 **Meristems**

A. Diagrammatic L.S. of shoot apex showing the positions of meristems. B. T.S. of shoot at A - A.

2. Apical meristems as the very name suggests are located at the apices of shoot, root and other organs.
3. They constitute the growing points of plants.
4. The growth in length of the plant and plant organs is due to apical meristems.
5. Growth is initiated by one or more cells of the apical meristems.
6. Apical meristem consists of two zones, promeristem and meristem.
7. Promeristem consists of actively dividing cells called apical initials.
8. The meristematic zone is classified into protoderm, procambium and ground meristems.
9. Intercalary meristems represent the detached portions of the apical meristems seperated due to the growth of an organ.
10. Intercalary meristems are always internodal.
11. In the early stages of development of the plant, whole or part of the internode is meristematic.
12. Subsequently a portion of it matures, as a result, patches of meristems are found embedded in permanent tissues.
13. Lateral meristems are present along the side of the stem.
14. The cells divide in only one plane and add to the girth of the tissues.
15. Typical example of a lateral meristem is the vascular cambium.
16. Vascular cambium produces secondary Xylem and Phloem thus bringing about secondary growth.

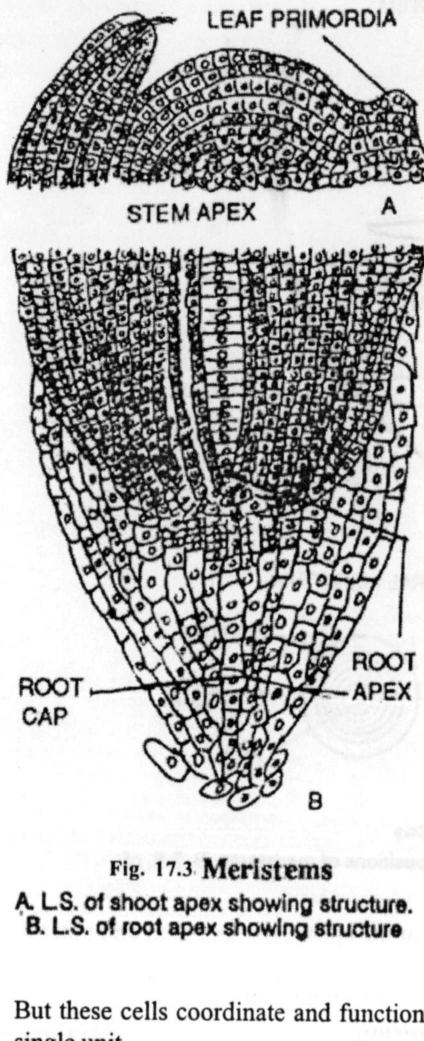

Fig. 17.3 **Meristems**

A. L.S. of shoot apex showing structure.
B. L.S. of root apex showing structure

PERMANENT TISSUES

1. As a contrast to the meristematic tissues, the cells of permanent tissues have no capacity of division.
2. They have attained their mature form and shape.
3. The cells of the permanent tissues may be living or dead, thin or thick walled.
4. Permanent tissues may be classified into simple and complex, based on their composition.
5. A simple tissue is homogenous i.e., it is made up of only one type of cells.
6. In a complex tissue there will be at least two different types of cells.
7. But these cells coordinate and function like a single unit.
8. Generally one or more types of simple tissues combine together to form a complex tissues.

SIMPLE PERMANENT TISSUES:

There are further classified into Parenchyma, Collenchyma and Sclerenchyma.

PARENCHYMA:

1. Parenchyma refers to a tissue composed of living cells variable in their morphology and physiology.
2. The word parenchyma is derived from the Greek *para*, beside and *en-chein*, to pour, meaning a semi-liquid substance poured beside other tissues.

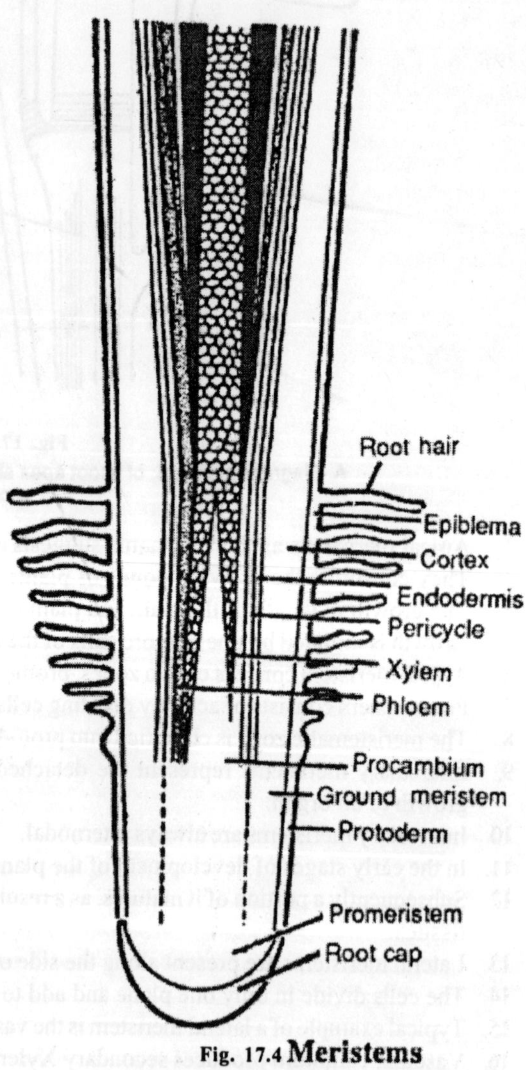

Fig. 17.4 **Meristems**

L.S. of root apex showing meristems

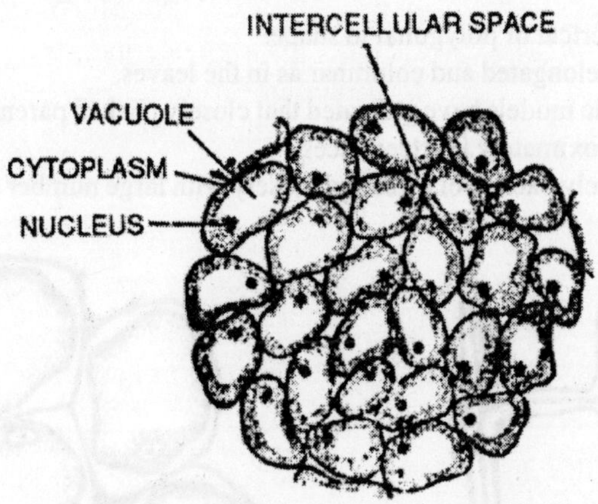

Fig. 17.5 **Permanent tissues.,**
Parenchyma

Distribution:

1. Parenchyma cells are present in almost all parts of the plant, like the pith and cortex of roots and stems, mesophyll of leaves, the flesh of succulent roots, the endosperm of seeds etc.
2. As components of heterogeneous tissues parenchyma cells form vascular rays.

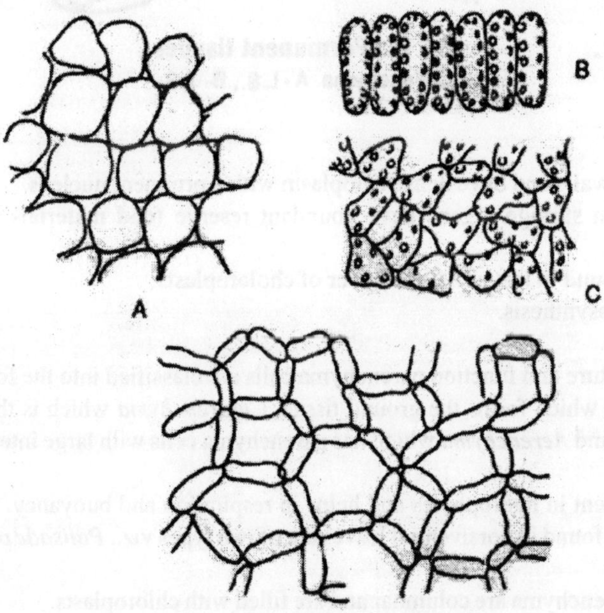

Fig. 17.6 **Permanent tissues.,**
Types of Parenchyma. A. Simple, B. Palisade, C. Spongy, D. Aerenchyma.

Cell shape:

1. Parenchyma cells are spherical or polygonal in shape.
2. Sometimes they may bee elongated and columnar as in the leaves.
3. Studies based on geometric models have indicated that closely packed parenchyma cells have the shape of a polyhedron with approximately fourteen faces.
4. Parenchyma cells are closely packed or arranged loosely with large number of intercellular spaces.

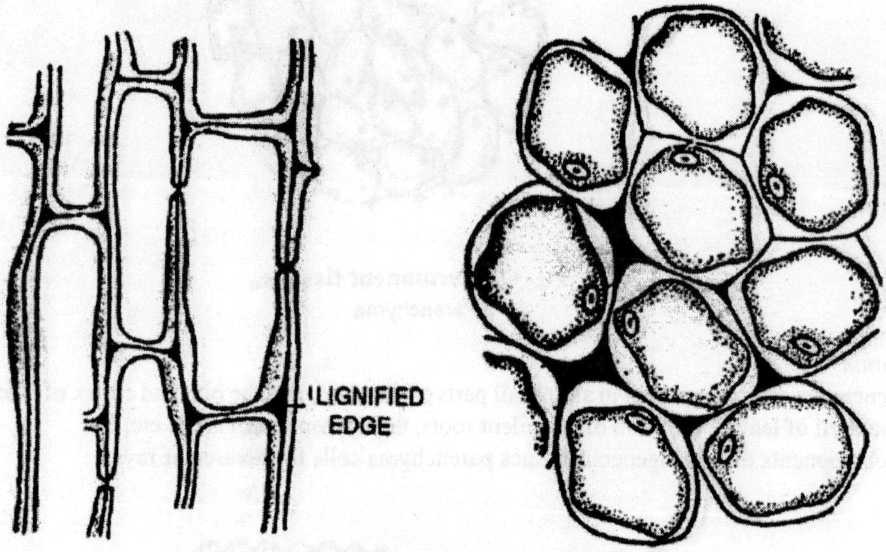

Fig. 17.7 Permanent tissues, Collenchyma. A - L.S., B- T.S.

Cell structure:

1. The cells have thin walls and have dense cytoplasm with a prminent nucleus.
2. Parenchyma cells in storage organs have abundant reserve food materials like proteins, fats and carbohyderates.
3. Parenchyma cells found in leaves have number of choloroplasts.
4. These help in photosynthesis.

Cell types:

1. Based on their structure and function parenchyma cells are classified into the following types.
2. *Simple parenchyma* which forms the ground tissue, *Chlorenchyma* which is the chief photosynthetic tissue in the plants, and *Aerenchyma* which has parenchyma cells with large intercellular spaces storing air.
3. Aerenchyma is present in hydrophytes and helps in respiration and buoyancy.
4. Chlorenchyma cells found in dorsiventral leaves are of two types viz., *Palisade parenchyma* and *spongy parenchyma*.
5. Cells of palisade parenchyma are columnar and are filled with chloroplasts.
6. Spongy parenchyma cells also have chloroplasts, but they are rich in starch grains.
7. Parenchyma cells found in Xerophytes (*Aloe, Agave* etc.) act as water storage tissues.

COLLENCHYMA:

1. Collenchyma is a living tissue and is composed of more or less thick primary non lignified walls.
2. Just as in parenchyma, even in collenchyma the protoplast is living.

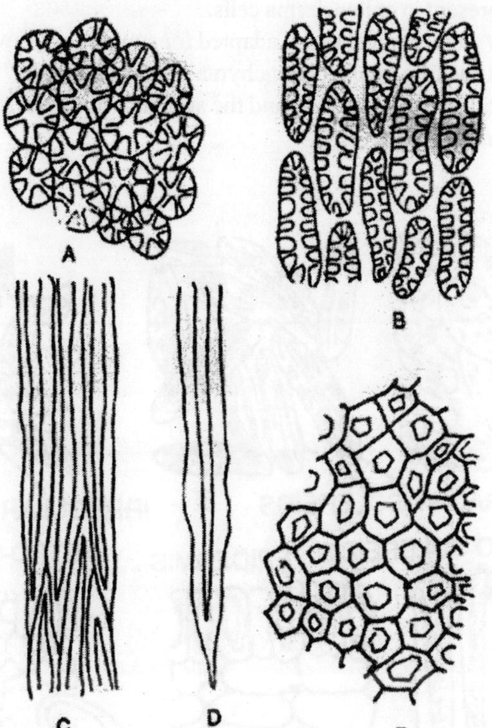

Fig. 17.8 **Permanent tissues.,**
Sclerenchyma. A., T.S. of sclereids. B., L.S. of the same. C. and D., L.S. of fibres. E. T.S. of Fibres

3. The word collenchyma is derived from the Greek word *colla*, meaning glue, which refers to the thick glistening wall

Distribution:

1. Collenchyma characterstically occurs in the pripheral position of stems, leaves, floral parts etc.
2. It may also occur in root cortex particularly if it is exposed to light.
3. In the stems and leaves of dicotyledons, collenchyma is hypodermal situated immediately below the epidermis or seperated from it by one or two layers of parenchyma.
4. Collenchyma cells are considerably elongated.
5. Sometimes, cells are 2 mm, long resembling fibres.
6. The shorter collenchyma cells are prismatic and appear polygonal in a transection.
7. In a long section the cells appear rectangular.

Cell structure and function:

1. The structure of the cell wall is the most characterstic feature of the collenchyma cells.
2. The thickenings are deposited unevenly, mainly deposited at the corners where several cells join together.
3. In long section, collenchyma cells show thin and thick wall portions dependng on the direction of the cut.
4. The walls containing mainly of cellulosic and pectic compounds and consist of much water.
5. Collenchyma cells have living protoplasts and may have chloroplastss.

6. Chloroplastss are less in number in long and narrow cells and more in cells, which approach parenchyma in form. Tannin may be present in collenchyma cells.
7. Collenchyma is primarily a mechanical tissue adapted for support of growing organs.
8. The thick walls and close packing, make collenchyma a strong tissue.
9. At the same time, the peculiarities of growth and the structure of its walls impart a considerable tensile strength with flexibility and plasticity.

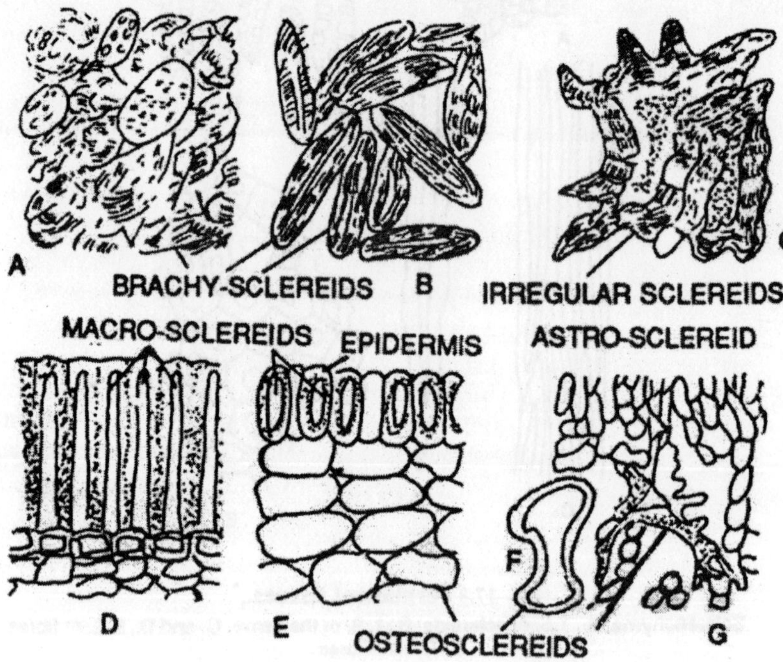

Fig. 17.9 **Permanent tissues.,**
Types of Sclereids. A and B. Brachysclereids, C. Irregular selereids, D and E. Macrosclereids, F. Osteosclereid, G. Astrosclereid.

SCLERENCHYMA:

1. This tissue consists of thick walled cells, (lignified), their main function being mechanical.
2. The word is derived from the Greek *Sclerous*, hard and *Enchyma* an infusion which emphasizes the hardness of the cell walls.
3. Sclerenchyma cells show variety in form, structure, origin and development , which the different types often integrating.
4. Generally, sclerenchyma cells are classified into *fibres* and *sclereids*.
5. Fibres are considered as long cells while sclereids, are relatively short cells.

Fibres:

1. Fibres are long narrow, thick, lignified cells with tapering ends.
2. The wall thickenings sometimes even block the lumen of the cells.
3. The pits of fibres are simple, small round and often oblique.
4. The middle lamella is conspicuous.

Sclereids:

1. These are short isodiametric cells but sometimes may be elongated also.
2. The cell walls are very thick and lignified but the cells do not have tapering ends.
3. The cells are dead and have a very narrow lumen.
4. The secondary walls show prominent pits which are simple.
5. Sometimes there is slight overarching of the secondary wall over the pit chamber.
6. The lignin deposition is uniform, but may be uneven as in the seed coats of some Leguminosae.
7. Sclereids are widely distributed in the plant body.
8. The cortex and the pith of gymnosperms and dicotyledons very often contain sclereids arranged singly or in groups.
9. Xylem and phloem may often contain sclereids besides fibres.
10. Sclereids may be present in leaves, seeds and fruits.
11. In fruits they are sometimes dispersed in the soft tissues.
12. Sclereids are classified into the following types viz., brachysclereids or stone cells (short isodiametric cells widely distributed in the flesh of fruits), macrosclereids long rod like cells found in the epidermis of leguminous seeds Osteosclereids (bone like cells) Astrosclereids (star like cells) Bizarre sclereids (Irregular sclereids), Filiform sclereids (long cells resembling fibres and trichosclereids branched, thin walled sclereids).

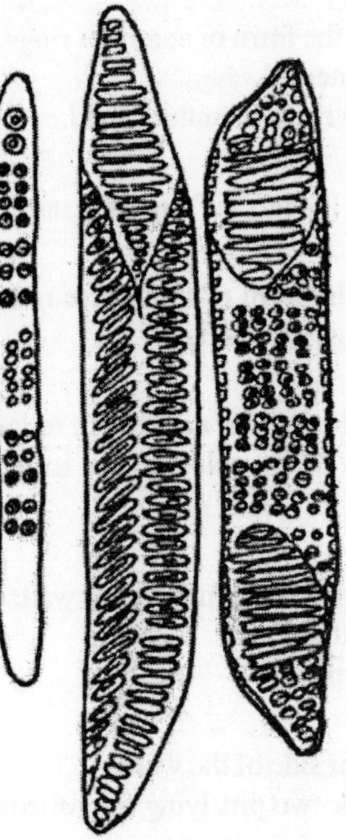

Fig. 17.10 **Permanent Tissues**
L.S. & T.S. of tracheids from Quercus alba

COMPLEX PERMANENT TISSUES

1. The vascular tissues are the most important among the complex tissues.
2. There are mainly two components in the vascular tissue viz., xylem and phloem.

XYLEM:

1. This is mainly a conducting tissue of water and mineral nutrients.
2. Structurally the xylem is most complex, because it consists of various types of living and non-living cells.
3. The xylem is mainly classified into primary and secondary xylem.
4. The former differentiates in the primary plant body and is derived from the procambium, while the latter originates from a fascicular cambium during secondary growth.

Elements of xylem:

1. The xylem is composed of different kinds of elements viz., tracheids, fibre tracheids, vessels or tracheae, wood fibres and wood parenchyma.

(a) Tracheids:

1. A tracheid is an elongated tube like cell with tapering rounded oval ends.
2. The tip of the tracheid of a secondary xylem may be chisel-like.
3. The walls are thickened due to the deposition of lignin, which are of various types.

Thickening of the cell wall:

1. The secondary wall acquires different types of thickenings.
2. This is specially seen in cells which are to differentiate into tracheary elements i.e, vessels and tracheids.
3. In these cells, the thickening is due to the deposition of lignin.
4. The deposition of lignin is not uniform, hence we find different types of thickenings.
5. Chiefly, the following types of thickening are seen in xylem.

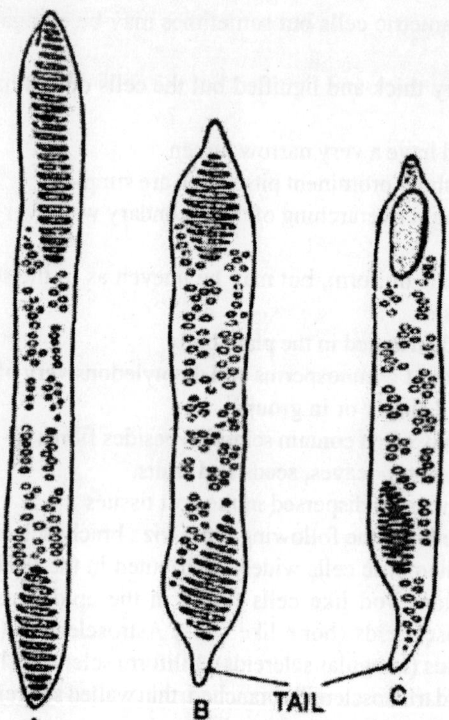

Fig. 17.11 **Permanent Tissues**

Type of vessels Simple C and Scalariform A-B

Annular:

1. Lignin deposition is in the form of annuli or rings.
2. These rings are unconnected.
3. The space between the rings is unthickened.

Spiral:

1. Lignification occurs in the form of a spiral band.

Scalariform (ladder like):

1. Wall thickening is in the form of transverse rods, very much resembling the rungs of a ladder.

Reticulate:

1. Lignin is deposited in the form of network or reticulum.
2. The unthickened areas of the cell wall are irregular in shape.

Pitted:

1. Lignin deposition is almost uniform over the wall, except for small pores here and there.
2. These pores are called pits.

Structure of pit:

1. Pits are formed on either side of the wall.
2. In any area there will be two pits lying opposite to each other.
3. These are usually termed as "pit pairs".
4. There is a space inside each pit called pit cavity or pit chamber.
5. There is a thin membrane which seperates the pit cavities of the opposite pits.
6. It is called pit membrane.
7. The opening to a pit is called the pit aperture.
8. A pit membrane is common to both the pits.
9. There are two types of pits viz. *simple pits* and *bordered pits*.
10. A simple pit is just an opening in the wall.
11. Whereas in a bordered pit, the pit aperture is overarched by the projection of the secondary wall.
12. In a pit pair both may be simple, both may be bordered or one may be simple and the other one bordered.
13. The lost mentioned one is generally called a half bordered pit pair.
14. In a bordered pit, the pit membrane becomes thick at the centre.
15. The thickening is known as the torus.
16. Tracheids are dead cells and like sclerenchyma have neither the nucleus nor the protoplast.
17. The lumen of the cell is large.
18. The wall of the tracheids is provided with a number of bordered pits.
19. In a cross section tracheids appear polygonal or rectangular.

(b) Fibre tracheids :

1. During the development of fibres the wall thickening increases whereas the lumen decreases.
2. In many cells, the lumen becomes so narrow that it is altogether obliterated.
3. In such cells there is little or no conduction and they may be said to have reached the stage of true fibre.

(c) Vessels:

1. These are evolved from tracheids and are present mainly in Angiosperms.
2. Vessels arise through the dissolution of the end walls of tracheids.
3. Vessels are composed of rows of cells placed one above the other with the transverse walls being eliminated.
4. This facilitates the easy flow of water.

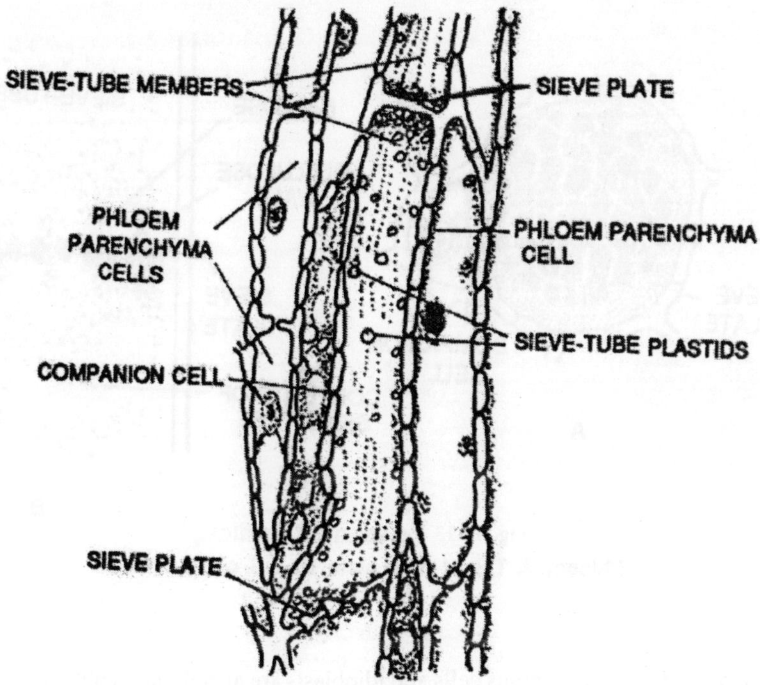

Fig. 17.12 **Permanent tissues., Phloem. Long sectional View.**

5. Vessels (tracheids also) may be of two types protoxylem and metaxylem.
6. The former matures a little early in the ontogeny and has a narrow lumen while the latter matures later and has a broad lumen.

(d) Wood fibres:

1. They are also called xylary fibres.
2. Sclerenchyma fibres associated with xylem are termed wood fibres.
3. They possess very thick walls and few small pits.
4. They originate from the derivatives of the cambium and have varied shapes.
5. In wood fibres, known as libriform fibres, the wall is extremely thick and lumen is almost absent.
6. Like the tracheids , wood fibres are also dead and their main function is mechanical support.

(e) Wood Parenchyma:

1. Parenchyma cells assicoated with xylem are called wood parenchyma.
2. These are living cells and remain so, as long as the tissue in which they lie conducts water.
3. The cells of wood parenchyma may be thin or thick walled have plenty of storage food and may help in conduction.

4. In secondary xylem wood parenchyma forms the rays which may help in radial conduction.

PHLOEM:

1. The phloem is the principal food conducting tissue in vascular plants.
2. Unlike xylem all the elements of phloem (except for phloem fibres) are living.
3. Like the xylem phloem is also of two types of viz., primary phloem (derived from procambium) and secondary phloem (derived from fascicular cambium).
4. The basic components of phloem are sieve elements, companion cells, phloem parenchyma and phloem

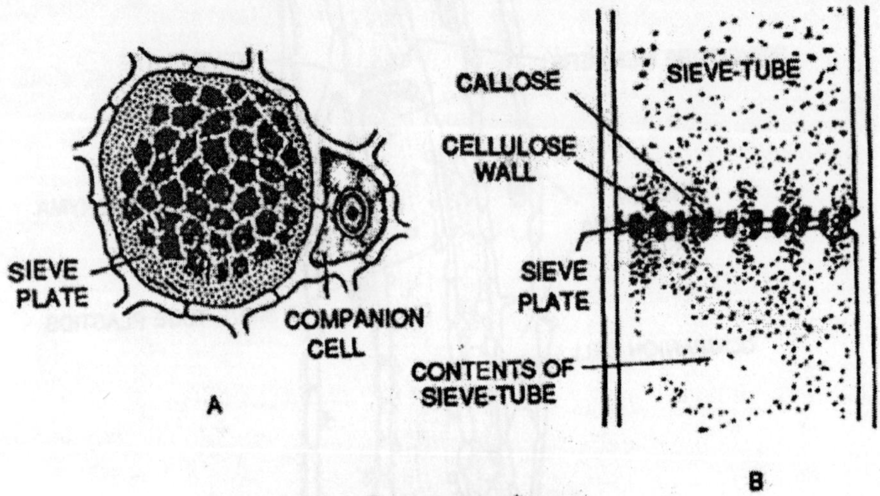

Fig. 17.13 **Permanent tissues., Phloem. A. T.S. of sieve plate, B., L.S. of sieve tube.**

fibres.

5. Occasionally sclereids, laticiferous cells and idioblasts are also found in phloem.

(a) Sieve elements:

1. What the tracheids are to xylem sieve elements are to phloem.
2. Sieve elements may be seggregated into relatively less specialized sieve cells (found in pteridophytes) and more specalized sieve tubes.
3. Seive tubes are living cells and have a thin cellulose wall.
4. In angiosperms they are enucleate, while they are nucleate in gymnosperms.
5. The protoplast of a sieve tube has a big central vacuole which is filled with viscous and albuminous proteins.
6. Leucoplasts are also reported in the vacuole.
7. The cross walls between the seieve tubes are porus and are called sieve plates as they resemble a sieve.
8. Sieve plates are usually of two types viz., simple and compound.
9. In simple sieve plates, the pores are large and the plate is horizontal or transverse in position.
10. The compound sieve plate is somewhat oblique.
11. The sieve pores are large and the plate is horizontal or transverse in position.
12. The compound sieve plate is somewhat oblique.
13. The sieve pores are very often plugged with a shining substance called *callus pad* which is usually deposited at the end of the spring season.
14. Chemically callus pads are made up of a carbohydrate called *callose*.

15. At the approach of the spring, the callus dissolves.
16. Through the sieve pores, thin cytoplasmic strands come out and enter into other cells.

(b) Companion cells:

1. These are specialised small cells usually associated with the sieve tubes.
2. These are living, nucleate cells and are supposed to compensate the absence of nuclei in sieve tubes.
3. Companion cells are absent in gymnosperms, where albuminous clls are present.
4. The cytoplasm of cmpanion cell is dense and granular with a prominent nucleus.
5. Ontogenetically both sieve tubes and companion cells are derived from the same mother cell.

(c) Phloem parenchyma :

1. Parenchyma cells in association with phloem are called phloem parenchyma.
2. They are no different from other ordinary parenchyma cells.
3. In shape, the cells may be elongated, pointed, broadly cylindrical or even subspherical.
4. The cells contain oil, starch mucilage, latex etc.
5. Phloem parenchyma is absent in most of the monocotyledons.

(d) Phloem fibres:

1. These are also called bast fibres.

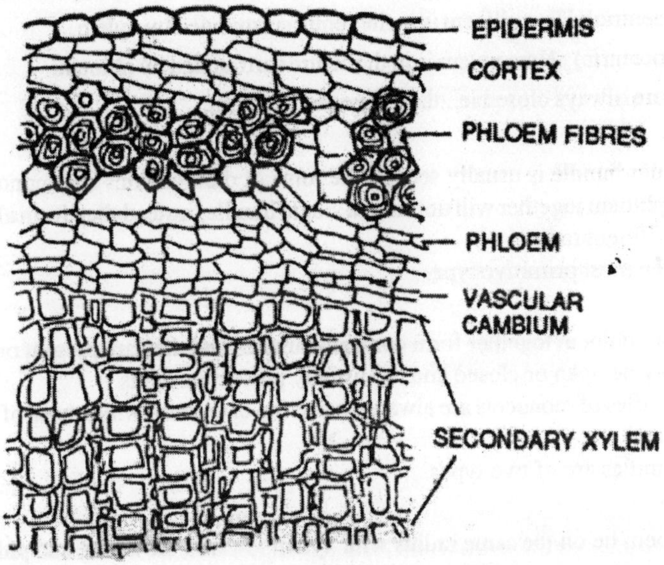

Fig. 17.14 **Permanent tissues.,**
Phloem fibres in the stem of Linum.

2. These are found both in primary and secondary phloem.
3. The walls have simple pits.
4. The bast fibres may be lignified or non lignified.

Types of Vascular bundles

1. A vascular bundle may be defined as grouping together of vascular elements namely xylem and phloem to form a functional unit.
2. Within a vascular bundle the two major components - xylem and phloem are arragned in various ways.
3. Depending on their relative position vascular bundles are categorized into the following types - cambium

which brings about secondary growth may or may not be present.

Open vascular bundles :

1. This is seen in dicotyledons.
2. Here between xylem and phloem, cambium (secondary meristem) is present.
3. As a result of this the vascular bundle is open to secondary growth.
4. There is no bundle sheath surrounding the vascular bundle.

Closed vascular bundle :

1. This is seen in monocotyledons.
2. Here there is no cambium in the vascular bundle.
3. Hence there is no secondary growth.
4. Further a bundle sheath surrounds the vascular bundle.
5. Based on the relative positions of xylem and phloem vascular bundles are classified into- *concentric*, *radial* and *conjoint* types.

Concentric vascular bundle:

1. In concentric vascular bundles xylem and phloem form concentric rings; either xylem surrounds the phloem or phloem surrounds the xylem.
2. Concentric vascular bundles are of two types-

Amphivasal (Leptocentric): Here phloem is in the centre surrounded by xylem.

Amphicribal (Hadrocentric) : Here xylem is in the centre surrounded by phloem.

Concentric bundles are always closed ie., they have no cambium.

Radial bundles:

1. This kind of a vascular bundle is usually seen in the roots of dicotyledons and monocotyledons.
2. Here the xylem and phloem together will not form a single bundle, instead they form alternating bundles and are present on different radii.
3. This is regarded as the most primitive type.

Conjoint bundles :

1. In this type xylem and phloem together form compact bundles and they are present on the same radius.
2. Conjoint bundles may be open or closed and are usually present in stems.
3. Conjoint vascular bundles of monocots are always closed (no cambium) while those of dicots are always open (cambium present).
4. Conjoint vascular bundles are of two types.

Collateral :

1. Here xylem and phloem lie on the same radius with xylem towards the centre and phloem towards the periphery.
2. In otherwords there is one group of phloem and one group of xylem.
3. In dicotyledons cambium is present in between xylem and phloem while it is absent in monocotyledons.
4. Collateral vascular bundles are found in stems of Angiosperms.

Bicollateral :

1. In these bundles there is one group of xylem surrounded by two groups of phloem - outer and inner.
2. Similarly two strips of cambia are also present.
3. Here the arrangement of various elements are as follows- inner phloem, inner cambium, xylem, outer cambium and outer phloem.
4. Bicollateral vascular bundles are usually seen in the members of the family Cucurbitaceae.

Chapter 18

ANATOMY OF ROOTS

Root of Helianthus

A transverse section of a young sunflower root shows the following features.

***Epidermis*:**

1. **This is the outermost layer.**
2. **This is also known as epiblema or piliferous layer and is made up of closely packed cells with no intercellular spaces.**
3. **Some of the epidermal cells produce unicellular extensions called root hairs.**
4. **These help in the absorption of nutrients from the soil.**

***Cortex*:**

1. **Internal to the epidermis is the cortex.**
2. **It is made up of several layers of thin walled spherical or oval parenchyma cells with large number of intercellular spaces.**
3. **Some of the outer layers of the cortex may become thick walled and function as exodermis.**

***Stele*:**

1. **Occupying the central region of the root is the vascular cylinder.**
2. **The outermost layer of the stele is the endodermis.**
3. **It is made up of a single layer of barrel shaped cells., closely packed without any intercellular spaces.**
4. **The endodermal cells have thickenings on their radial walls called casparian thickenings.**
5. **In between endodermal cells thin walled cells are present.**
6. **They are called the passage cells.**
7. **Usually they are present opposite the protoxylem.**
8. **Internal to the endodermis is the pericycle.**
9. **It is made up of a single layer of rough, thin walled parenchyma cells.**
10. **Surrounded by the pericycle are the vascular bundles.**

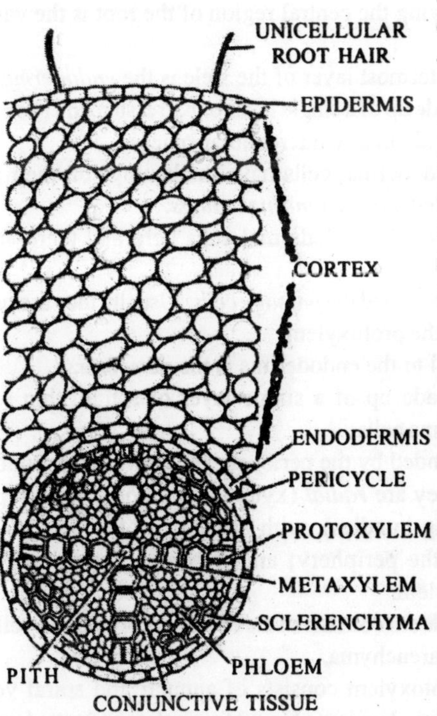

Fig. 18.1 Anatomy of root. T.S. of dicot root. Detail of a sector.

11. They are radial (Xylem and Phloem form separate alternating bundles) with *exarch* (protoxylem pointing towards the periphery) and *tetrarch* (four protoxylem points) xylem.
12. Xylem consists of annular and sprial vessels with metaxylem being made up of reticulate and pitted vessels.
13. Occupying the central region of the stele is a narrow parenchymatous pitch.
14. Sometimes the metaxylem groups join together so that pith is obliterated.
15. Between the radial vascular bundles is present parenchyma constituting the conjuctive tissue.

BEAN

A transverse section of a young bean root shows the following features.

Epidermis:

1. This is the outermost layer.
2. This is also known as *Epiblema* or *Piliferous layer* and is made of closely packed cells with no intercellular spaces.
3. Some of the epidermal cells produce unicellular extensions called root hairs.
4. These help in the absorptionof nutrients from the soil.

Cortex:

1. Internal to the epidermis is the cortex.
2. It is made up of several layers of thin walled spherical or oval parenchyma cells with large number of intercellular spaces.
3. Some of the outer layers of the cortex may become thick walled and function as **exodermis**.

Stele:

1. Occupying the central region of the root is the vascular cylinder.
2. The outermost layer of the stele is the *endodermis.*
3. It is made up of a single layer of barrel shaped cells closely packed without any intercellular spaces.
4. The endodermal cells have thickenings on their radial walls called *Casparian thickenings*.
5. A few of the endodermal cells here and there are not thickened.
6. They are called the *passage cells*. Usually they are present opposite the protoxylem.
7. Internal to the endodermis is the *pericycle.*
8. It is made up of a single layer of rough, thin walled parenchyma cells.
9. Surrounded by the pericycle are the vascular bundles.
10. They are *radial* (xylem and phloem form separate alternating bundles), with *exarch* (protoxylem pointing towards the periphery) and *tetrarch* (Four protoxylem points) xylem.
11. Phloem consists of sieve tubes, companion cells and phloem parenchyma.
12. Protoxylem consists of annular and spiral vessels while Metaxylem is made up of reticulate and pitted vessels.
13. Occupying the central region of the stele is a narrow parenchymatous pith.

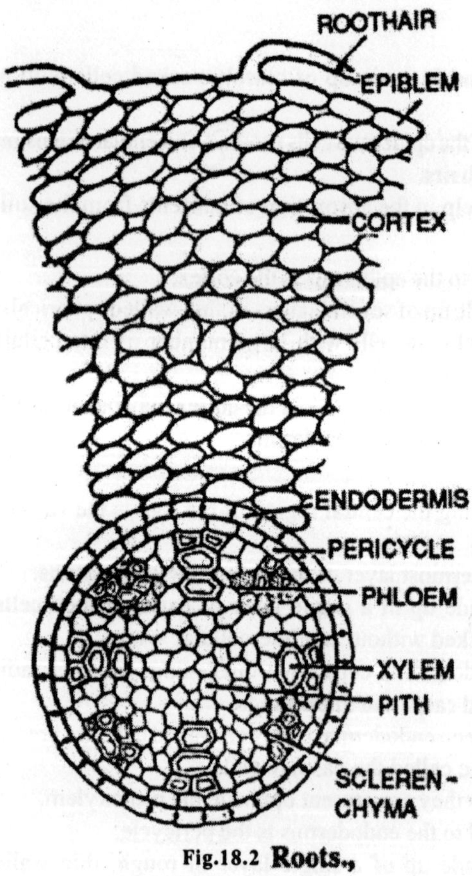

Fig.18.2 **Roots.,**
T.S. of Bean root.

14. Sometimes the metaxylem groups join together so that pith is obliterated.
15. Between the radial vascular bundles is present parenchyma, constituting the conjunctive tissue.

RANUNCULUS

1. A T.S shows three regions - Epiblema, cortex and stele.
2. Epiblema is made up of a single layer of cells.
3. Some cells of epiblema produce unicellular root hairs.
4. Cortex is made up of many layers of rounded or polygonal parenchyma cells.
5. Some of the layers of cells immediately below the epiblema constitute exodermis.
6. The cells of the cortex is loosely arranged with intercellular spaces.
7. The cells of the cortex are rich in starch grains.
8. The stele is externally bounded by endodermis, the cells of which have casparian thickenings.
9. Next to the endodermis internally is the single layered peri cycle.
10. The vascular bundles are radial.
11. The xylem is tetrarch (four strands).
12. The xylem is exarch (protxylem points towards the perrphery).
13. Alternating with the xylem are found bundles of pholem.
14. Phloem consists of seive tubes, companion cells and phloem parenchyma.
15. Conjunctive parenchyma is present between the vascular bundles.
16. There is no pith.

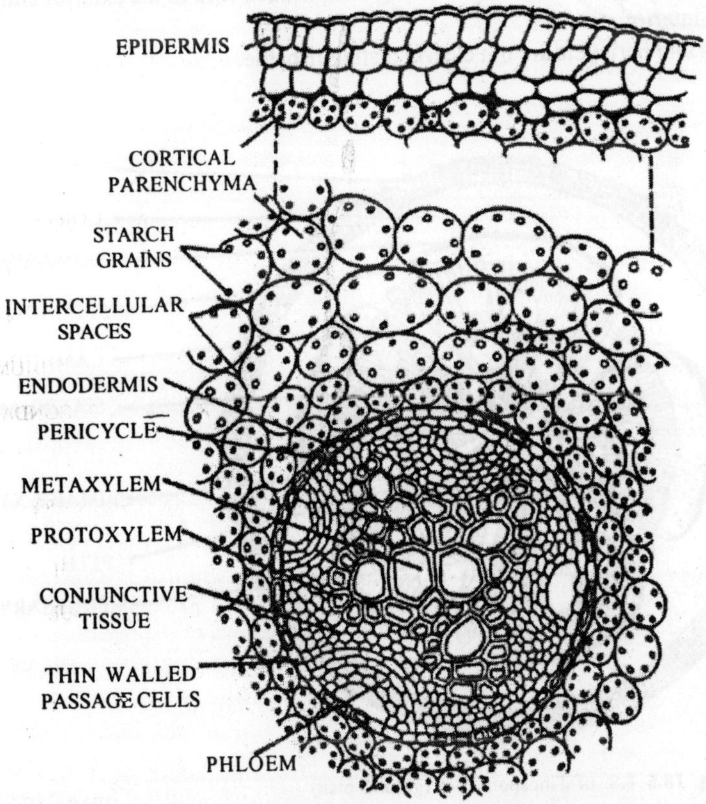

Fig. 18.3 T.S. of Root of Ranunculus - A part enlarged

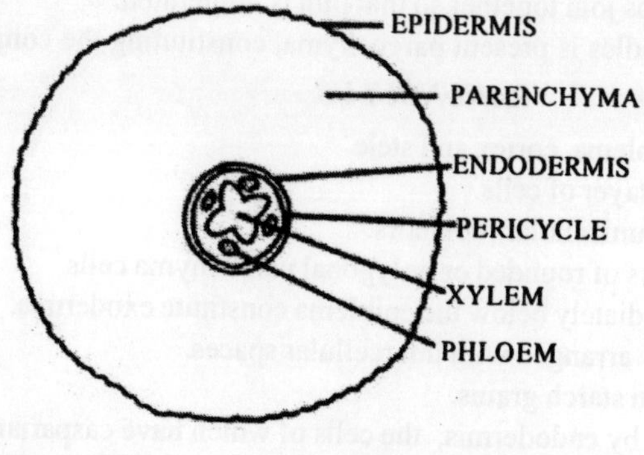

Fig. 18.4 T.S. of Root of Ranunculus Ground plan

TINOSPORA

1. A transfer section of the root shows rhizodermis, cork, cork cambiam and vasculature.
2. The rhizodermis is the outermost layer and is made up of rectangular cells.
3. Immediately below the rhizodermis is found the cork which is made up of few layers of rectangular cells.
4. The cells of the cork are dead.
5. Internal to the cork is found the cork cambium or phellogen.
6. The cells of phellogen are actively dividing and produce cork to the exterior and secondary cortex towards the interior.
7. Endodermis and pericycle are not clearly distinguishable.

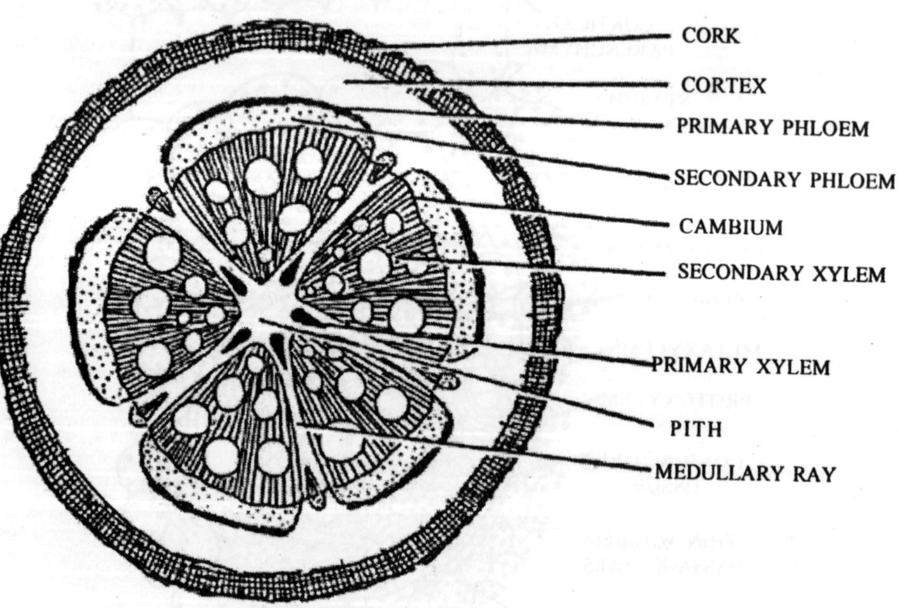

Fig. 18.5 T.S. of Tinospora root (ground plan)

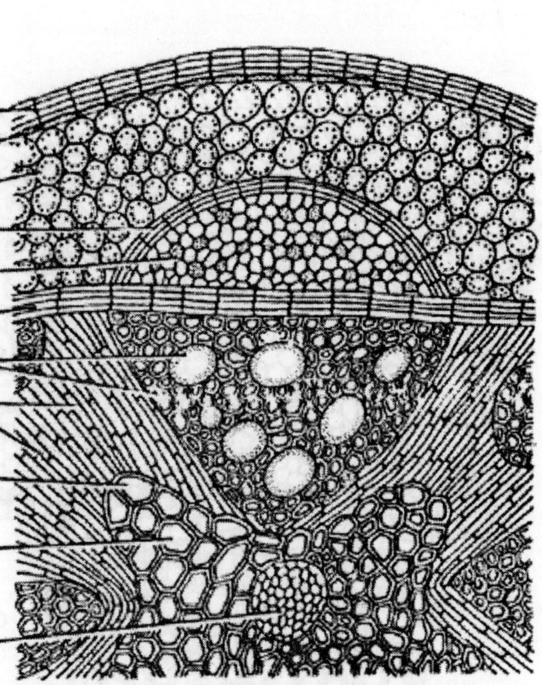

Fig. 18.6 T.S. of Tinospora root (A sector enlarged)

8. The vascular consists of primary phloem secondary phloem and secondary xylem.
9. The primary phloem is crushed and is found next to the secondary cortex.
10. Next to the primary phloem is secondary phloem consisting of seive tubes companion cells and phloem parenchyma.
11. Next to the secondary phloem is the fascicular cambium.
12. Internal to the cambium is the secondary xylem consisting of large vessels.
13. Very close to the centre are found primary xylem elements.
14. The primary xylem is exarch and tetrarch.
15. Pith is very narrow and found in centre.
16. Between the vascular bundles are found medullary rays.

PROP ROOT OF BANYAN (*Ficus bengalensis*)

1. A transverse section of the aerial root of *Ficus* in the primary structure shows an outermost epliblema followed by cortex and stele.
2. The cortex consists of parenchymatous cells which may have chloroplasts.
3. The stele is surrounded by endodermis and pericycle.
4. The vascular bundles are *radial* and six to nine in number (*polyarch*).
5. In the central region there is a well developed parenchymatous pith.
6. After the initiation of secondary growth which takes place as in other dicotyledonous roots, the following features are seen

Cork or phellem:

1. This replaces the epibelma and consists of 4-6 layers of suberiseed cells.
2. Certain cells of the cork contain tannin. Lenticels are also present.

Cork cambium:

1. This is present below the cork. It is this which produces the cork tissue.
2. The cells of the cork cambium are rectangular and actively dividing.

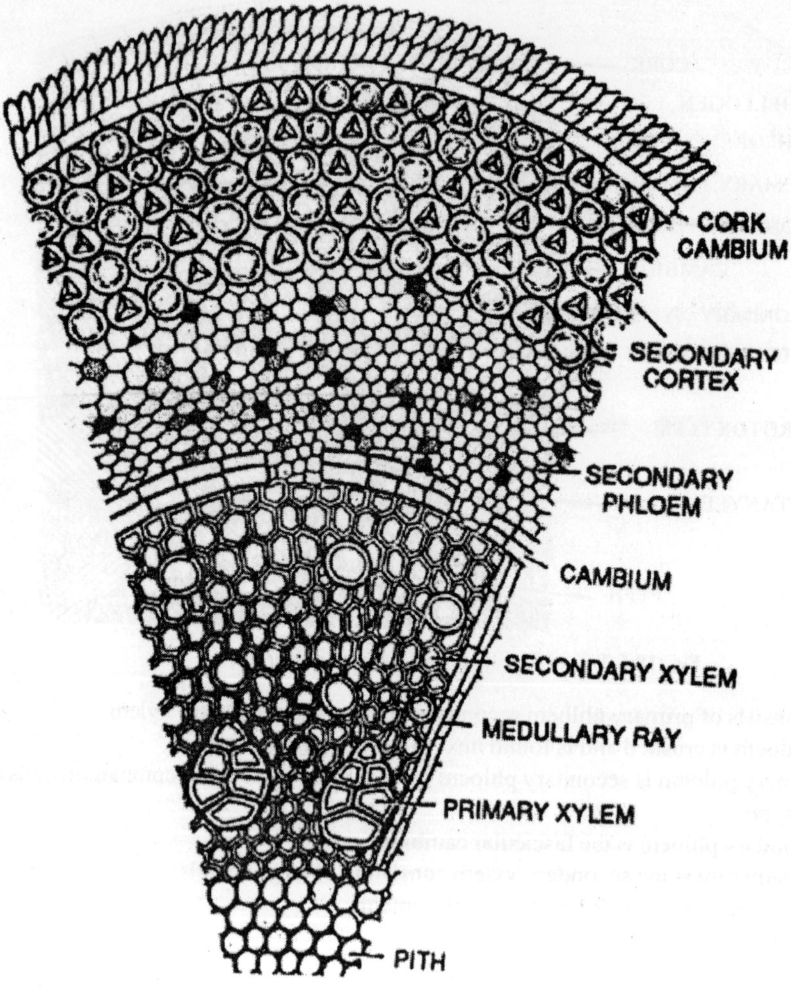

Fig. 18.7 T.S. of Banyan root (A part enlarged)

Secondary cortex:

1. This is also produced by the cork cambium on its inner face.
2. The cells are thin walled parenchyma, containing chloroplasts.

Vascular tissues:

1. After the initiation of secondary growth also, both primary and secondary vasculature can be seen.
2. The primary vascular bundles are radial with exarch, polyarch xylem.
3. Secondary vasculature forms a complete ring and is produced due to the activity of cambium.
4. In a transverse section, cambium is seen between secondary xylem and secondary phloem.
5. Fibres may be seen in secondary phloem.
6. Secondar xylem consists of vessels, tracheids and wood parenchyma.
7. In the centre there is a parenchymatous pith.
8. Medullary rays are present traversing across the secondary vascular tissues.

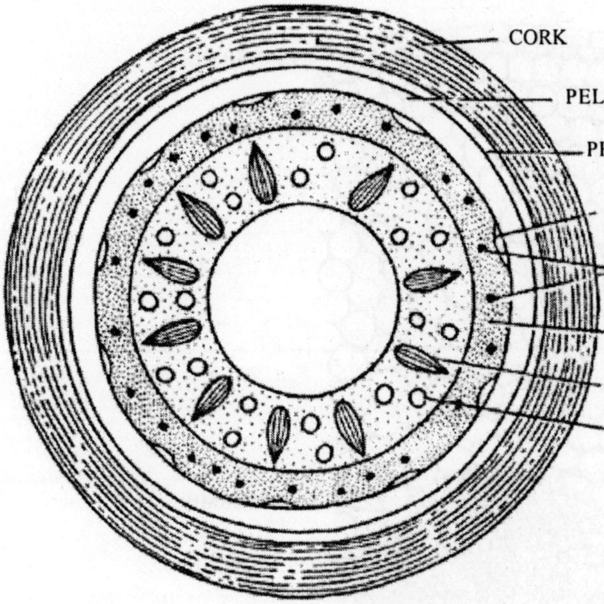

Fig. 18.8 T.S. of Prop root of Banyan (ground plan)

SMILAX

1. A transfer section shows epiblema, cortex, and stele.
2. Epibelma is single layered and it produces a number of unicellular root hairs.
3. Next to the epibelma is exodermis consisting of a single layer of thick walled cells.
4. The cortex is many layered consisting of loosely arranged thin walled parenchyma cells.
5. The cells of cortex are oval in shape and have a number of starch grains.
6. The stele is externally bounded by endodermis.
7. The cells of endodermis are barrel shaped and have casparian thickenings on their radial and tangential walls.
8. Next to the endodermis is single layered peri cycle.
9. Vascular bundles are radial and they are many in number.
10. Xylem and phloem form seperate alternative bundles.
11. The xylem is polyarch (many bundles) and exarch.
12. Phloem consists of seive tubes companion cells and phloem parenchyma.
13. Occupying the central region of the root is the pith consisting of loosely arranged parenchyma cells.

CANNA

1. A transverse section shows epibelma, cortex and stele.
2. Epibelma is single layered and it produces a number of unicellular root hairs.
3. Next to the epibelma is exodermis consisting of a single layer of thick walled cells.
4. The cortex is many layered consisting of loosely arranged thin walled parenchyma cells.
5. The cells of cortex are oval in shape and have a number of starch grains.
6. The stele is externally bounded by endodermis.
7. The cells of endodermis are barrel shaped and have casparian thickenings on their radial and tangential walls.
8. Next to the endodermis is single layered peri cycle.
9. Vascular bundles are radial and they are many in number.

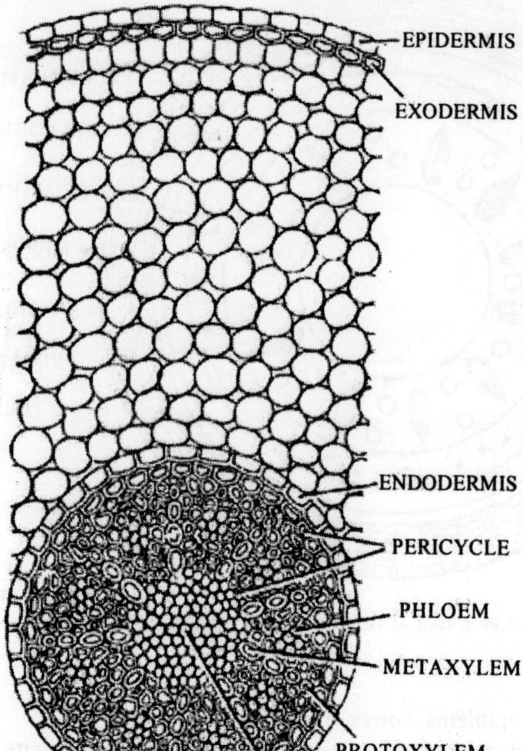

Fig. 18.9 Anatomy of Smilax root (A sector enlarged)

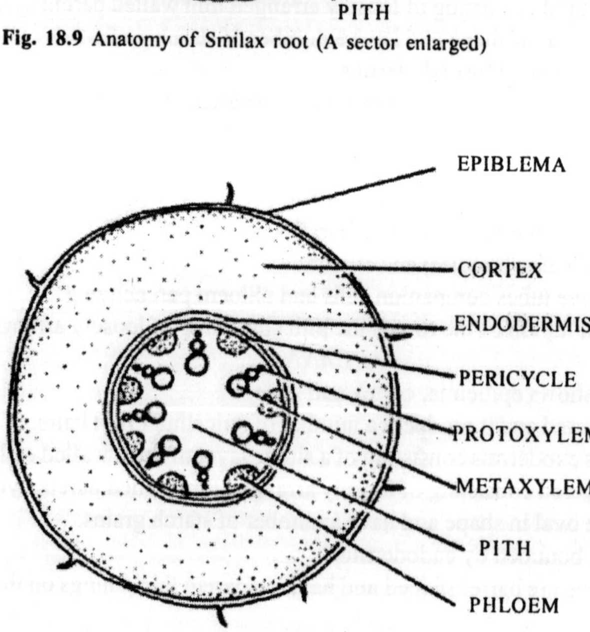

Fig. 18.10 T.S of Cana root (diagrammatic)

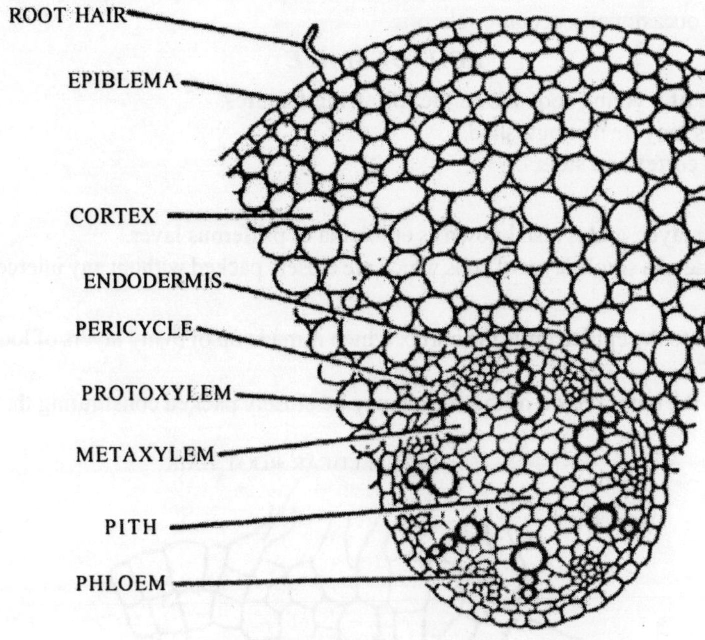

Fig. 18.11 T.S of Cana root (A part cellular)

10. Xylem and phloem form seperate alternative bundles.
11. The xylem is polyarch (many bundles) and exarch.
12. Phloem consists of seive tubes companion cells and phloem parenchyma.

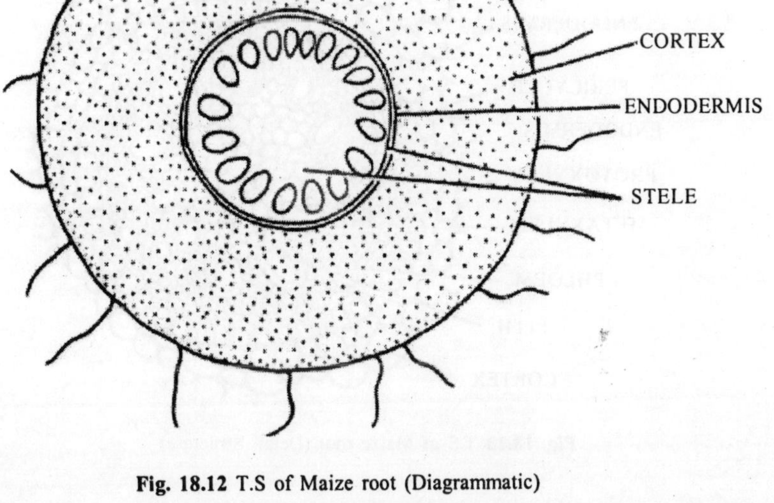

Fig. 18.12 T.S of Maize root (Diagrammatic)

13. Occupying the central region of the root is the pith consisting of loosely arranged parenchyma cells.
14. The cells of the pith occasionally become sclerotic.

ROOT OF MAIZE

1. A transverse section of a young root shows the following features.
2. Mainly three regions may be distinguished.
3. These are epiderms, cortex and stele.

Epidermis:

1. This is the outermost layer, and is also known as epiblema or piliferous layer.
2. The epidermis is made of a single layer of cells which are closely packed without any intercellular spaces.

Cortex:

1. Immediately internal to the epidermis is the cortex which is made up of many layers of loosely arranged parenchyma cells.
2. Some of the cells in the outer region of the cortex may be closely packed constituting the hypodermis.

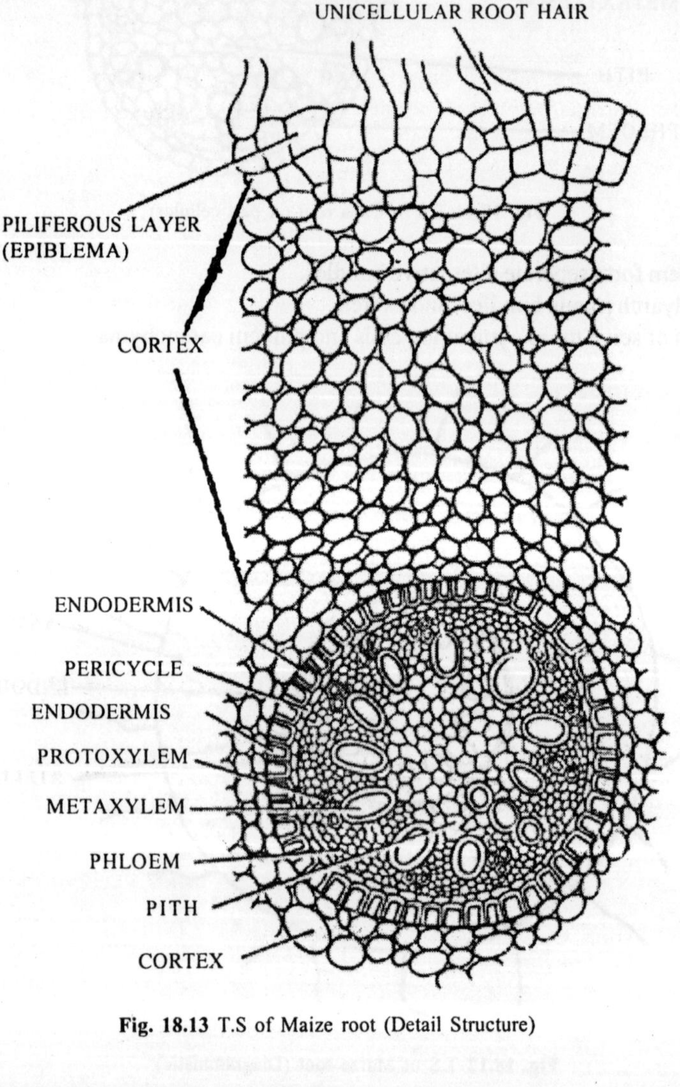

Fig. 18.13 T.S of Maize root (Detail Structure)

Stele:

1. The stele is bounded by an endodermis and pericycle.
2. Endodermal cells are barrel shaped and have casparian thickenings on their radial and tangential walls.
3. Pericycle is represented by a single layer of thin walled parenchyma cells.
4. The vascular bundles are many in number (polyarch) and radial with exarch xylem.
5. In the centre of the stele is a parenchymatous pith.
6. The main points of difference between the dicotyledonous root and the monocotyledonous root are the number of xylem groups and the presence or absence of pith.

MUSA ROOT

1. A transverse section of the root shows as usual epidermis, cortex and stele.
2. Epidermis or epiblema is single layered.
3. Internal to the epidermis is the cortex which is made up of many layers of loosely arranged parenchyma cells.
4. The stele is bounded by an endodermis and pericycle.
5. The vascular bundles are radial, with exarch and polyarch xylem.
6. There is a central pith.
7. In between the vascular bundles is the conjunctive tissue.
8. The conjunctive tissue as well as pith are lignified and have sclerotic cells.
9. Lying scattered in the pith are many isolated metaxylem vessels.
10. Additional phloem groups are also found scattered in the pith.

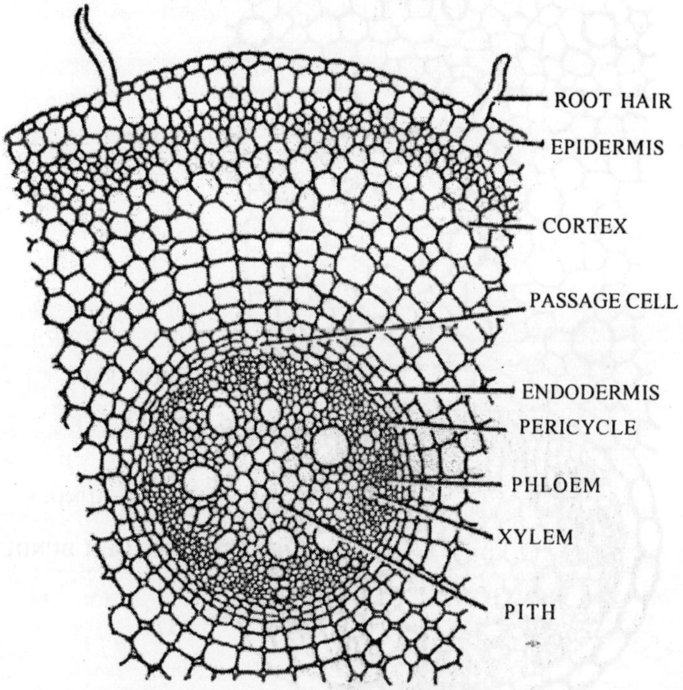

Fig. 18.14 T.S of Musa root (A portion enlarged)

ORCHID ROOT (*Vanda*)

1. The epiphytic roots of orchids have certain special features.
2. This is associated with their special function.
3. The epiphytic roots are also known as assimilatory roots isnce they help in photosynthesis.
4. A transverse section of the root of *Vanda* shows the following details:

Velamen:

1. This is the outermost region, and may also be called a multilayered epidermis.
2. The cells here are non living and elongated radially.
3. Arrangement of the cells is compact.
4. Velamen cells have the capacity to retain air and water.
5. Sometimes the cell walls develop fibrous thickenings.

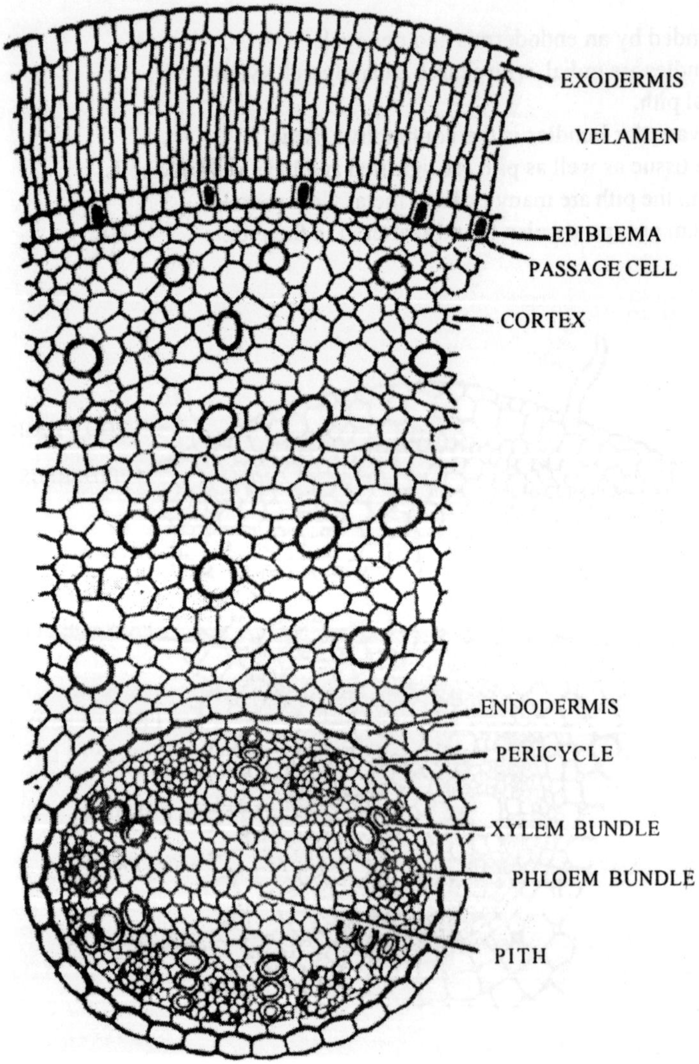

Fig. 18.15 T.S of Orchid root (A portion enlarged)

Exodermis:

1. The outermost layer of the cortex is called exodermis.
2. It is made up of single layer of cells with thickened outer and lateral walls.
3. Here and there some thin walled cells are present called passage cells.

Cortex:

1. Internal to the exodermis is the multilayered cortex.
2. It consists of many layers of thin walled loosely arranged parenchyma cells.

Stele:

1. Endodermis and pericycle are present.
2. There structure is as in other roots.
3. The vascular bundles are radial with *exarch* and *polyarch* xylem.
4. There is a well developed parenchymatous pith in the centre.
5. Conjuctive parenchyma is present between the vascular bundles.

Chapter 19

ANATOMY OF STEMS

***HELIANTHUS* (Sunflower):**

A transverse section of young sunflower stem shows an outmost epidermis, a middle cortex and a central stele.

Epidermis:

1. This is single layered with closely packed rectangular cells.
2. Above the epidermis there is a well defined cuticle.
3. Some of the epidermal cells produce multicellular hairs.
4. Occasionally stomata may be seen in the epidermis.

Cortex:

1. There are two regions in the cortex.

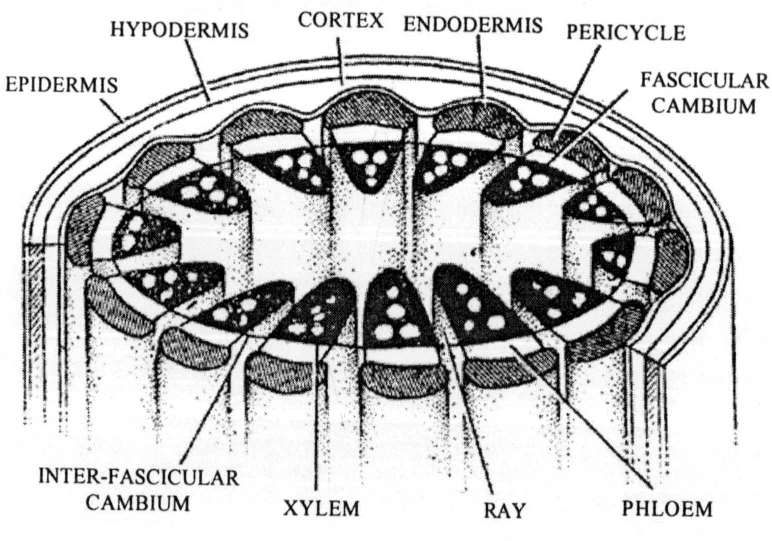

Fig. 19.1 Stems. T.S. of Sunflower stem (Ground plan)

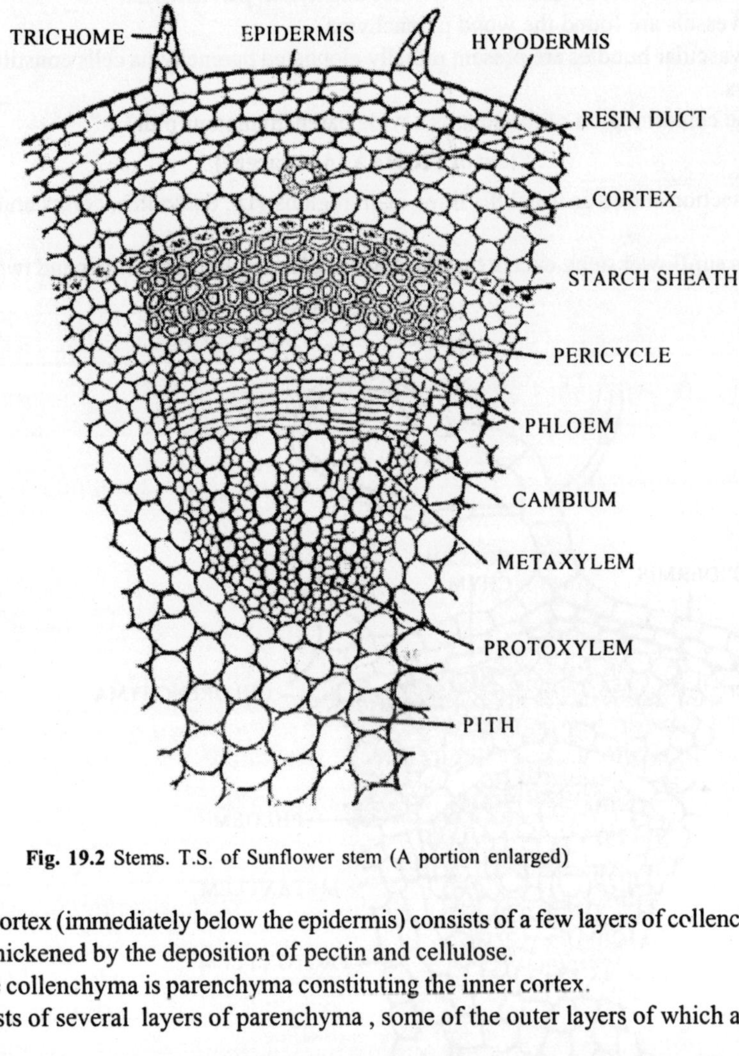

Fig. 19.2 Stems. T.S. of Sunflower stem (A portion enlarged)

2. The outer cortex (immediately below the epidermis) consists of a few layers of collenchyma whose corner walls are thickened by the deposition of pectin and cellulose.
3. Next to the collenchyma is parenchyma constituting the inner cortex.
4. This consists of several layers of parenchyma , some of the outer layers of which are chlorophyllous.

Stele:

1. The outermost layer is the endodermis whose cells are rich in starch grains; hence also known as the starch sheath.
2. It has the thickening on radial walls called casparian thickenings.
3. Next to the endodermis is the pericycle.
4. It is represented by patches (cap like) of sclerenchyma present immediately above each vascular bundle.
5. The vascular bundles are many in number, arranged in the form of a ring surrounding a central pith.
6. Each vascular bundle is conjoint (xylem and phloem combining to form a single bundle), collateral (xylem and pholem on the same radius with x/p arrangement), open (open to secondary growth i.e. cabium is present) with endarch (protoxylem pointing towards the pith) xylem.
7. Phloem consists of seive tubes companion cells and phloem parenchyma .
8. In between phloem and xylem is present cambium which is said to be intrafascicular (present within a vascular bundle).
9. The cambial cells are thin walled, rectangular and actively dividing.
10. Next to the cambium, internally is the xylem, with the protoxylem facing the pith.

11. Xylem consists of vessels, trachieds, wood fibres and xylem parenchyma.
12. Between the vessels are found the wood parenchyma.
13. Between the vascular bundles are present radially elongated parenchyma cells constituting the primary medullary rays.
14. Occupying the central region of the stem is a wide parenchymatous pith.

Stem of TRIDAX (Asteraceae)

A transverse section shows as usual the three main regions viz., epidermis, cortex and stele.

Epidermis:

1. It is same as in sunflower stem, except that the epidermal hairs are much longer and two cells wide at the base.

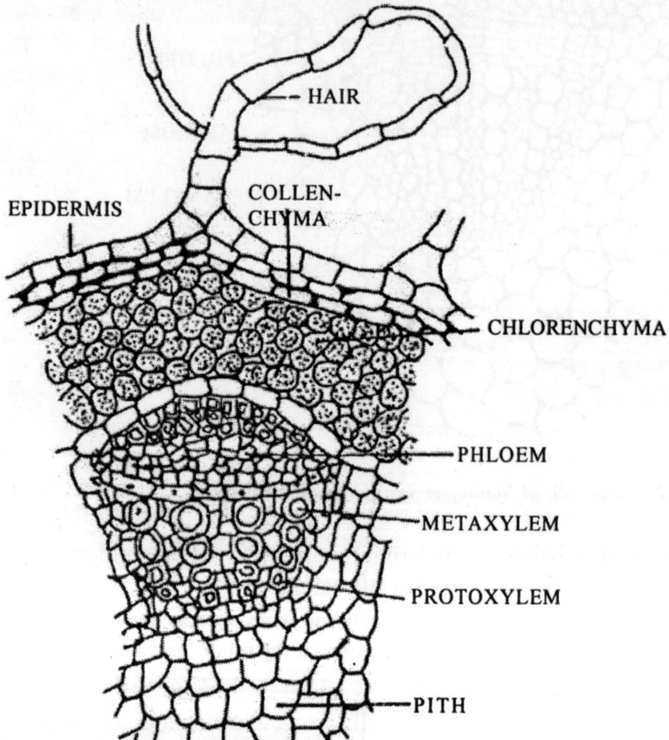

Fig. 19.3 Stems. T.S. of Tridax stem (A portion enlarged)

Cortex:

1. It is also same as in sunflower stem.
2. All the cells of parenchyma are chlorophyllous.
3. Endodermis, pericycle, vascular bundles, pith and medullary rays are same as in sunflower stem.

XANTHIUM

1. A transverse section of a young stem shows epidermis, cortex and stele.
2. Epidermis is made up of a single layer of closely packed cells.
3. Some of the cells of epidermis produce multicellular hairs.

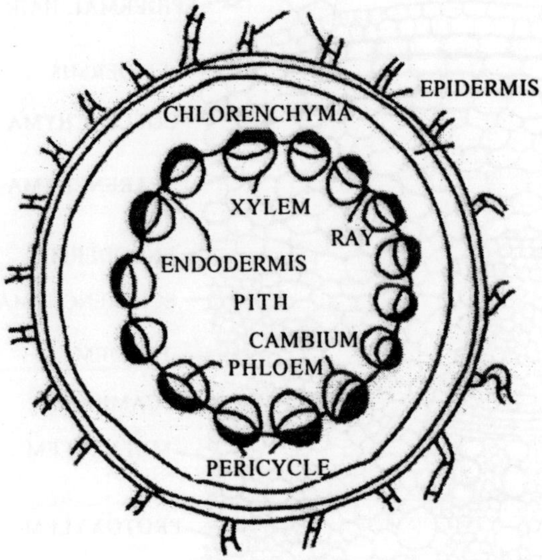

Fig. 19.4 Stems. T.S. of Tridax stem (ground plan)

4. Some times there is deposition of cuticle above the epidermis.
5. Cortex is very broad and is made up of two zones.
6. The outer zone of the cortex consists of two to three layers of collenchyma cells.
7. The collenchyma cells are closely packed and have thickened corner walls.
8. The inner zone of the cortex consists of many layers of parenchyma cells.
9. The parenchyma cells are loosely arranged and oval in shape.
10. Bounding the stele externally is the endodermis which is not as conspicuous as in the root.
11. Next to the endodermis is the pericycle which is discontinous and found in the form of patches of sclerenchyma immediately above the vascular bundles.
12. The vascular bundles are many in number arranged in the form of a ring surrounding a central pith.
13. The vascular bundles are collateral, conjoint, and open.
14. Between xylem and phloem is present the fascicular cambium.
15. The cambial cells are narrow rectangular, and actively dividing.
16. The cambium produces secondary xylem to the interior and secondary phloem to the exterior.
17. Medullary rays are found between the vascular bundles.
18. The pith is broad and made up of loosely arranged rounded or oval parenchyma cells.

RICINUS

1. A transverse section of a young stem shows epidermis, cortex and stele.
2. Epidermis is made up of a single layer of closely packed cells.
3. Some of the cells of epidermis produce multicellular hairs.
4. Some times there is deposition of cuticle above the epidermis.
5. Cortex is very broad and is made up of two zones.
6. The outer zone of the cortex consists of two to three layers of collenchyma cells.
7. The collenchyma cells are closely packed and have thickened corner walls.
8. The inner zone of the cortex consists of many layers of parenchyma cells.
9. The parenchyma cells are loosely arranged and oval in shape.

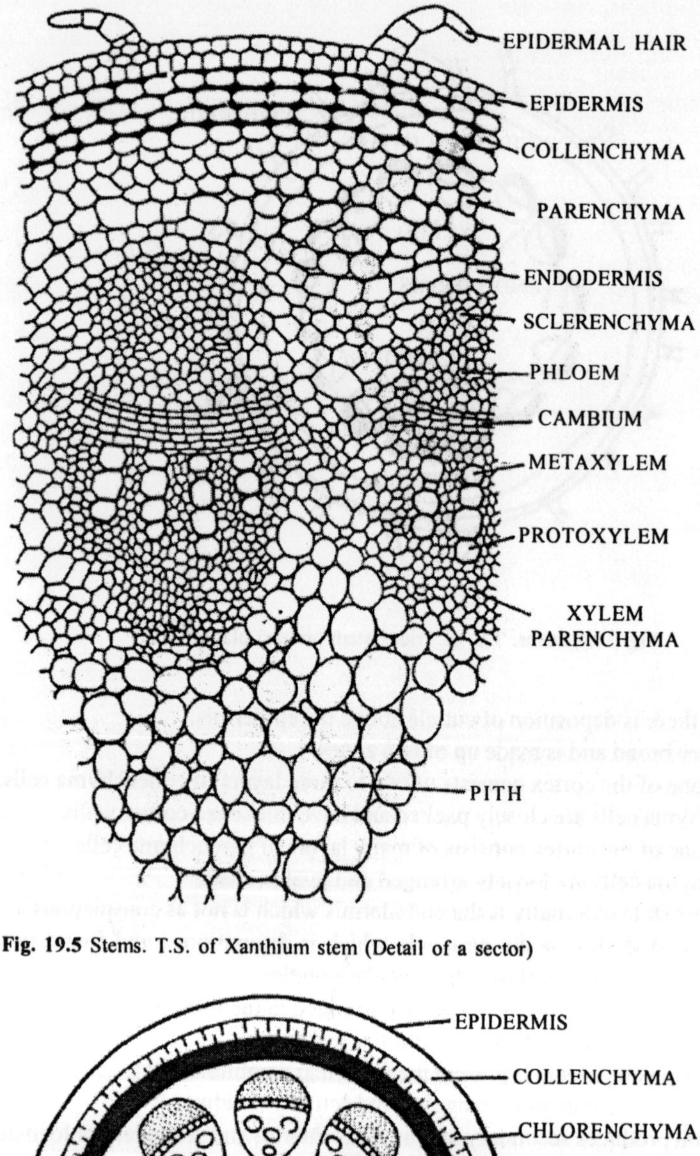

Fig. 19.5 Stems. T.S. of Xanthium stem (Detail of a sector)

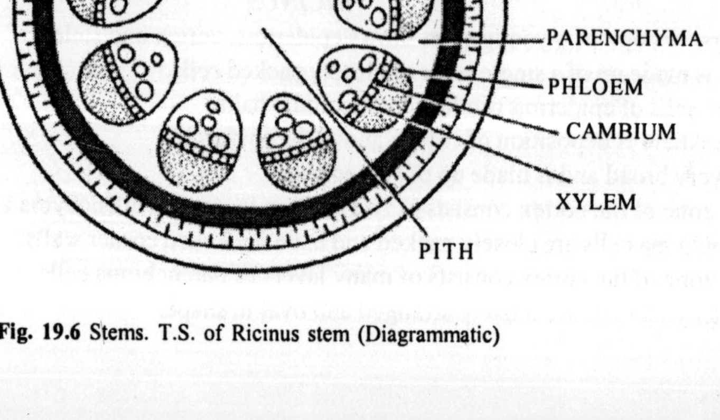

Fig. 19.6 Stems. T.S. of Ricinus stem (Diagrammatic)

10. Bounding the stele externally is the endodermis which is not as conspicuous as in the root.
11. Next to the endodermis is the pericycle which is discontinous and found in the form of patches of sclerenchyma immediately above the vascular bundles.
12. The vascular bundles are many in number arranged in the form of a ring surrounding a central pith.
13. The vascular bundles are collateral, conjoint, and open.
14. Between xylem and phloem is present the fascicular cambium.
15. The cambial cells are narrow rectangular, and actively dividing.
16. The cambium produces secondary xylem to the interior and secondary phloem to the exterior.
17. Medullary rays are found between the vascular bundles.
18. The pith is broad and made up of loosely arranged rounded or oval parenchyma cells.

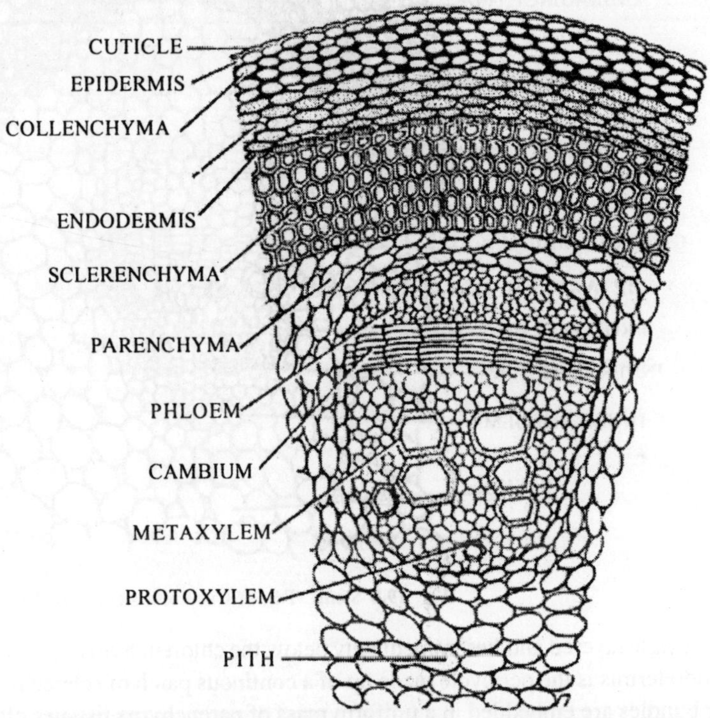

Fig. 19.7 Stems. T.S. of Ricinus stem (A portion of sector)

CUCURBITA

1. A transverse section of a young stem shows epidermis, cortex and stele.
2. Epidermis is made up of a single layer of closely packed cells.
3. Some of the cells of epidermis produce multicellular hairs.
4. Some times there is deposition of cuticle above the epidermis.
5. Cortex is very broad and is made up of two zones.
6. The outer zone of cortex consists of collenchyma cells.
7. The collenchyma is many layered near the ridges while near the forus they are either absent or very narrow.
8. The inner zone cortex consists of chlorenchyma cells consisting of chloroplasts.
9. The cells of chlorenchyma help in photo synthesis.

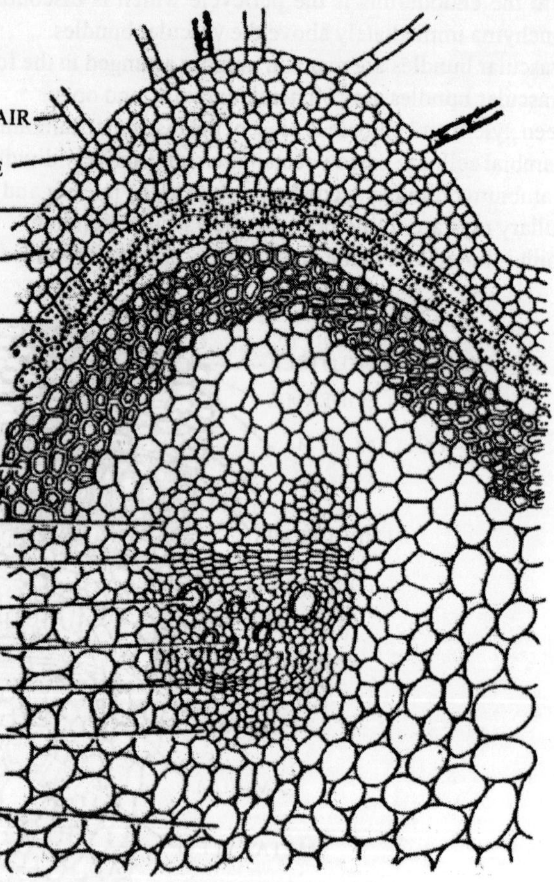

Fig. 19.8 Stems. T.S. of Cucurbita stem (Detail of a sector)

10. Endodermis is single layered and lies immediately below the chlorenchyma.
11. Next to the endodermis is the pericycle made up of a continous patch of sclerechyma.
12. The vascular bundles are embedded in a uniform mass of parenchyma tisssues often referred to as the ground tissue.
13. The ground tissue extends from the pericycle to the central pith cavity.
14. The vascular bundles are usually ten in number and arranged in two rows.
15. Outer row corresponds to the riges and inner row corresponds to the furrows.
16. The vascular bundles are conjoint, bicollateral, and open.
17. The bicollateral vascular bundles consists of outer phloem, outer xylem, outer cambium, xylem, inner cambium, and inner phloem.
18. The xylem consists of both proto and metaxylem elements.
19. There are two strips of cambia in each vascular bundle and its activity is confined to the vascular bundles.
20. Normally the outer cambial strip in the vascular bundle is the only one which shows activity.
21. Phloem occurs in two patches, the inner one towards the pith while the outer one is towards the pericycle.
22. There is a central pith cavity.

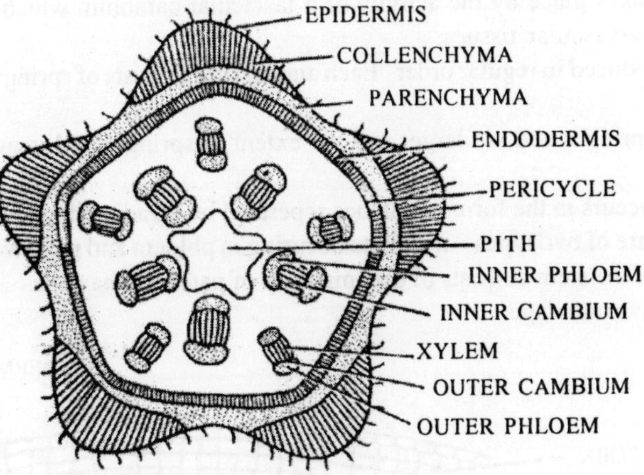

Fig. 19.9 Stems. T.S. of Cucurbita stem (Diagrammatic)

TILIA

1. The primary structure of the stem is typical to a dicotyledonous stem.
2. In the secondary structure the stem shows a rupture epidermis with cuticle deposited above it.
3. In very old stems the epidermis is completely replaced by the periderm.
4. The periderm consists of outer cork tissue, middle cork cambium, and inner secondary cortex.
5. Next to the secondary cortex is the primary cortex made up of a few layers of parenchyma cells.
6. In the secondary structure of the stem endodermis and pericycle cannot be distinguished.

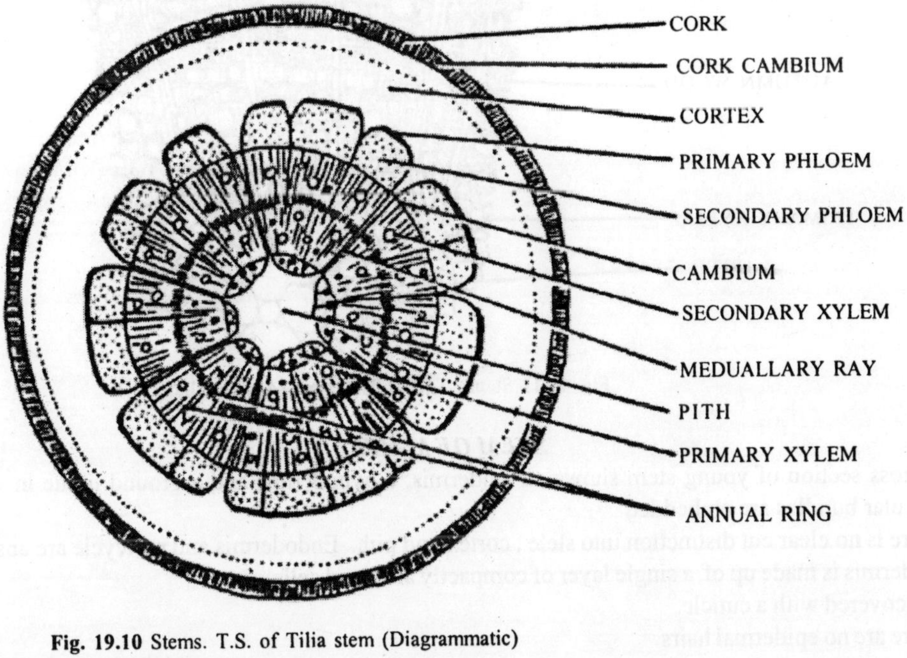

Fig. 19.10 Stems. T.S. of Tilia stem (Diagrammatic)

7. Secondary growth takes place by the activity of a fascicular cambium which forms a continous ring producing secondary vascular tissues.
8. Annular rings are produced in regular order. Each annual ring consists of spring wood and autumn wood which alternate.
9. The vessels of the spring wood are larger and the extent of spring wood is more in comparison with autumn wood.
10. Secondary phloem occurs in the form of patches seperated by triradiated medullary rays.
11. The medullary rays are of two types - triradiate occuring in phloem and narrow rays occuring in xylem.
12. The central region of the stem consists of pith made up of parenchyma cells.
13. Some cells of pith have mucilage.

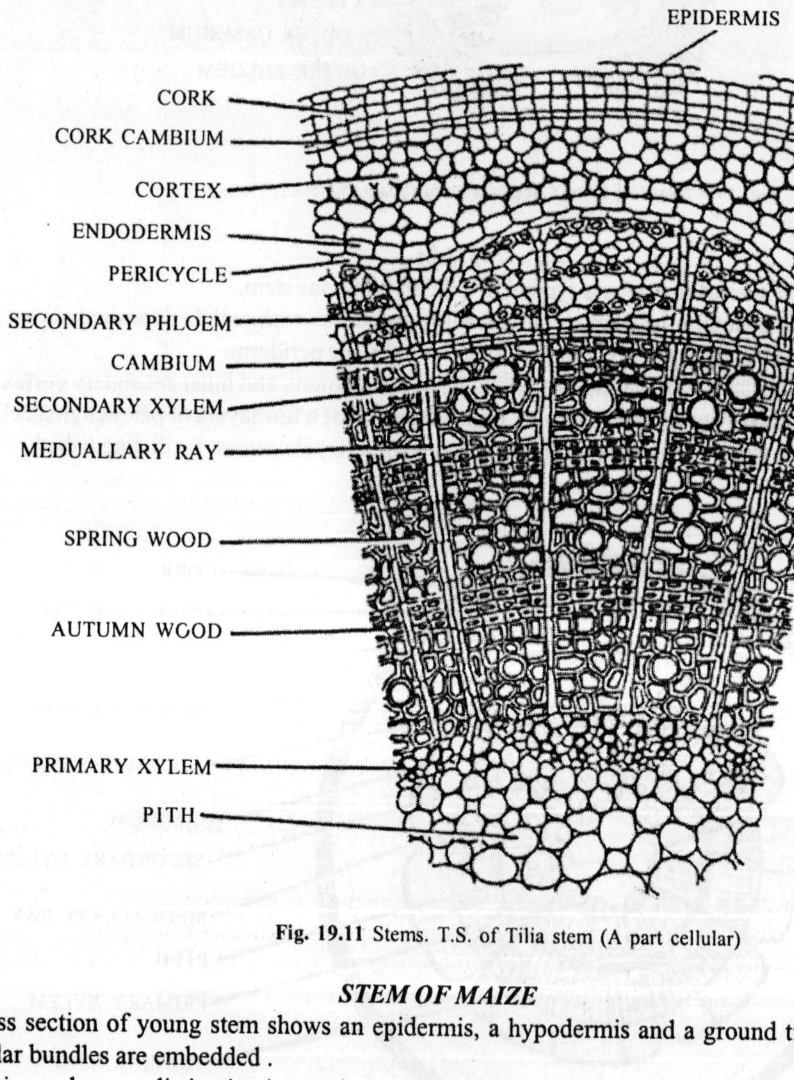

Fig. 19.11 Stems. T.S. of Tilia stem (A part cellular)

STEM OF MAIZE

1. A cross section of young stem shows an epidermis, a hypodermis and a ground tissue in which the vascular bundles are embedded .
2. There is no clear cut distinction into stele , cortex and pith. Endodermis and pericycle are absent.
3. Epidermis is made up of a single layer of compactly arranged cells.
4. It is covered with a cuticle.
5. There are no epidermal hairs.

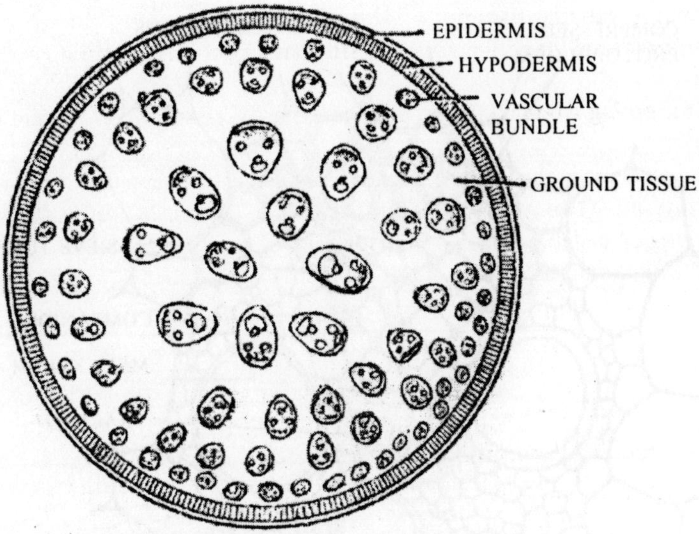

Fig. 19.12 Stems. T.S. of Maize stem (Ground plan)

6. Hypodermis is internal to the epidermis.
7. It consists of few layers of sclerenchyma cells.
8. Ground tissue is represented by loosely arranged, thin walled parenchyma cells.
9. Embedded in this tissue are the vascular bundles.
10. Vascular bundles are scattered in the ground tissue.
11. They are crowded towards the margin but smaller in size, while towards the centre they are big in size and sparsely distributed.

Structure of a vascular bundle:

1. Each bundle is oval in shape, conjoint, collateral endarch and closed (cambium is absent hence no secondary growth).
2. Surrounding the bundle there is a sclernchymatous bundle sheath.
3. The phloem is above the xylem.
4. It consists of sieve tubes and companion cells.
5. There is no phloem parenchyma.
6. The outer cells of phloem (protophloem) are crushed while the inner cells (metaphloem) are healthy.
7. Xylem is reduced; it consists of four (sometimes three) vessels (two meta and two proto) arranged in the form of 'Y'.
8. Metaxylem vessels are pitted, while protoxylem vessels are annular and spiral.
9. Sometimes in between the metaxylem vessels, small, pitted tracheas are present.
10. Below the protoxylem a lysigneous cavity is present.
11. It consists water and is known as the water containing cavity.
12. Surrounding this are present cells of wood parenchyma .

STEM OF *VANDA:*

In a cross section the following regions are seen:

Epidermis:

1. It is single layered.

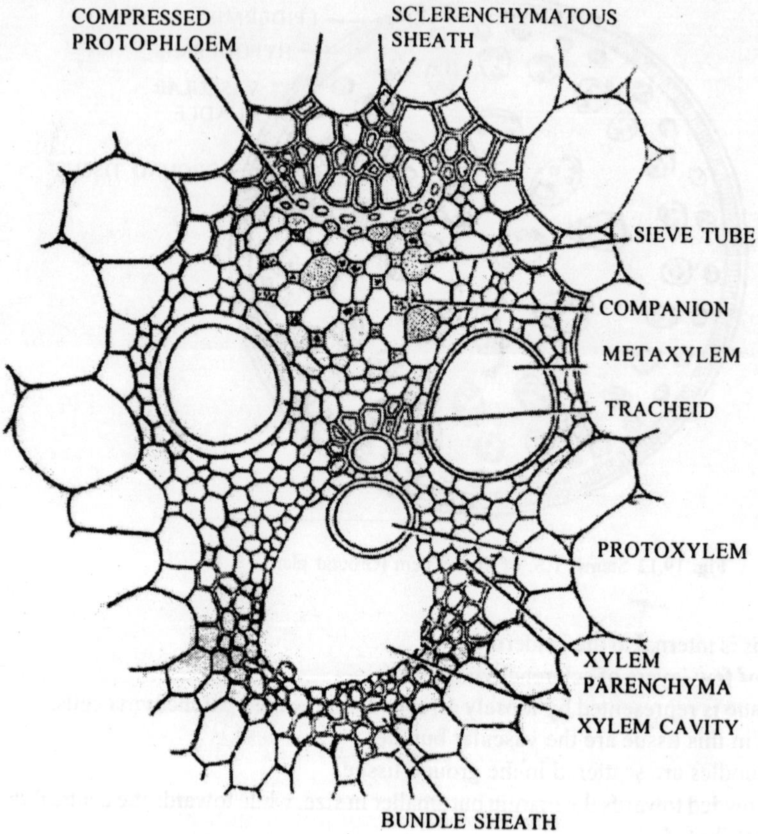

Fig. 19.13 Stems. T.S. of Maize stem (Magnified)

2. Cuticle is deposited above the epidermis.
3. There is no hypodermis.

Ground tissue:

1. Below the epidermis there are one or two layers of chlorenchyma, below which is the parenchyma consisting of thin walled, loosely arranged, oval cells.
2. This zone of the ground tissue may be called the cortex.
3. The cortex is demarcated from the rest of the central parenchymatous ground tissue by means of few layers of sclerenchyma which may be called pericycle.
4. The vascular bundles are scattered in the central ground tissue.
5. A few of the bundles at the centre are larger than the rest.
6. Vascular bundles are conjoint, collateral and closed, with endarch xylem.
7. Each vascular bundle is surrounded by a sclerenchymatous bundle sheath.
8. Structure of the phloem is same as in maize stem.
9. Xylem is reduced. There is no water containing cavity.
10. Each vascular bundle has a large vessel besides tracheids.

STEM OF *ASPARAGUS:*

In a cross section, the following regions are seen.

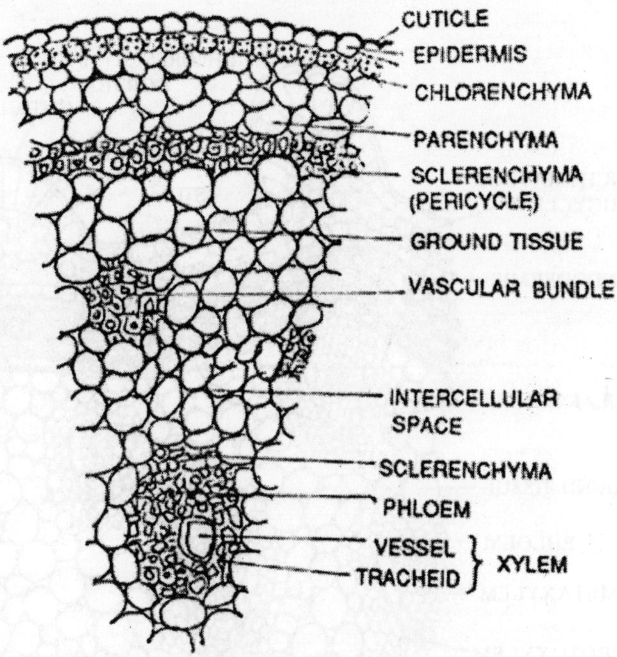

Fig. 19.14 **Stems.**
T.S. of Vanda stem (A portion enlarged).

Epidermis:

1. It is single layered and is covered by cuticle.
2. There are no epidermal hairs. There is no hypodermis.

Ground tissue:

1. It extends from the periphery to the centre of the axis and may be divided into four zones.
2. The outer zone which is immediately below the epidermis consists of a few layers of chlorenchyma with little intercellular spaces.

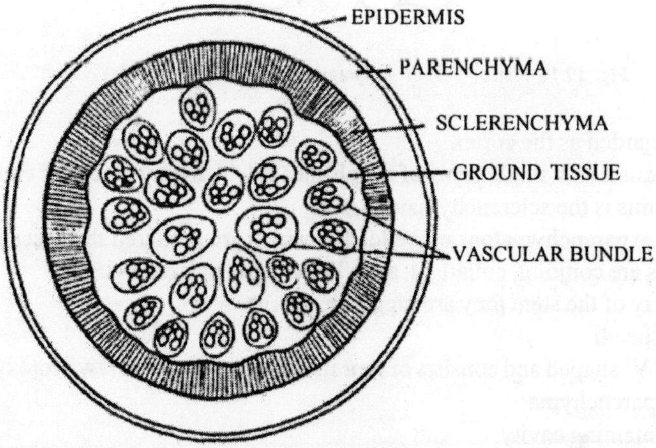

Fig. 19.15 Stems. T.S. of Asparagus stem

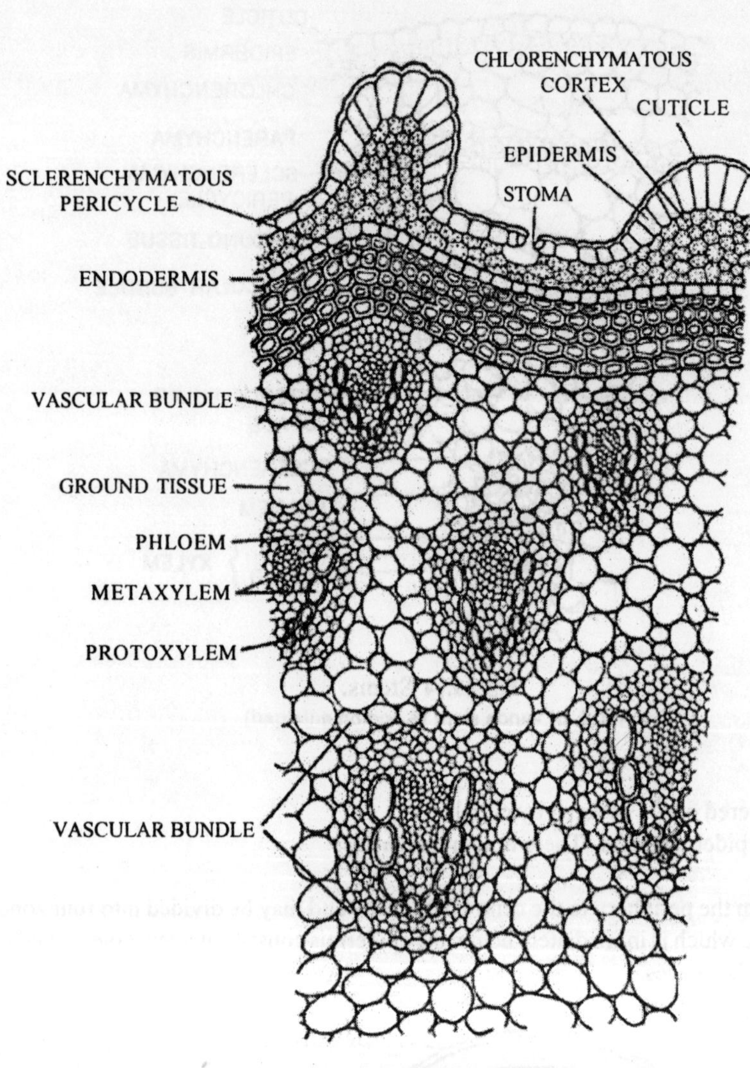

Fig. 19.16 Stems. T.S. of Asparagus stem (vascular bundle)

3. This zone may be regarded as the cortex.
4. The second zone consists of a single layer of starch sheath which may be called endodermis.
5. Next to this endodermis is the sclerenchymatous zone.
6. The innermost zone is parenchymatous embedded in which are scattered the vascular bundles.
7. The vascular bundles are conjoint, collateral, and closed, with endarch xylem.
8. Towards the periphery of the stem they are bigger in size.
9. There is no bundle sheath.
10. The xylem strand is 'V' shaped and consists of four metaxylem vessels, a few protoxylem vessels, a few tracheids and wood parenchyma.
11. There is no water containing cavity.
12. Phloem is similar in structure to that of other monocotyledons.

CANNA

1. A transverse section shows three regions namely epidermis, cortex and Stele.
2. Epidermis is the outermost layer. It is singled layered and made up of small polygonal cells.
3. The outer walls of epidermal cells are cutinized.
4. Next to the epidermis is the ground tissue.
5. Immediately below the epidermis is found the cortex.
6. The cortex consists of a few layers of parenchyma cells. These are large and polygonal.
7. Immediately below the cortex is found a single layer of chlorenchyma.
8. Next to the chlorenchyma are found patches of sclerenchyma which usually abutt the vascular bundles.
9. The entire inner region of the stem next to this consists off a uniform mass off parenchyma cells constituting the ground tissue.

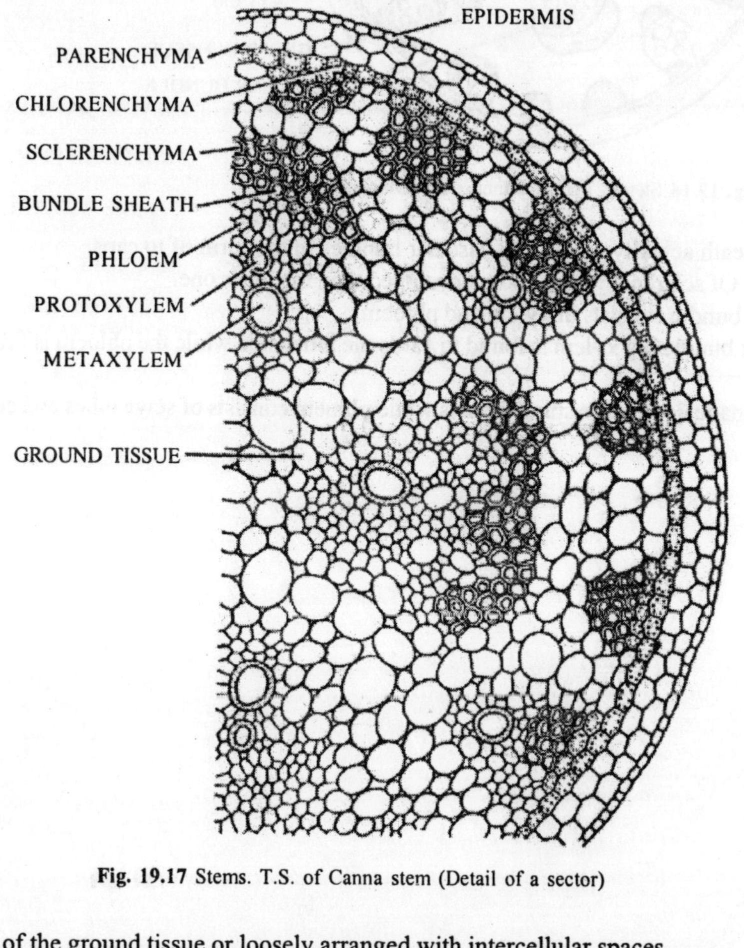

Fig. 19.17 Stems. T.S. of Canna stem (Detail of a sector)

10. The cells of the ground tissue or loosely arranged with intercellular spaces.
11. Embedded in the mass of ground tissue are found a number of vascular bundles.
12. The vascular bundles are scattered without any arrangement.
13. The vascular bundles towards the periphery are smaller in size when compared to those in the center.
14. The vascular bundles are collateral and closed.
15. There is no cambium hence there is no secondary growth.
16. Each vascular bundle is incompletely surrounded by a bundle sheath made up of sclerenchyma.

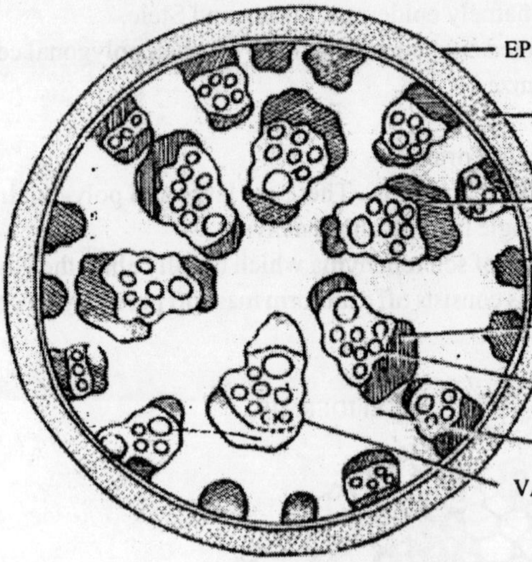

Fig. 19.14 Stems. T.S. of Canna stem (Diagrammatic)

17. The bundle sheath actually covers the vascular bundles in the form of to caps.
18. The outer cap Of sclerenchyma is more developed than the inner one.
19. Each vascular bundle consists of xylem and phloem.
20. In the vascular bundle big xylem is found towards the inner side while the phloem is found towards the outside.
21. The xylem is made up of large spiral vessels while phloem consists of seive tubes and companion cells.

Chapter 20

ANATOMY OF LEAVES

LEAF OF HELIANTHUS

The leaf of sunflower shows the features of a typical dicotyledonous leaf. In a transverse section it shows three well defined regions viz. epidermis, mesophyll and veins.

Epidermis:

1. There are two epidermal layers covering either surfaces of the leaf.
2. The epidermis on the upper surface is called the upper epidermis and the one on the lower surface is called the lower epidermis.
3. In some cases, covering the epidermal layers will be a layer of cuticle.
4. In the leaf of sunflower the upper epidermis is continous while the continuity of the lower epidermis is interrupted by the presence of stomata.
5. There are small pores helping in gaseous exchange and the loss of water during transpiration.
6. Each stoma is covered by two guard cells which regulate the opening and closing of the stomatal aperture.

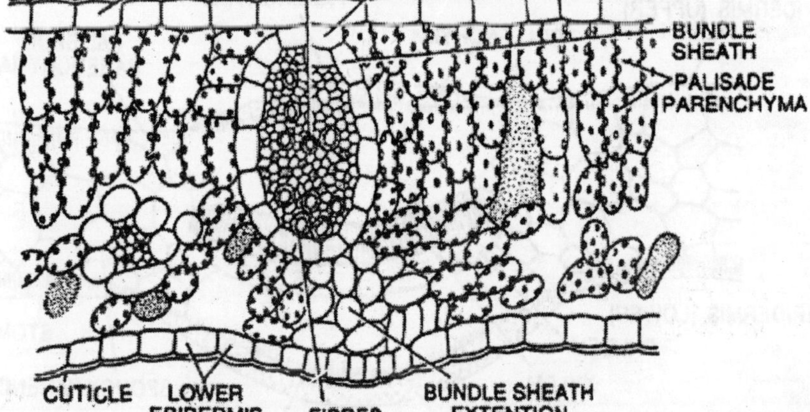

Fig. 20.1 Anatomy of Sunflower leaf (A portion enlarged)

7. The guard cells are crescent shaped with thick inner walls.
8. Internally each stoma leads into an air space.

Mesophyll:

1. In between the two epidermis layers is the mesophyll.
2. It is differentiated and has well developed palisade tissue below the upper epidermis.
3. Above the lower epidermis and below the palisade tissue is the spongy parenchyma which has loosely arranged cells with large number of air chambers.
4. Palisade tissue is chlorophyllous and it is the chief photosyntietic tissue of the plant.
5. There are a few chloroplasts in the spongy parenchyma cells also but their main function is storage.
6. There are rich in starch grains.

Vein:

1. The veins are arranged in the form of a network and in a cross section they show the vascular bundles.
2. Each vein representing a vascular bundle is surrounded by a parenchymatous envelope called the bundle sheath.
3. Within the bundle sheath are the vascular elements viz, the xylem and the phloem.
4. Unlike in stems the xylem is above and the phloem is below.

LEAF OF *DALBERGIA:*

The leaf of *Dalbergia* may be studied as an example of a typica dicot leaf. A cross section of the leaf shows the following features.

Epidermis:

1. There are two epidermal layers viz., upper and lower.
2. Both the layers are uniseriate and composed of barrel shaped cells with thick outer walls.
3. There is a well developed cuticle above the upper epidermis.
4. The leaves are hypostomatic (stomata found only on the lower surface).
5. The stomata internally lead to air cavities.
6. In between the two epidermal layers is the mesophyll.
7. It is differentiated and has a well developed palisade tissue below the upper epidermis.
8. Above the lower epidermis and below the palisade tissue is the spongy parenchyma which has loosely arranged cells with large number of air chambers.

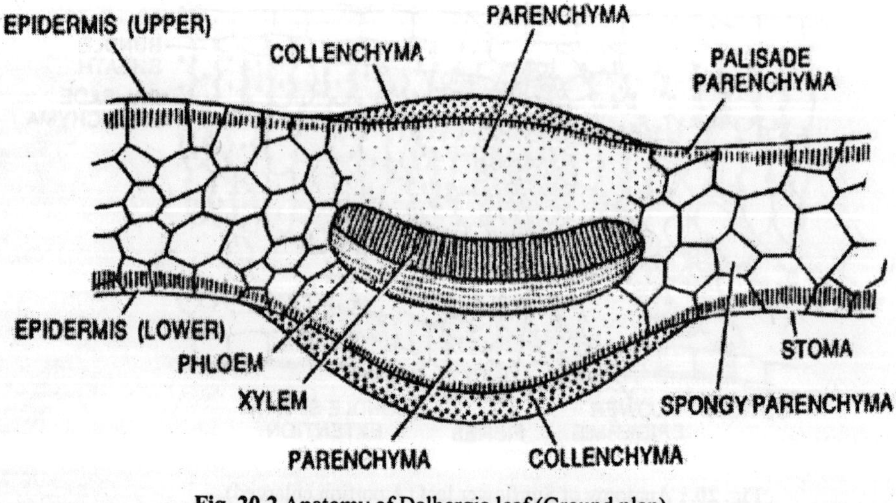

Fig. 20.2 Anatomy of Dalbergia leaf (Ground plan)

9. Palisade tissue is chlorophyllous and it is the chief photosynthetic tissue of the plant.
10. There are a few chloroplasts in the spongy parenchyma cells also but their main function is storage.
11. They are rich in starch grains.
12. Near the midrib region, internal to the epidermis is collenchyma.
13. Collenchyma cells are more towards the lower surface.
14. Immediately surrounding the vein is a bundle sheath which is parenchymatous.
15. It is very distinct in *Dalbergia* leaf.
16. The vascular bundle is *cresent* shaped and shows x/p arrangement.

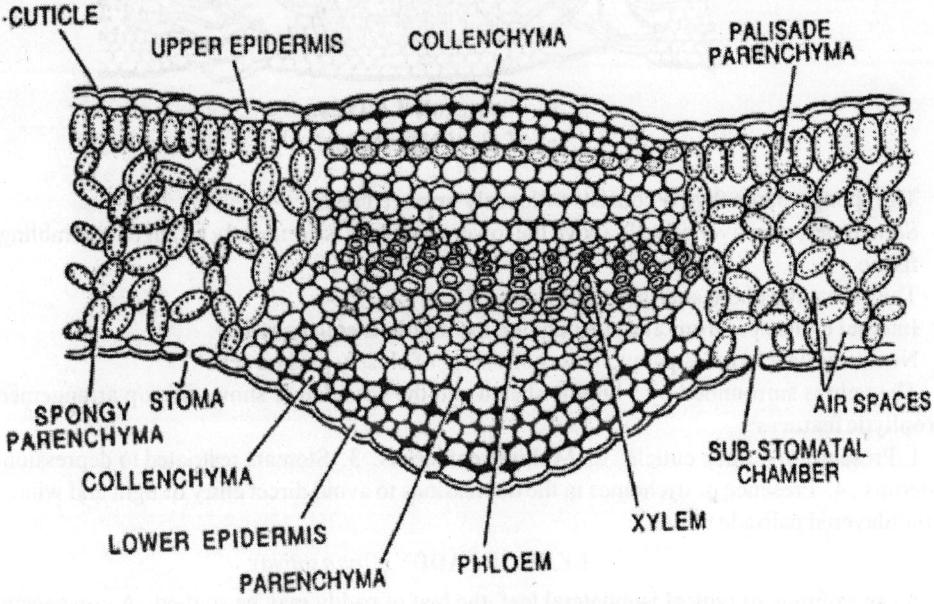

Fig. 20.3 Anatomy of Dalbergia leaf (A portion enlarged)

LEAF OF *NERIUM* (Oleander)

Nerium belongs to the family *Apocynaceae* of dicotyledons. It shows many xerophytic adaptations. A cross section of the leaf shows the following:

Epidermis:

1. The leaf is dorsiventral and shows two epidermal regions viz., upper and lower.
2. The upper epidermis is multilayered and it consists of two to three layers of barrel shaped cells arranged compactly.
3. The outermost epidermal layer is covered with a thick cuticle.
4. The structure of the lower epidermis differs from that of the upper in that it is single layered.
5. There is the deposition of a cuticle even above the lower epidermis.
6. At certain places the epidermal cells incurve forming depressions of pits.
7. The stomata are found in these depressions (hypostomatic).
8. The pits are protected with a number of *trichomes* (hairs) growing from the sides.
9. The presence of the trichomes retards the direct entry of the light and hence checks transpiration.

Mesophyll:

1. It is differentiated into palisade and spongy parenchyma.
2. Usually two or three layers of palisade tissue is present below the upper epidermis.

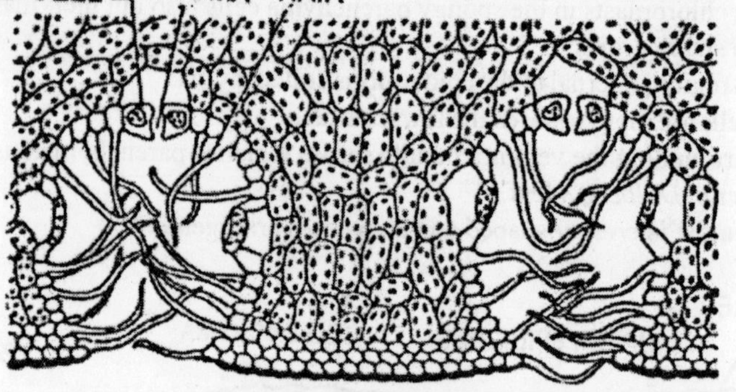

Fig. 20.4 **Leaves.,**
T.S. of Nerium leaf.

3. The spongy parenchyma consists of loosely arranged cells.
4. Sometimes, one layer of cells above the lower epidermis is compactly arranged resembling the palisade tissue.
5. The air cavities here and there show druses.
6. Internal to the epidermis and surrounding the midrib is collenchyma.
7. Next to collenchyma is parenchyma consisting of closely packed cells.
8. The vein is surrounded by a bundle sheath and the vasculature shows the x/p arrangement.

Xerophytic features:

1. Presence of a thick cuticle. 2. Multiple epidermis. 3. Stomata restricted to depression of the lower epidermis. 4. Presence of trichomes in the depressions to avoid direct entry of light and wind. 5. Presence of multilayered palisade tissue.

LEAF OF PADDY (*Oryza sativa*):

As an example of typical isobilateral leaf, the leaf of paddy may be studied. A cross section shows the following:

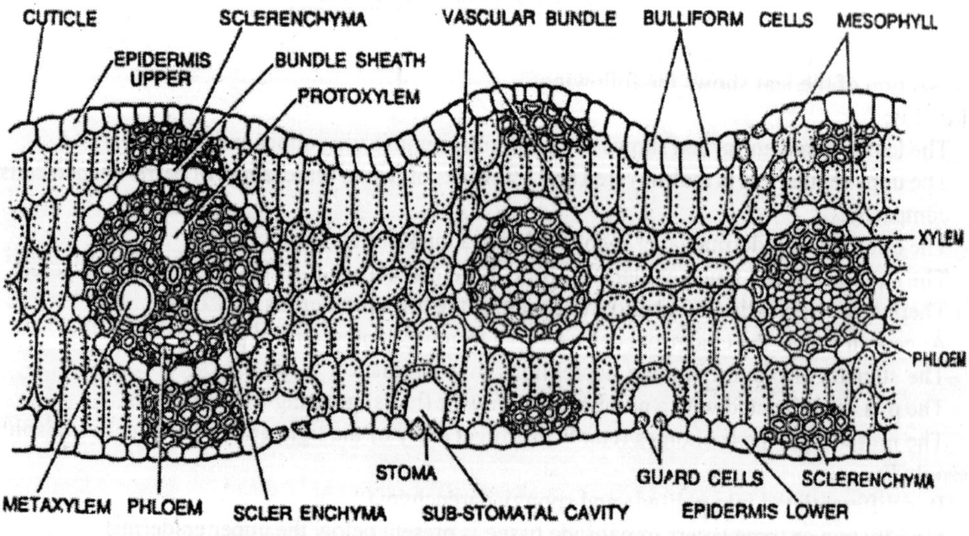

Fig. 20.5 Leaves. T.S. of Paddy leaf

Epidermis:

1. There will be two epidermal layers - upper and lower.
2. The epidermal layers are uniseriate and composed of more or less oval cells arranged compactly.
3. There is the deposition of a cuticle both one the upper epidermis as well as on the lower epidermis.
4. Besides the normal cells, the upper epidermis consists of special cells called *bulliform cells* or *motor cells*.
5. These cells are larger than the other epidermal cells and are thin walled.
6. These help in the rolling of the leaves during dry weather.
7. Leaves are *amphistomatic* (stomata are present on both the epidermal layers).
8. Each stoma internally leads into a large substomatal air chamber.

Mesophyll:

1. It is undifferentiated.
2. But the mesophyll cells below the upper epidermis may show a compact arrangement.

Vascular bundles:

1. The veins are arranged parallely and are not interconnected.
2. There are two pads of sclerenchyma supporting the vein.
3. The vein itself is surrounded by a bundle sheath made up of parenchyma cells.
4. Within the bundle sheath are present collateral vascular elements showing x/p arrangement.

LEAF OF MAIZE

A cross section shows the following.

EPIDERMIS :

1. There will be two epidermal layers upper and lower.
2. The epidermal layers are uniseriate and composed of more or less oval cells arranged compactly.
3. There is the deposition of a cuticle both on the upper epidermis as well on the lower epidermis .
4. Besides the normal cells, the upper epidermis consists of special cells called bulliform cells or motor cells.
5. These cells are larger than the other epidermal cells and are thin walled.
6. These help in the rolling of the leaves during dry weather .

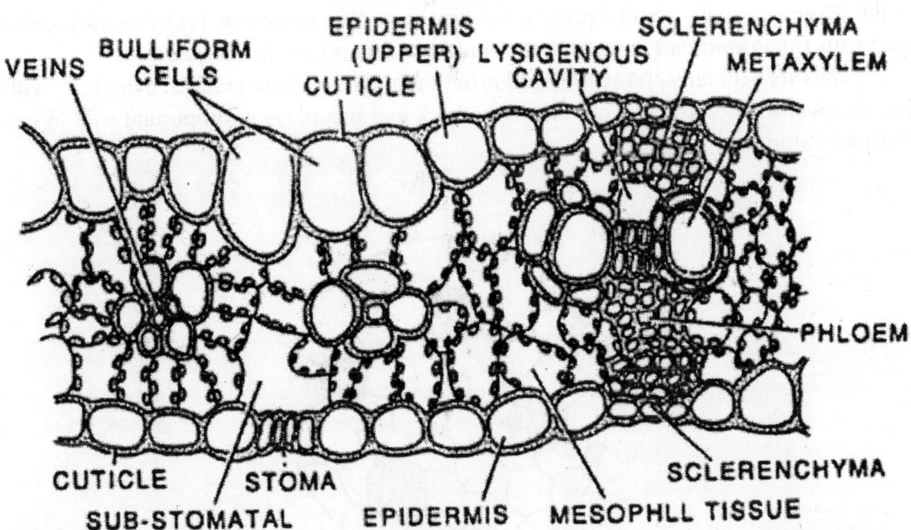

Fig. 20.6 Leaves. T.S. of Maize leaf

7. Leaves are amphistomatic (stomata are present on both the epidermal layers), Each stoma internally leads into a large substomatal air chamber.

Mesophyll:

1. It is undifferentiated.
2. But the mesophyll cells below the upper epidermis may show a compact arrangement.

Vascular bundles:

1. The veins are arranged parallely and are not interconnected.
2. There are two pads of sclerenchyma supporting the vein.
3. The vein itself is surrounded by the bundle sheath made up of parenchyma cells.
4. Within the bundle sheath are present collateral vascular elements showing X/P arrangement.

STOMATA

1. Stomata are microscopic openings present in the epidermis.
2. Each stoma has a stomatal aperture or opening surrounded by two guard cells.

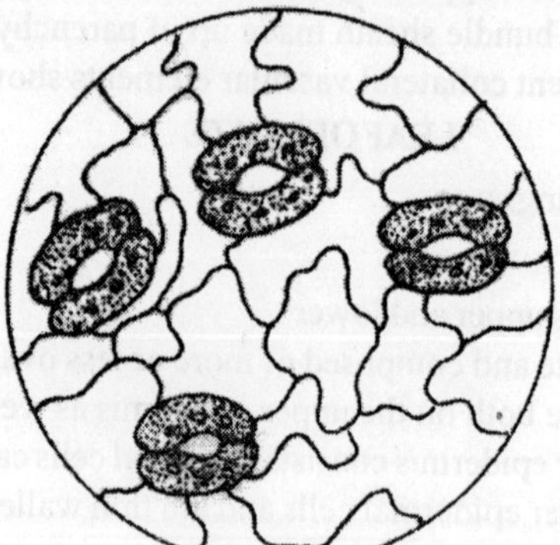

Fig. 20.7 Anomocytic stomata (Cucurbita)

3. In the leaves, stomata may be present only on the upper epidermis (*epistomatic*), only on the lower epidermis (*hypostomatic*) or on both the epidermal layers (*amphistomatic*).
4. Guard cells are generally crescent shaped in outline and have dense granular cytoplasm with chloroplasts.
5. The cell walls of guard cells are unevenly thick and this plays an important role in the opening and closing of stomata.

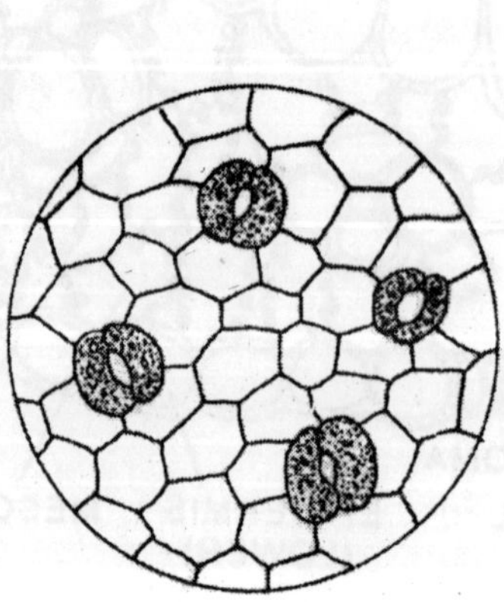

Fig. 20.8 Anomocytic stomata (Diospyros)

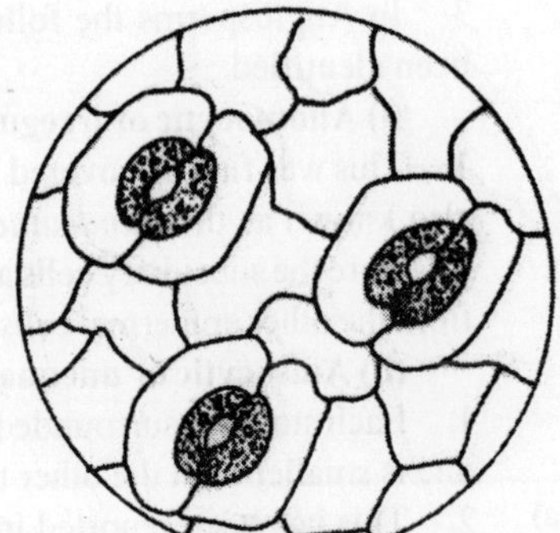

Fig. 20.9 Paracytic stomata (Ixora)

6. **Usually the wall of guard cells facing the pore is thicker than the one away from the pore.**
7. **During day time there is increased turgor pressure in the guard cells, as a result the thin walls are pulled away so that the concavity of the front (thick) wall increases, and the aperture becomes wider.**

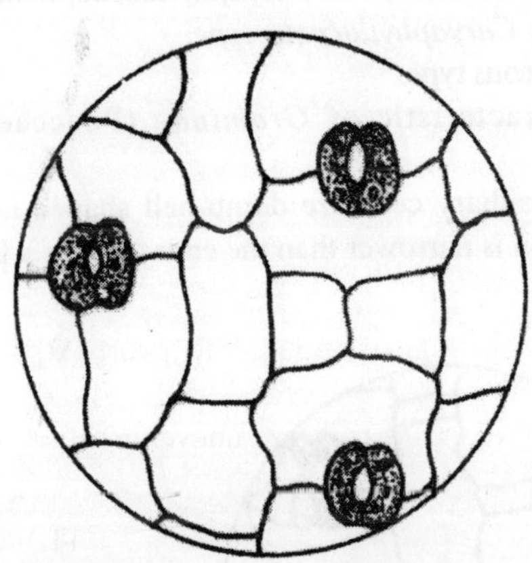

Fig. 20.10 Diacytic stomata (Dianthus)

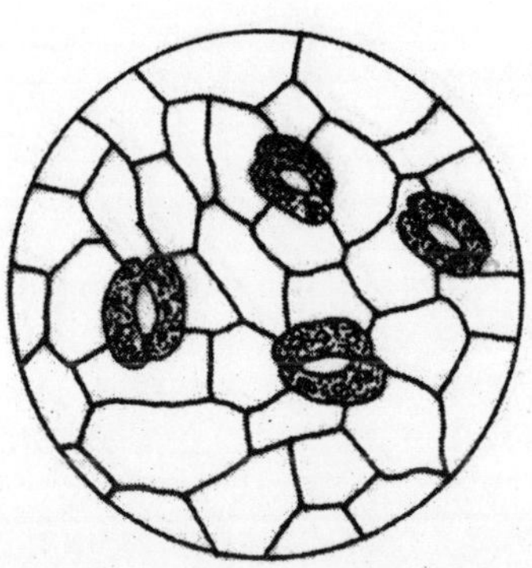

Fig. 20.11 Anisocytic stomata (Petunia)

8. Stomata are present on the aerial organs of plants like leaf, stem, sepals etc.

9. The chief distribution of stomata however is on the leaves.

10. Rarely stomata are enocuntered on petals, stamens and ovules also, but these are non functional.

TYPES OF STOMATA:

1. Classification of the types of stomata is mainly based on the nature and number of *subsidiary cells* associated with them (stomata).

2. Subsidiary cells are epidermal cells that are morphologically

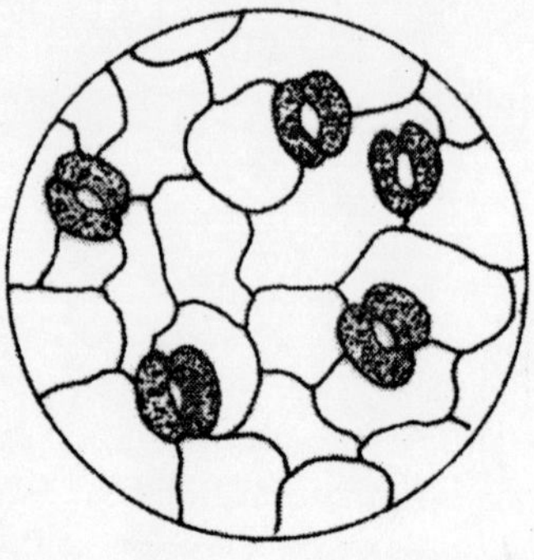

Fig. 20.12 Anisocytic stomata (Brassica)

distinct from other epidermal cells are are in close association with the stomata.

3. In Angiosperms the following main types of stomata have been identified.

(i) Anomocytic or irregular celled type:

1. This was first discovered in the family *Ranunculaceae*, hence also known as the *Ranuculaceous type*:
2. Here the subsidiary cells are absent in that they are not distinct from the other epidermal cells.

(ii) Anisocytic or unequal celled type:

1. Each stoma is surrounded by three subsidiary cells, of which one is smaller than the other two.
2. This has been reported in *Cruciferae*.

(iii) Paracytic or parallel celled type:

1. Stoma is surrounded by two equal parallely arranged subsidiary cells.
2. This is of frequent occurrence in *Rubiaceae*, hence also known as the *Rubiaceous type*.

(iv) Diacytic or cross celled type:

1. The stoma is surrounded by two subsidiary cells, which are at right angles to the guard cells unlike in the paracytic type.
2. This was first discovered in *Caryophyllaceae*, hence also known as the *Caryophyllaceous type*.

(v) Graminaceous type:

1. This is characteristic of *Graminae* (Poaceae) and *Cyperaceae*.
2. Here the subsidiary cells are dumb-bell shaped i.e., the middle portion is narrower than the ends and has a thicker wall.

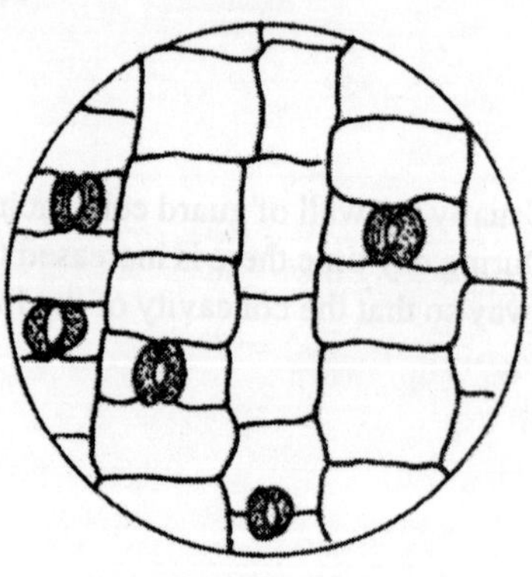

Fig. 20.13 Stomata of Rheo

Fig. 20.14 Graminaceous stomata

Chapter 21

ANOMALOUS STRUCTURE OF STEMS

STEM OF *ARISTOLOCHIA* (Aristolochiaceae)

1. The young stem has a typical dicotyledonous structure.
2. Pericycle is continous and is represented by sclerenchyma.
3. The vasculature consists of collateral bundles separated by wide interfascicular area.
4. During secondary growth inter and intrafascicular cambia become active and form a ring.
5. But while the intrafascicular cambium forms the vascular tissues, the segments of interfascicular cambium form only ray like parenchyma with the result the vascular bundles though increased in size, remain discrete.
6. With the increase in diameter the pith and rays get crushed.

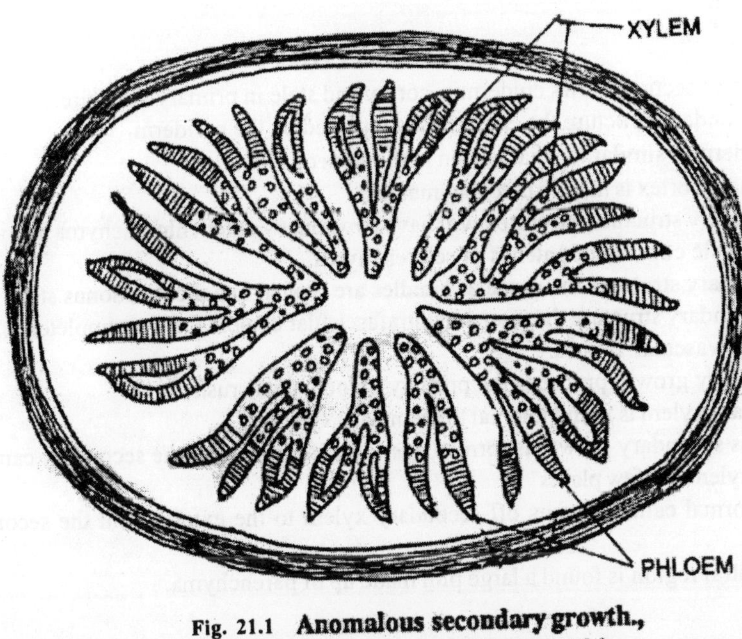

Fig. 21.1 **Anomalous secondary growth., T.S. of Aristolochia stem (Diagrammatic).**

7. Sometimes, portions of intrafascicular cambia also form parenchymatous rays, with the result the vascular bundles become bifurcated.
8. A stongly fluted vascular cylinder is thus formed.
9. This is an instance of the abnormal behaviour of the normal cambium.

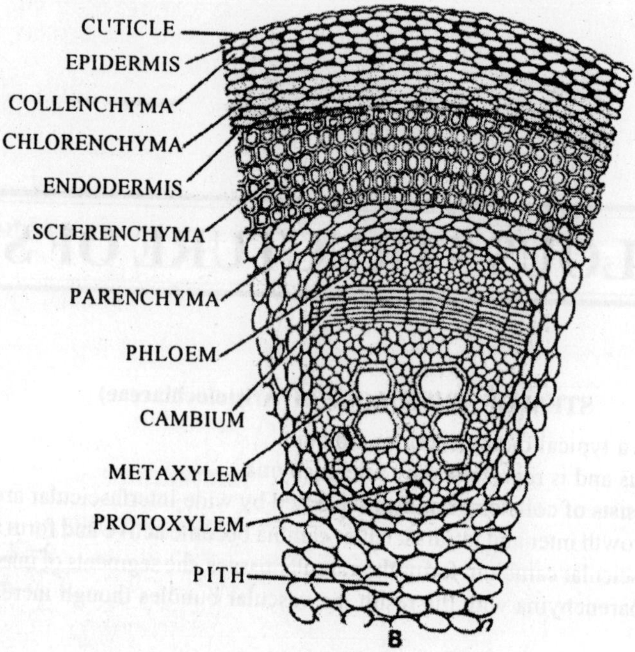

Fig. 21.2 Anatomy of stem. T.S. of young stem of Aristolochia (dicot). detail of a sector

TECOMA

1. A transverse section shows epidermis, cortex and stele in primary structure.
2. In the secondary structure the epidermis is replaced by the periderm.
3. The periderm is similar to what is seen in the stem of *Tilia*.
4. The primary cortex is mainly parenchymatous.
5. In the primary structure the cortex will have two zones - outer chlorenchyma and inner parenchyma.
6. The pericycle consists of patches of sclerenchyma.
7. In the primary structure the vascular bundles are typical to a dicotyledonus stem.
8. In the secondary structure the inter and intrafascicular cambia form a completely closed ring and cut off secondary vascular tissues.
9. As secondary growth proceeds the primary phloem gets crushed.
10. The primary xylem is found internal to secondary xylem.
11. Anomalus secondary growth is formed due to the formation of the secondary cambium internal to the primary xylem in a few places.
12. This abnormal cambium cuts off secondary xylem to the exterior and the secondary phloem to the interior.
13. In the central region is found a large pith made up of parenchyma.

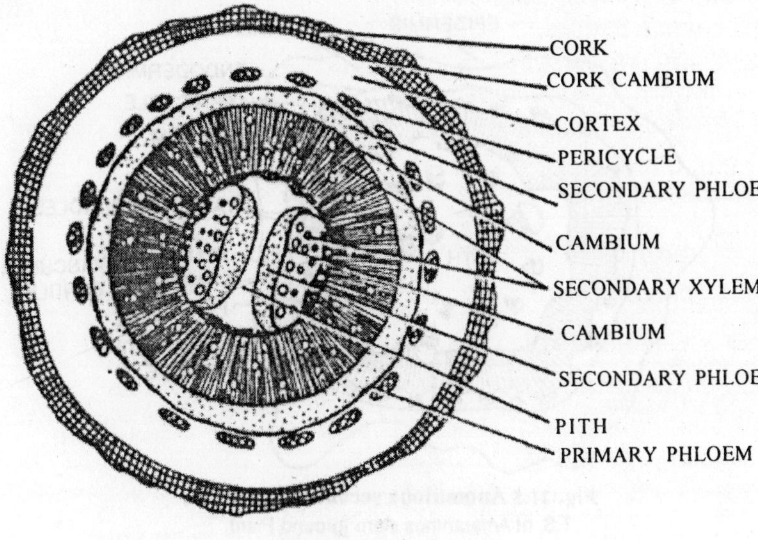

Fig. 21.3 T.S. of Tecoma stem (diagrammetic)

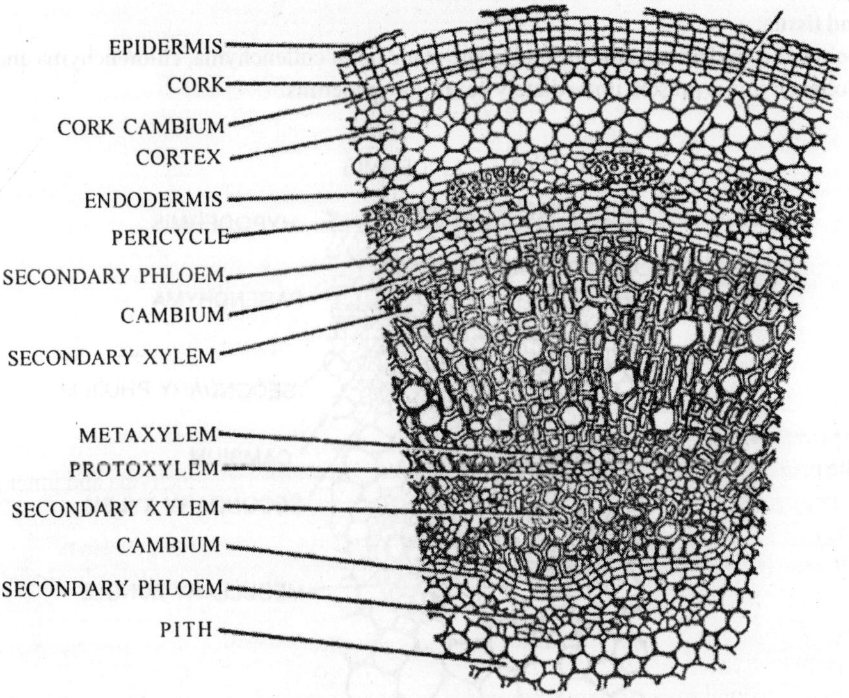

Fig. 21.4 T.S. of Tecoma stem (a part cellular)

STEM OF *AMARANTHUS* (Amaranthaceae)

The stem of *Amaranthus* shows certain anomalous structures, both in the primary structure as well as in the secondary thickening.

A cross section of the stem shows the following features.

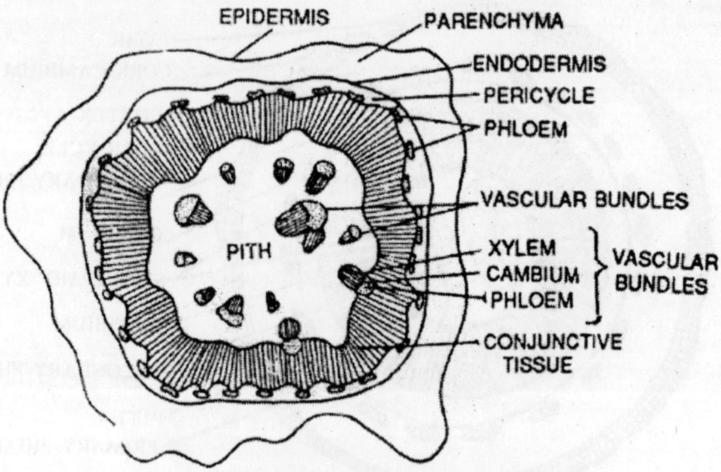

Fig. 21.5 Anomalous secondary growth.,
T.S. of Amaranthus stem (ground Plan).

Epidermis:

1. This is the outermost layer and is uniseriate.

Ground tissue:

1. Below the epidermis is the ground tissue made up of collenchyma, chlorenchyma and parenchyma.
2. Collenchyma is present immediately below the epidermis.

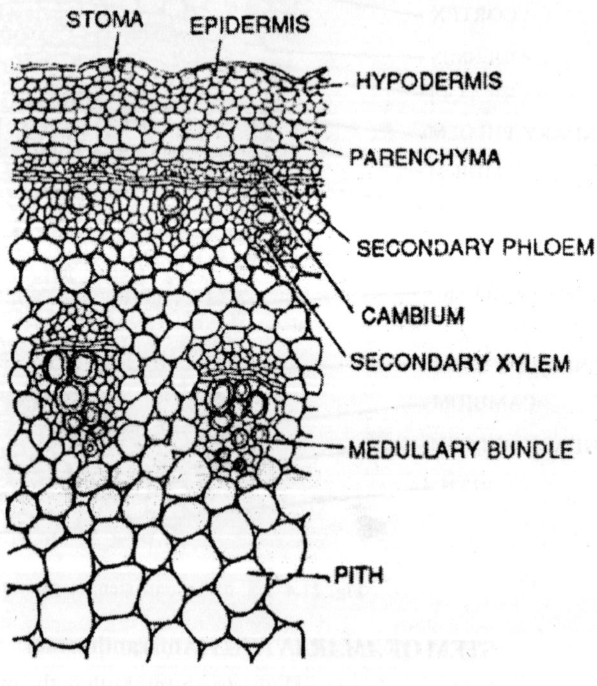

Fig. 21.6 Anomalous secondary growth.,
T.S. of Amaranthus stem (a portion enlarged).

3. Here and there it is interrupted by patches of chlorenchyma.
4. Below the collenchyma is a multilayered parenchymatous zone constituting the inner cortex.
5. The endodermis is not very conspicous.
6. In the primary structure the entire region inner to the endodermis is filled by loosely arranged, thin walled parenchyma cells.
7. Embedded in this, are found the scattered primary vascular bundles.
8. These are the medullary bundles.
9. There is no ring of vascular bundles in the primary structure as in a normal dicot stem.
10. The medullary bundles are conjoint, collateral, and open with endarch xylem.
11. Vascular bundles towards the periphery are smaller than those towards the centre.
12. The cambia of these bundles produce a little secondary vascular tissue (which is limited to the bundles) and very soon cease to function.
13. Secondary growth is brought about by a cambium originating outside the stele.
14. This cuts off secondary vascular bundles, which lie embedded in the conjunctive tissue around the pith region.
15. This is an example of *medullary bundles,* and *replacement of normal cambium by abnormal accessory cambium.*
16. There is anomaly both in the primary structure as well as in the secondary growth.

STEM OF *BOUGAINVILLEA* (Nyctaginaceae)

The stem in its primary structure shows an epidermis, cortex and a vascular system. The epidermis is single layered and has many multicellular hairs.

1. **Cortex** is made up of collenchyma and chlorenchyma.
2. Collenchyma is present in the form of patches.
3. Chlorenchyma is internal to collenchyma and extends between the patches of the latter.

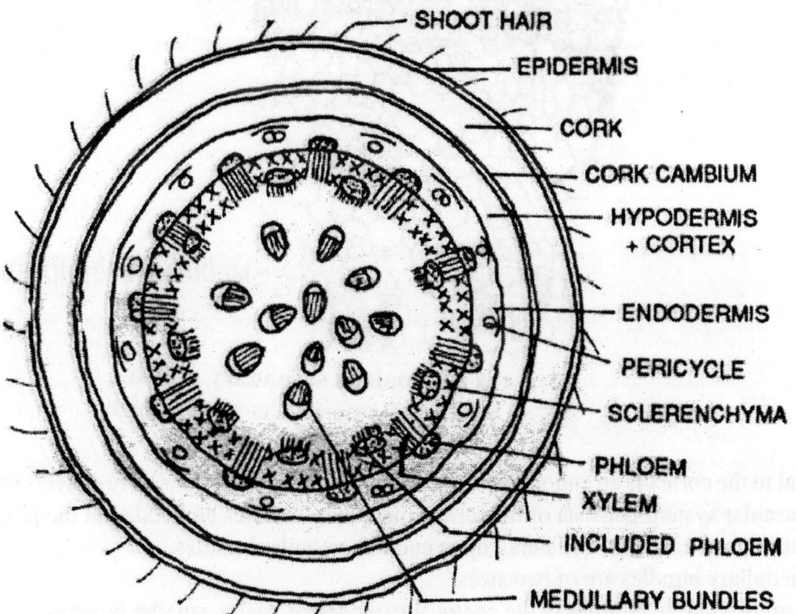

Fig. 21.7 Anomalous secondary growth., T.S. of Bougainvillea stem (ground plan).

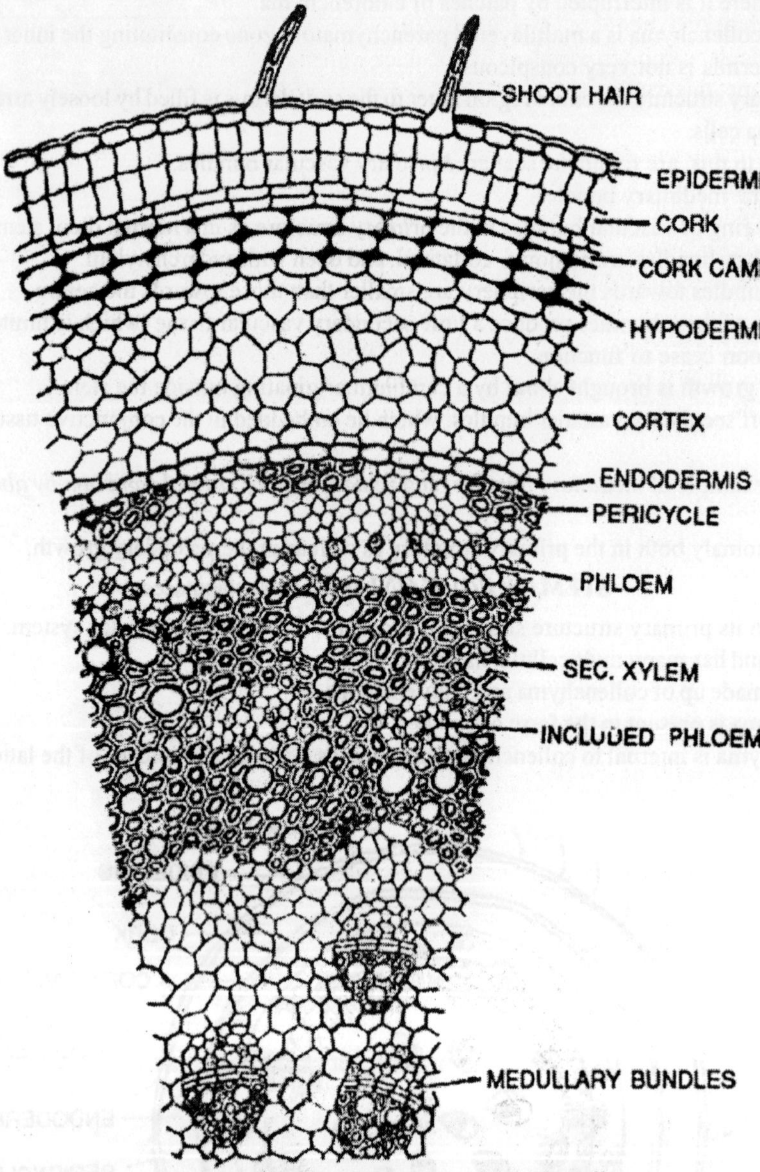

Fig. 21.8 **Anomalous secondary growth.,**
T.S. of Bougainvillea stem (A portion enlarged).

4. Internal to the cortex is an endodermis and a pericycle made up of one or two layers of parenchyma cells.
5. The vascular system consists of numerous medullary bundles embedded in the pith, in addition to the normal rings of conjoint, collateral, open endarch vascular bundles.
6. The medullary bundles are of two sizes.
7. There are four large bundles in the centre surrounded by many, smaller bundles.
8. Medullary bundles are similar in structure to that of normal bundles.
9. Fascicular cambium in these bundles has a limited activity.
10. During secondary growth, more than one cambial ring is formed.

11. The first cambial ring produces vasculature, only where the primary vasculature is present.
12. The interfascicular cambium produces living, thick walled conjunctive tissue.
13. After some time, this cambium ceases to function and a second cambial ring arises external to the first one.
14. The second cambial ring functions in the same way as the first. producing vasculature only in strips.
15. Successive cambial rings arise and function in the same way producing discrete secondary vascular bundles embedded in a sclerenchymatous conjunctive tissue.
16. The type of secondary growth here is same as in *Amaranthus* i.e., *medullary vascular bundles* and abnormal behaviour of *accessory cambia.*

PEPEROMIA

1. A transverse of the section shows the epidermis, cortex and the vasculature.
2. Anomalous secondary growth is seen in the stem.
3. Epidermis is made up of a single layer of compactly arranged cells.
4. There is the deposition of cuticle above the epidermis.
5. Next to the epidermis is found the cortex consisting of few layers of collenchyma cells.
6. Next to the cortex is found a uniform mass of parenchyma constituting the ground tissue.
7. Embedded in the ground tissue are found a number of scattered vascular bundles.
8. This arrangement is unusual for dicotyledons.

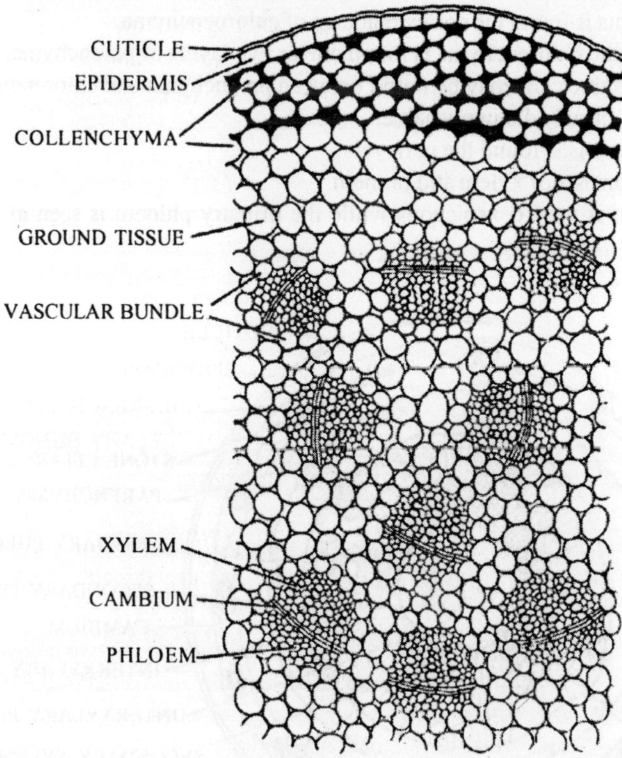

Fig. 21.9 Anatomy of anomalour stem T.S. of Peperomia (dicot stem;

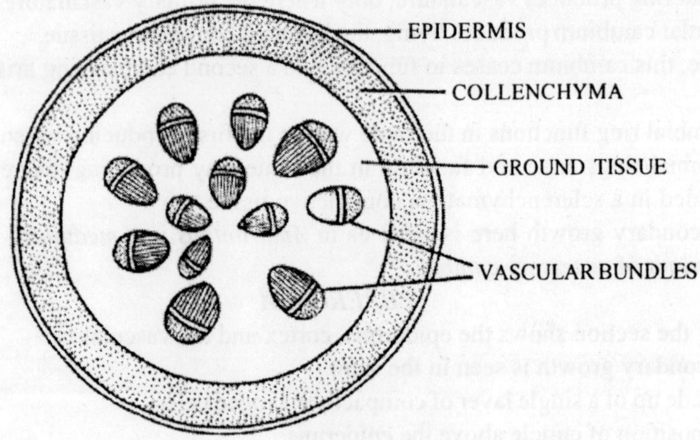

Fig. 21.10 Anatomy of anomalous stem T.S. of Peperomia (dicot stem; detail of a sector)

LEPTADEMIA

1. A transverse section of the stem show epidermis, cortex and stele.
2. Anomalous vascular structure are seen in the stem.
3. Epidermis is made up of a single layer of compactly arranged cells.
4. There is the deposition of cuticle above the epidermis.
5. Below the epidermis is found the cortex made up of chloroenchyma.
6. Sometimes below the chlorenchyma is found one or two layers of parenchyma.
7. Occasionally few stone cells may be found between parenchyma and chlorenchyma.
8. The endodermis is clearly distinguishable.
9. Next to the endodermis is found the pericycle.
10. The vasculature consists of xylem and phloem.
11. Secondary phloem is very conspicuous while the primary phloem is seen at regions opposite to the primary phloem.

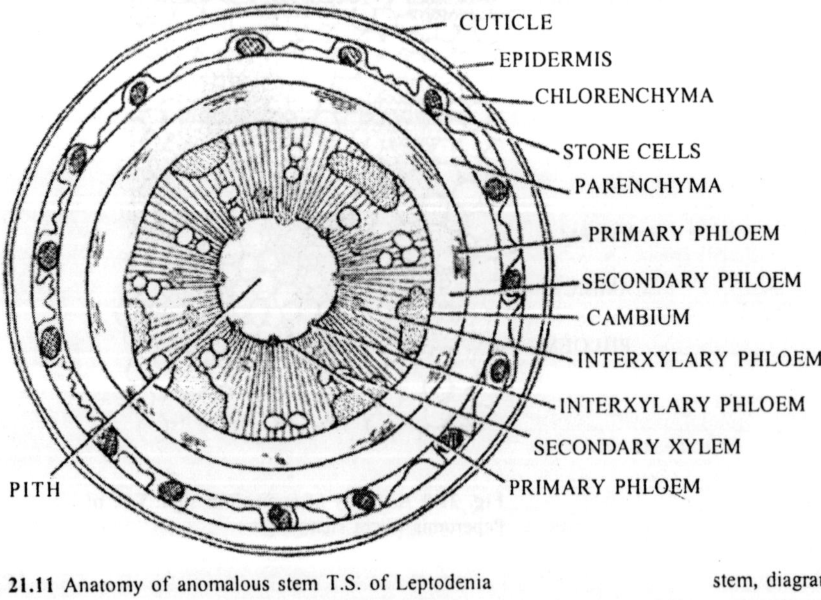

Fig. 21.11 Anatomy of anomalous stem T.S. of Leptodenia stem, diagrammatic

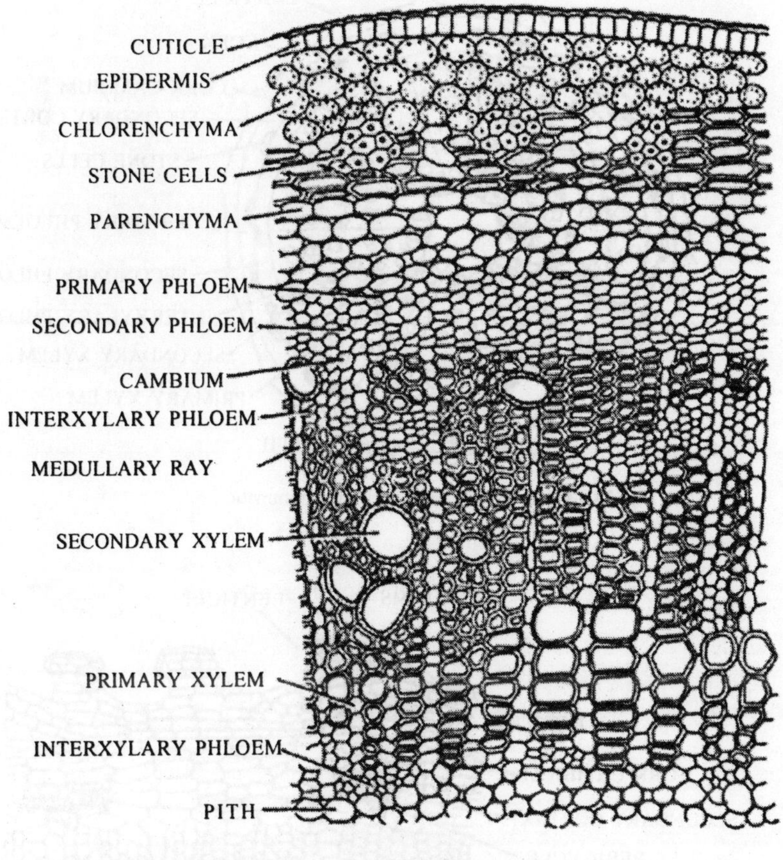

Fig. 21.12 Anatomy of anomalous stem T.S. of Leptodenia . stem, detail of a sector

12. Next to the secondary phloem is the cambium made up of many layers of rectuangular cells.
13. The xylem is found next to the phloem.
14. The primary xylem is found to the interior.
15. The secondary xylem consists of large vessels.
16. The secondary vasculature has unusual feature such as interxylary and intraxylary phloem.
17. The inter xylary phloem is found between the strands of the secondary xylem.
18. The intra xylary phloem is found at the border of the pith.
19. Uni or multi seritae medullary rays are found between the vascular bundles.
20. Latex in the cells are found embedded in the parenchymatous cells.

SALVADORA

1. A transverse section of the stem shows epidermis, cortex and stele.
2. Anomalous vascular structure are seen in the stem.
3. Epidermis is made up of a single layer of compactly arranged cells.
4. There is the deposition of cuticle above the epidermis.
5. Next to the epidermis is found a few layers of parenchyma.
6. Next to the parenchyma is found three to four layers of chlorenchyma.
7. In the secondary structure the epidermis is replaced by the cork tissue.

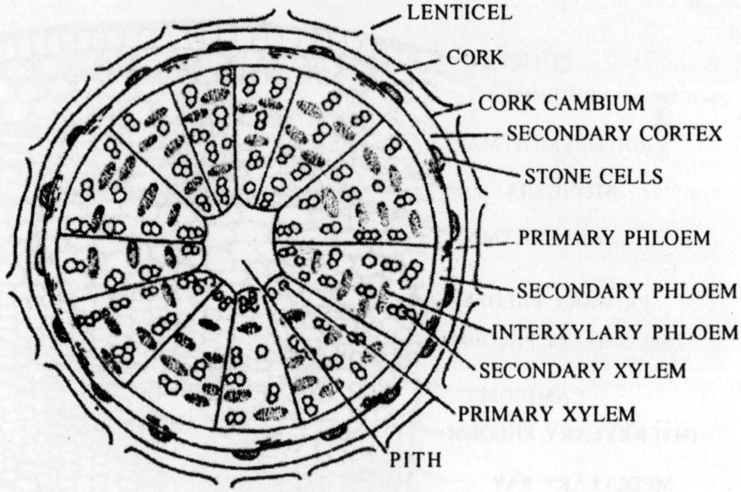

Fig. 21.13 T.S. of Salvadora stem, diagrammatic

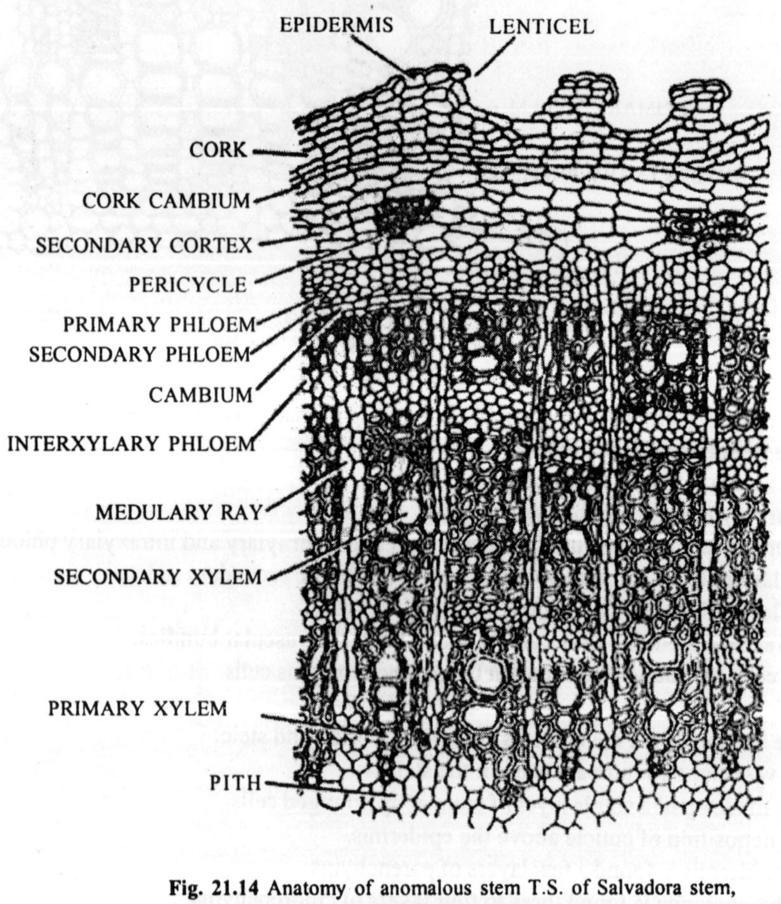

Fig. 21.14 Anatomy of anomalous stem T.S. of Salvadora stem,
Detail of a sector

8. Internal to the cork is found the cork cambium and the secondary cortex.
9. Outlining the stele is found the endodermis.
10. Pericycle is found next to the endodermis and is represented by the patches of fibers opposite the vascular bundles.
11. Internal to the pericycle is found the secondary phloem and sum amount of primary phloem.
12. The phloem bundles is seperated by uni or multiseriate medullary rays.
13. Next to the phloem is found the cambial zone.
14. The cambium is seperated into patches by the medullary rays.
15. Next to the cambium is found the secondary xylem.
16. Towards the pith is found primary xylem.
17. The secondary xylem consists of patches of phloem constituting the interxylary phloem.
18. Towards the center is found the pith made up of parenchyma.
19. The main anomalous feature in the stem is the presence of the patches of interxylary phloem arising on the inner side of the cambium.

STRYCHNOS

1. A transverse section of the stem shows epidermis, cortex and stele.
2. Anomalous vascular structure are seen in the stem.
3. Epidermis is made up of a single layer of compactly arranged cells.
4. There is the deposition of cuticle above the epidermis.

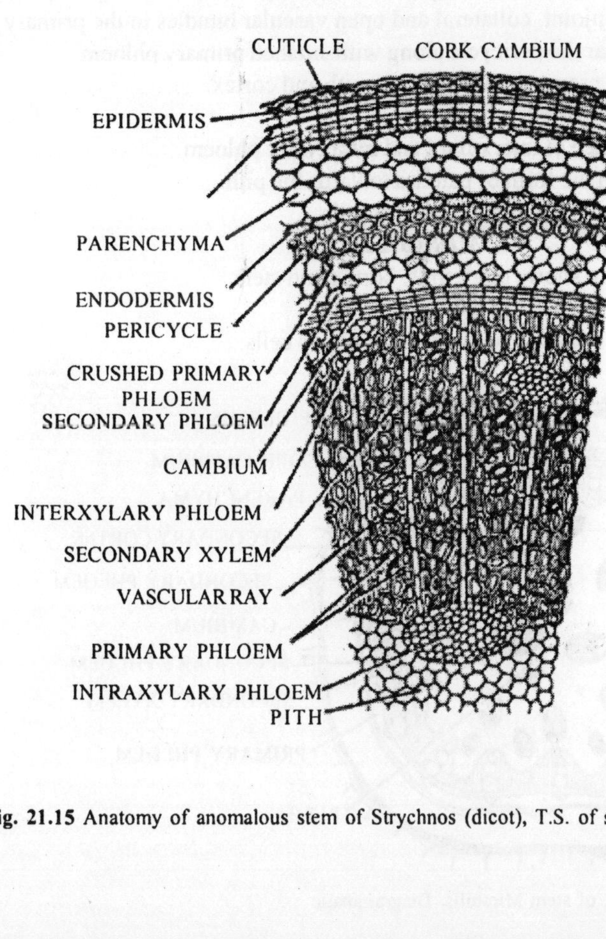

Fig. 21.15 Anatomy of anomalous stem of Strychnos (dicot), T.S. of stem, detal of sector

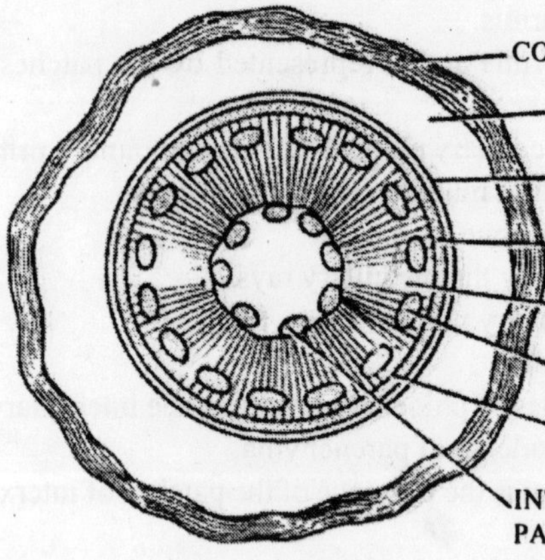

Fig. 21.16 T.S. of Strychnos stem (dicot), diagrammatic

5. Cortex consists of a few layers of chloencyma followed by parenchyma.
6. In the secondary structure the epidermis is replaced by the cork.
7. Internal to cortex is the endodermis.
8. Next to the endodermis is found a continous layer of sclerenchyma representing the pericycle.
9. The vasculature consists of conjoint, collateral and open vascular bundles in the primary structure.
10. Secondary phloem is found near the pericycle along with curshed primary phloem.
11. Uniseriate medullary rays are present connecting the pith and cortex.
12. Secondary xylem is found next to the cambium.
13. The unusual features are presence of interxylary and intraxylary phloem.
14. Intraxylary phloem is found in the form of patches towards the pith.
15. The central region consits of a wide parenchymatous pith.

MIRABILIS

1. A transverse section of the stem shows epidermis, cortex and stele.
2. Anomalous vascular structure are seen in the stem.
3. Epidermis is made up of a single layer of compactly arranged cells.

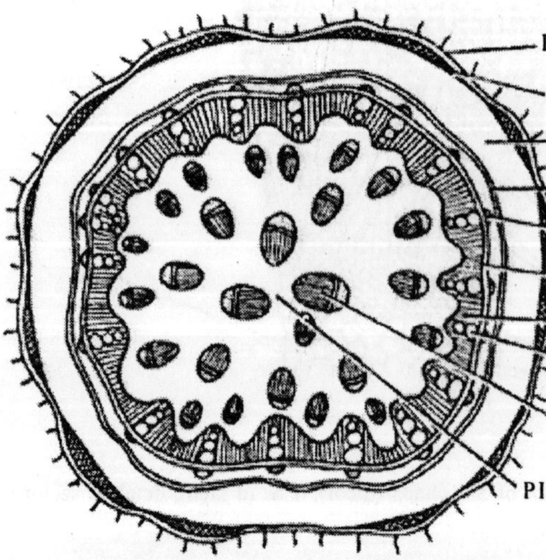

Fig. 21.11 T.S. of stem Mirabilis, Diagrammatic

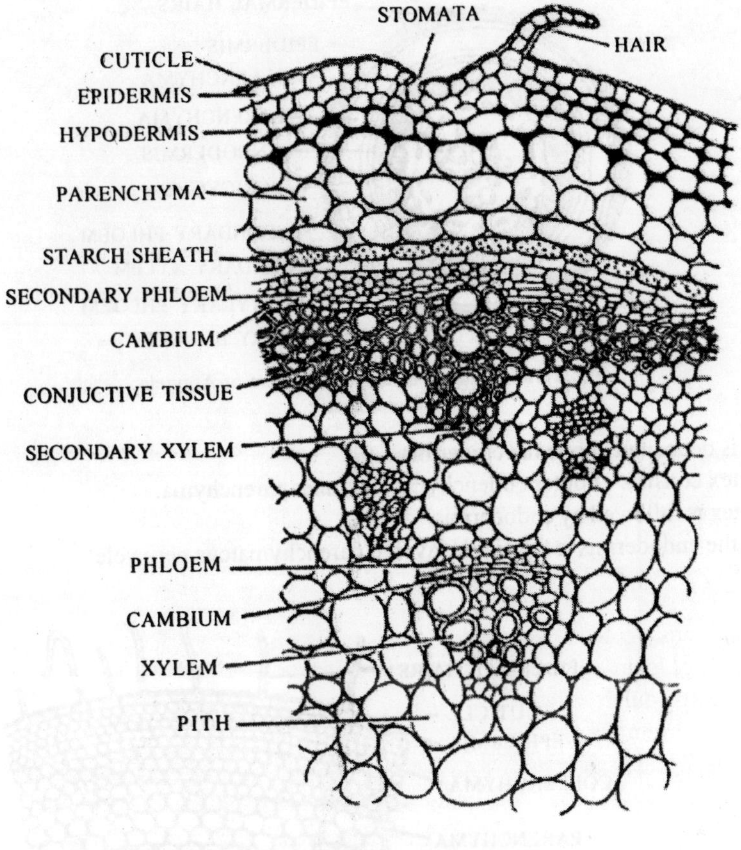

Fig. 21.18 Anatomy of anomalous stem of Mirabilis, Detail of sector of same

4. There is the deposition of cuticle above the epidermis.
5. Cortex consists of few layers of chloenchyma followed by parenchyma.
6. Internal to the cortex is the endodermis.
7. Internal to the endodermis is found pericycle made up of a single layer of parenchyma.
8. Next to the pericyle is the secondary phloem along with patches of primary phloem.
9. Next to the phloem is the cambium.
10. Internal to the cambium is the secondary xylem.
11. The central region of the stem consists of a uniform mass of parenchyma constituting the ground tissue.
12. Embedded in this ground tissue are embedded a number of vascular bundles which are reffered to as the medullary vascular cells.
13. Anomalous structures are seen in the form of a number of rings of vascular bundles occuring in the medullary region.
14. These vascular bundles are irregular arrangement.

ARGYREIA

1. A cross section of the stem is some what rectangular with four projections at the corner.
2. In the primary structure the usual epidermis, cortex and stele are seen.
3. Epidermis is single layered made up of compactly arranged cells.

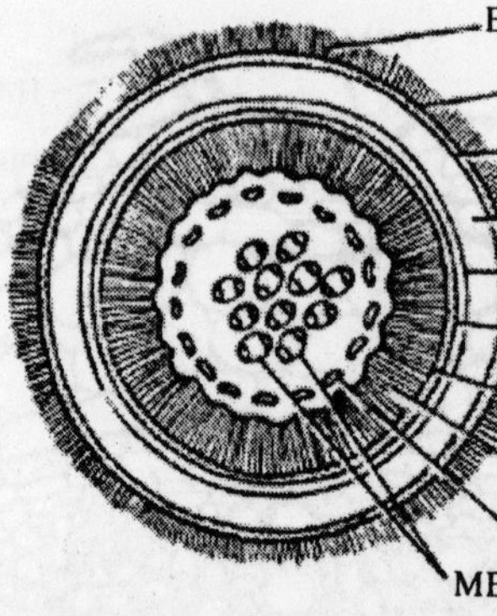

Fig. 21.19 Anatomy of anomalous stem of Argyreia Diagrammatic

4. Cuticle is deposited above the epidermis.
5. The cortex consists of outer collenchyma and inner parenchyma.
6. The cortex is followed by endodermis.
7. Next to the endodermis is the single layered parenchymatous pericycle.

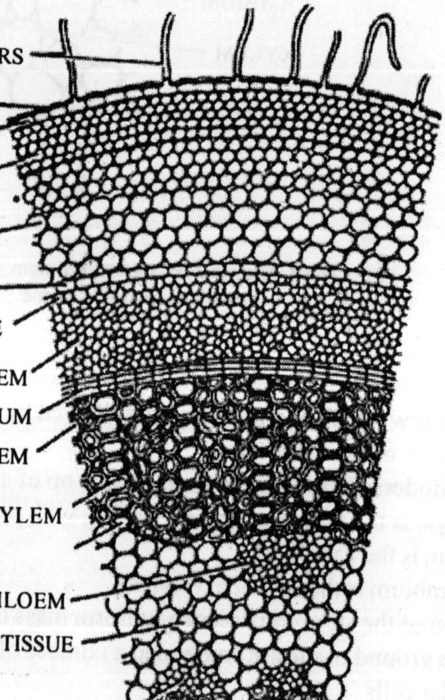

Fig. 21.20 Anatomy of anomalous stem of Argyreia, Detail of a sector of same

8. The vasculature consists of conjoint, colletral and open vascular bundles in the primary structure.
9. Next to the pericycle is the secondary vasculature consisting of secondary phloem, cambium followed by secondary xylem.
10. Central region is occupied by the pith.
11. The unusal feature is the presence of several discrete vascular bundles found embedded in the pith.
12. Another unusal feature is the presence of patches of interxylary pholem.

CHENOPODIUM

1. A cross section of the stem is some what rectangular with four projections at the corner.
2. In the primary structure the usual epidermis, cortex and stele are seen.
3. Epidermis is single layered made up of compactly arranged cells.
4. Cuticle is deposited above the epidermis.
5. The cortex is divided into two zones - outer collenchymatous and inner parenchymatous.
6. Internal to the cortex is the endodermis.
7. A parenchymatous pericylce is found internal to the endodermis.
8. The primary vascular bundles are typical to a dicotyledonous stem.
9. In the region of the pericycle develops an unusual cambium external to primary phloem.
10. It is this which bings about the abnormal secondary growth.
11. Normal seconday also occurs in the primary vascular bundles due to the activity of fasicular cambium.
12. This cambium however is active only for a short period.

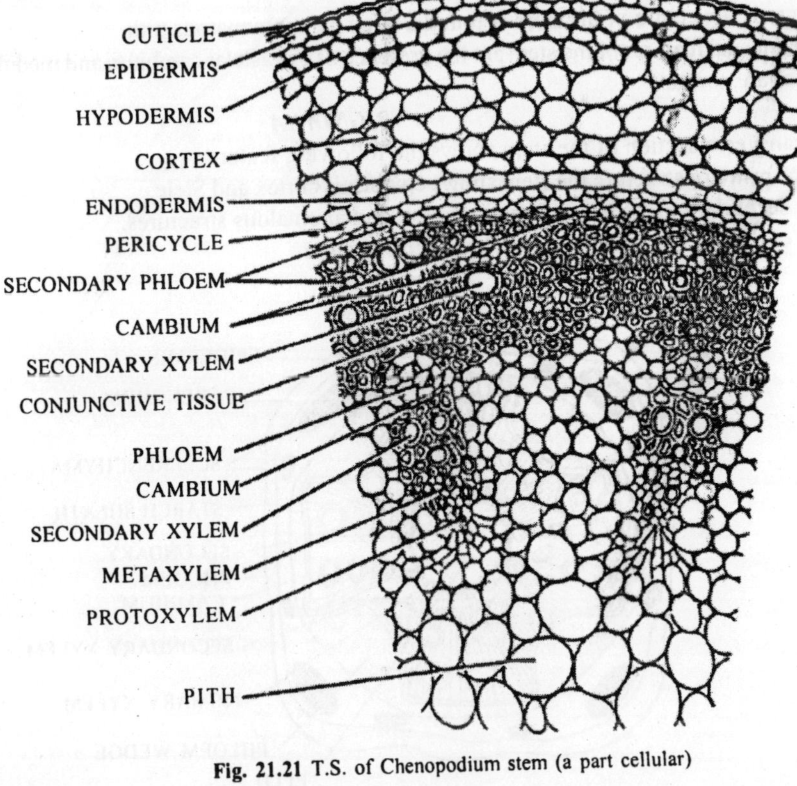

Fig. 21.21 T.S. of Chenopodium stem (a part cellular)

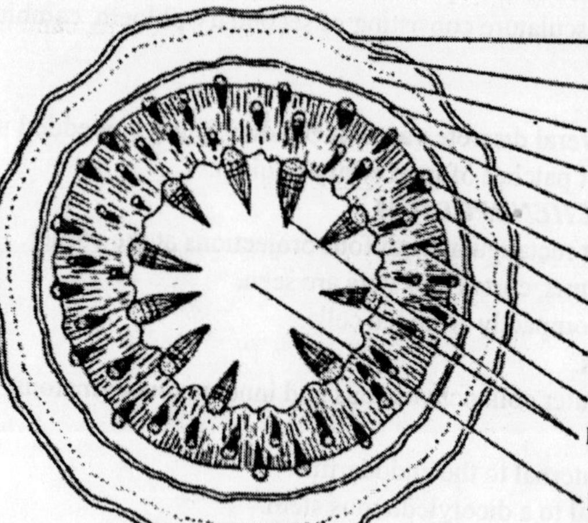

Fig. 21.22 T.S. of Chenopodium stem (diagrammetic)

13. The abnormal cambium ring produces thin walled cells towards the inner face increasing the quantity of ground tissue.
14. As a result of this the primary vascular bundles are pushed into the centre appearing like medullary vascular bundles.
15. Towards the centre of the stem is found a large parenchymatous pith.
16. The unusual features of the stem are the presence of extrastelar cambium and medullary vascular bundles (these are normal).

BIGNONIA

1. A transverse section of the stem shows the following features.
2. In the primary structure the stem shows epidermis cortex and Stele.
3. The secondary growth shows both normal and anomalous structures.
4. The outline of the section will not be exactly circular.

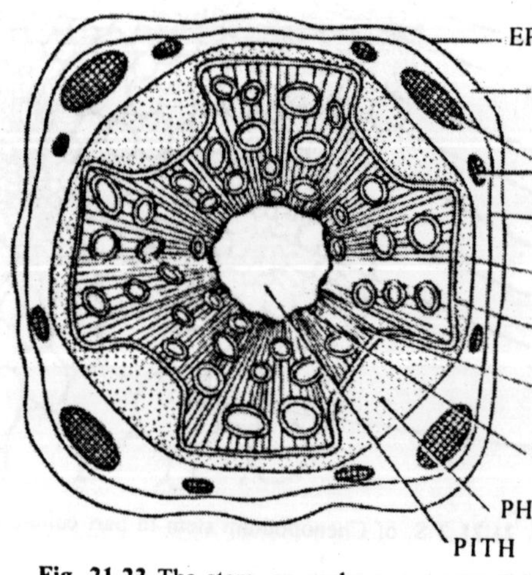

Fig. 21.23 The stem, anomalous structure. T.S. of Bignonia (dicot). detail of the structure

5. Epidermis is made up off a single layer of cells and is covered by a thick cuticle.
6. In older stems however the epidermis is replaced by the periderm.
7. The periderm is made up of Cork, cork cambium and secondary cortex.
8. The cortex is made up of the a few layers of parenchyma.
9. The endodermis is not distinct and is made up of barrel shaped cells.
10. Pericycle is represented by patches of sclerenchyma.
11. Secondary growth takes place by the activity of normal fascicular cambium.
12. The normal cambium behaves abnormally in that it cuts off more secondary phloem to the exterior and less secondary xylem towards the interior in three or four places while in other regions it cuts off more xylem towards the interior and less phloem towards the exterior.
13. As a result the secondary vascular region becomes wavey.
14. Due to this abnormal activity in many places the secondary phloem becomes 'U' shaped and projects into secondary xylem.
15. Some of the cells in the phloem become fibrous.
16. Occupying the centre region is found the pith.

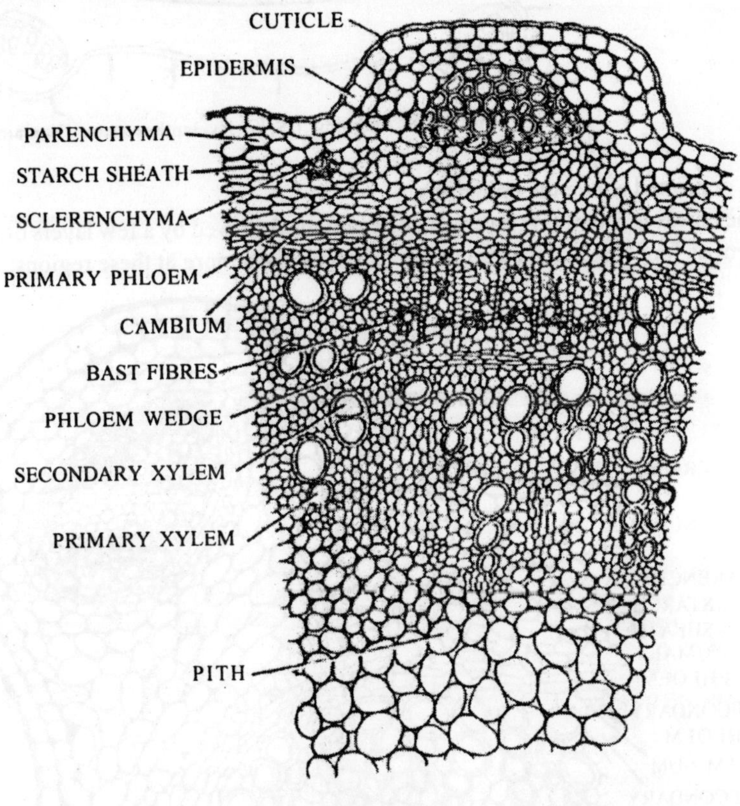

Fig. 21.24 The stem, anomalous structure. T.S. of Bignonia (dicot). diagrammetic

NYCTANTHES

1. A cross section of the stem is some what rectangular with four projections at the corner.
2. In the primary structure the usual epidermis, cortex and stele are seen.
3. Epidermis is single layered made up of compactly arranged cells.

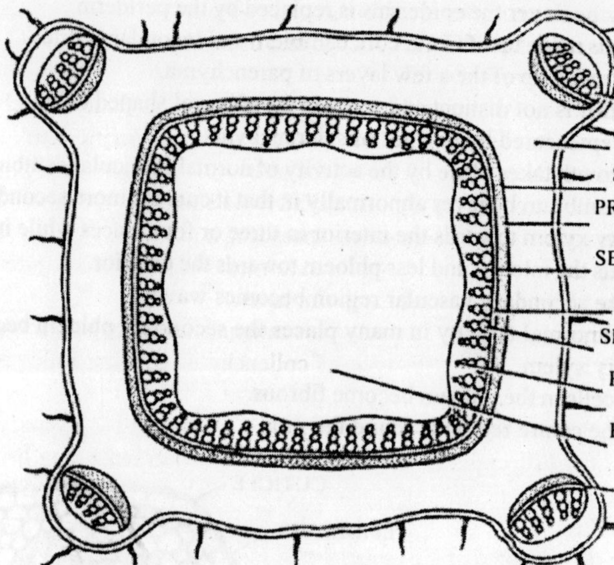

Fig. 21.25 Anatomy T.S. of Nyctanthes stem, Diagrammatic

4. Cuticle is deposited above the epidermis.
5. Next to the cortex is found few layers of collenchyma followed by a few layers of parenchyma.
6. As the stem as four ridges the exten of the parenchyma is more at these regions.

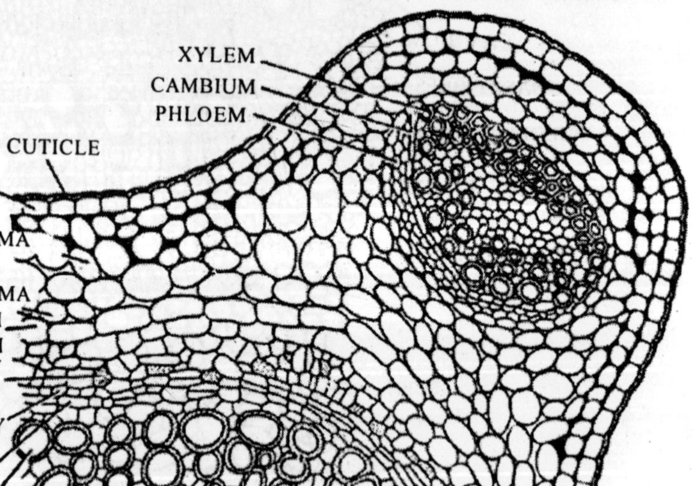

Fig. 21.26 Anatomy. T.S. of Nyctanthes stem, sector

7. Embedded in the parenchyma at the ridges are found four cortical vascular bundles.
8. These cortical vascular bundles are anomalous in origin.
9. These cortical vascular bundles are colletral and open.
10. Next to the cortex is the endodermis.
11. Endodermis is followed by pericycle and secondary vascular tissues which are normal.
12. Cambium is found between the secondary pholem and secondary xylem.
13. Occupying the secondary region is found the pith.

BOERHAVIA STEM

1. *Boerhavia diffusa* is a small semiprostrate spreading herb.
2. A transverse section of the stem shows an outermost single layered epidermis.
3. The epidermis bears a number of multicellular hairs.
4. In the primary structure, next to the epidermis internally is the hypodermis.
5. The hypodermis is made up of 2-3 layers of collenchyma which is followed by parenchymatous cortex having few chlorophyllous cells.
6. Next to the cortex is an indistinct endodermis lining externally 1-2 layered pericycle.
7. The stele consists of 3 rings of vascular bundles of which inner ring has two large centrally located medullary bundles, the middle ring is made up of 6-14 loosely arranged bundles and the outer ring has small 15-20 bundles.
8. Secondary growth is limited in the inner and middle rings.
9. As a result they slightly increase their size.
10. The inter and intrafascicular cambial strips join together to form the cambial ring.

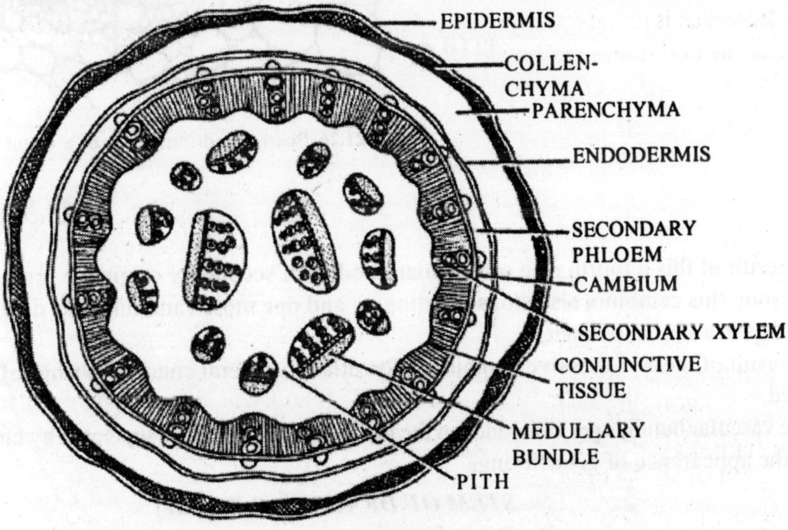

Fig. 21.25 Boerhavia diffusa. T.S. stem-diagrammetic

11. The cambial cells produce secondary xylem in the infrafascicular region and lignified conjunctive tissue in the interfascicular region on the inner side and on the outer side secondary phloem facing the secondary xylem and parenchyma opposite the conjunctive tissue.
12. After the formation of secondary tissue, the cambium ceases its activity and a new cambial ring gets differentiated by combining the secondary parenchyma cells opposite to the conjunctive tissue and the cells of pericycle outside the phloem.
13. This accessory cambium functions in a manner similar to the fascicular cambium and produces similar tissues.

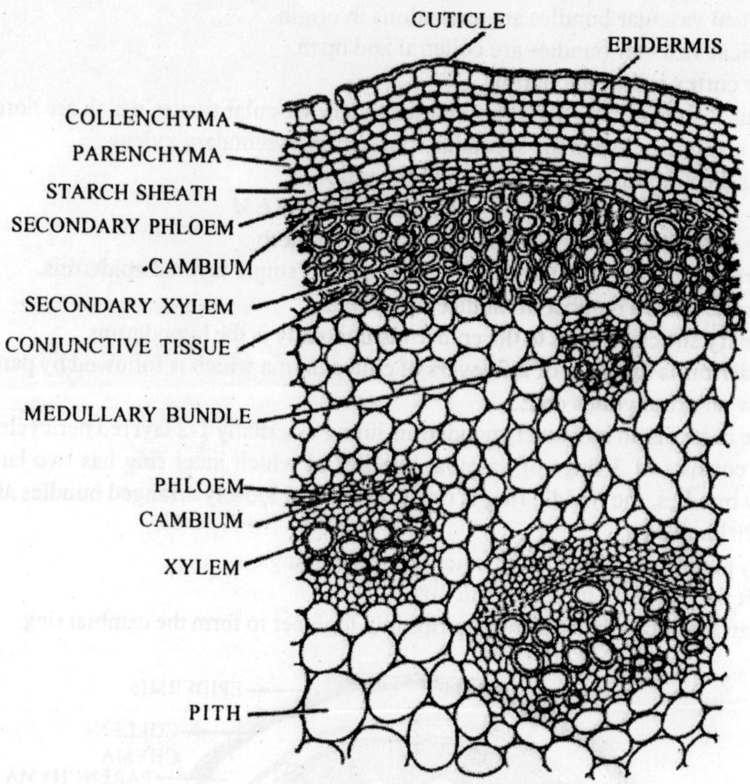

Fig. 21.26 Boerhavia diffusa. T.S. of a sector of enlarge

14. As a result of this a fourth ring of vascular bundles of secondary origin are produced.
15. Very soon this cambium also stops functioning and one more cambium gets differentiated, which also functions in a similar manner.
16. As a result of these successive cambial differentiation several concentric rings of vascular bundles are formed.
17. These vascular bundles get embedded in the thick conjunctive tissue seperated by him walled parenchyma give the appearance of growth rings.

STEM OF DRACENA (Liliaceae)

1. Generally, in monocots ther will be no secondary growth due to the lack of cambium.
2. Abnormally, secondary growth occurs in members like *Dracena, Yucca* etc.
3. The stem of *Dracena* in its primary structure consists of a normal monocotyledonous structure.
4. It has a number of conjoint, collateral and closed vascular bundles scattered in the ground tissue.
5. Secondary growth is initiated by a multilayered cambium arising outside the primary vascular bundles.
6. This abnormal cambium produces more cells towards its inner face than towards the outer face.
7. The outer derivatives develop into thin walled parenchyma whereas the inner derivatives develop partly into lignified conjunctive parenchyma and partly into secondary vascular tissues.
8. Each cambial cell divides anticlinally to form a row of four cells.

9. Further division are irregular.
10. The cells towards the periphery develop into xylem and those towards the centre develop into phloem.
11. After secondary growth, the stem of *Dracena* shows the following features.
12. Outermost layer is the epidermis.
13. Beneath the epidermis is the cork tissue.
14. Internal to this is the parenchymatous cortex which is secondary in origin.

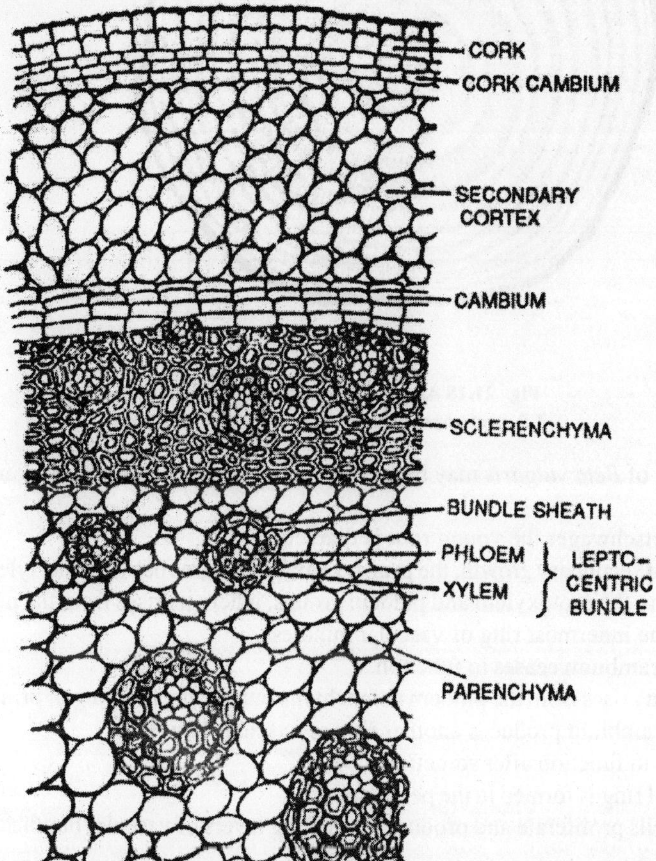

Fig. 21.20 Anomalous secondary growth.
T.S. of Dracena stem (A portion enlarged).

15. It is derived from the cambium.
16. Next to the secondary cortex is the cambium.
17. Internal to the cambium are found the secondary vascular bundles embedded in the thick walled conjunctive tissue.
18. The secondary vascular bundles are *amphivasal* (phloem surrounded by xylem on all the sides).
19. Towards the centre of the stem are found the normal primary vascular bundles.

STORAGE ROOT OF BETA *VULGARIS* (Beet root)

1. The storage roots generally represent the combination of hypocotyl and root and exhibit a great variation in their secondary structure.

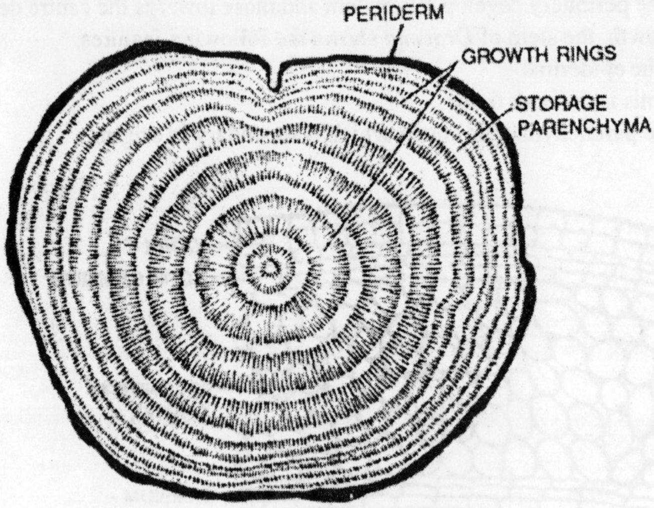

Fig. 21.28 Anomalous secondary growth.,
T.S. of storage root of Beta vulgaris (ground plan).

2. The storage root of *Beta vulgaris* may be taken as an example of abnormal secondary thickening in the storage roots.
3. According to Artschwager the young root is diarach.
4. At the intiation of secondary growth, the primary cambium arises between the xylem and phloem groups, except opposite to the protoxylem and phloem groups, where it arises from the pericycle.
5. This produces the innermost ring of vascular bundles.
6. Very soon, this cambium ceases to function.
7. Second cambium arises from the phloem parenchyma outside the first ring of primary cambium.
8. The secondary cambium produces another ring of vascular bundles.
9. This also ceases to function after sometime.
10. The next cambial ring is formed in the pericycle.
11. The pericycle cells proliferate and produce alternating layers of vascular bundles and parenchyma cells of pericycle.
12. In this way a number of vascular rings are formed.
13. The vascular bundles are collateral and are separated by narrow bands of radial parenchyma, rich in reserve food.

PART FOUR
EMBRYOLOGY OF ANGIOSPERMS

Chapter 22

Flower and Organogeny

Structure of a typical flower: Basically, a flower consists of three parts viz., *receptacle, sterile appendages* and *fertile appendages*.

Receptacle represents the expanded or swollen tip of the stalk(pedicel) of the flower. It is on this that the appendages of the flower are borne. The receptacle may be flat, globose or concial in shape.

The sterile appendages (accessory organs) are of two types viz., *calyx* and *corolla*. Sometimes the sterile appendages cannot be distinguished into calyx and corolla as in *Michelia*. In such cases the sterile appendages are termed *perianth lobes* or *tepals*.

Calyx is the outermost or lower most of the sterile appendages. The individual members are called sepals and they are generally green in colour. The sepals in flower may be fused or free.

Corolla is next to calyx. It is the most attractive of the floral organs. The various colours of the flowers are due to corolla. The individual members are called *Petals*. The petals in the corolla may be fused or free.

In some flowers, the sterile appendages are entirely absent. Such flowers posses only the fertile appendages and are called *achlamydeous*.

The fertile appendages are also of two types viz., *Androecium* and *Gynoecium*.

Androecium represents the collection of male reproductive structures. In arrangement, it is found next to corolla. The individual members are called *stamens*. Each stamen has three parts viz., *filament, anther* and *connective*. Filament is stalk like and it bears the anther. The portion of the filament that joins the anther is called the connective. It is in the anther (Microsporangium) that the pollen grains are formed which ultimately develop into the male gametophyte.

Gynoecium represents the collection of female reproductive structures. It is also called pistil and has three parts viz., *ovary, style* and *stigma*. Ovary is the basal swollen part of the gynoecium and is made up of *carpels* (megasporophylls). Within the carpels and attached to the plaenta are found the *ovules* (megasporangia). It is within the ovule, the embryo sac (female gametophyte) develops. Further developments i.e., fertilization, endosperm and embryo also take place in the ovule. Style is a stalk that arises from the ovary and it facilitates the growth of the pollen tube. The style terminates in a stigma which serves as a lodging for pollen grains to settle and germinate.

Organogeny: The development of floral parts is called organogeny. The floral primordium arises on the inflorescence axis (peduncle). First, bract arises as a lateral projection in the axil of which the flower originates. The floral shoot subsequently becomes flattened and forms the receptacle. The receptacle at this stage shows a group of meristematic cells with dense cytoplasm and conspicuous nucleus. The various floral organs originate from the receptacle either *spirally* or *cyclically*. In the spiral type, the recptacle shows a tendency towards indeterminate growth in a linear fashion and assumes a dome like shape. This type is regarded primitive and is seen in *Magnoliaceae*. In the cyclic type, the floral parts arise in distinct whorls. This is seen in the majority of flowering plants. An intermediate condition where the calyx and corolla are spiral, with cyclic androecium and gynoecium is seen in the *amentiferae* group.

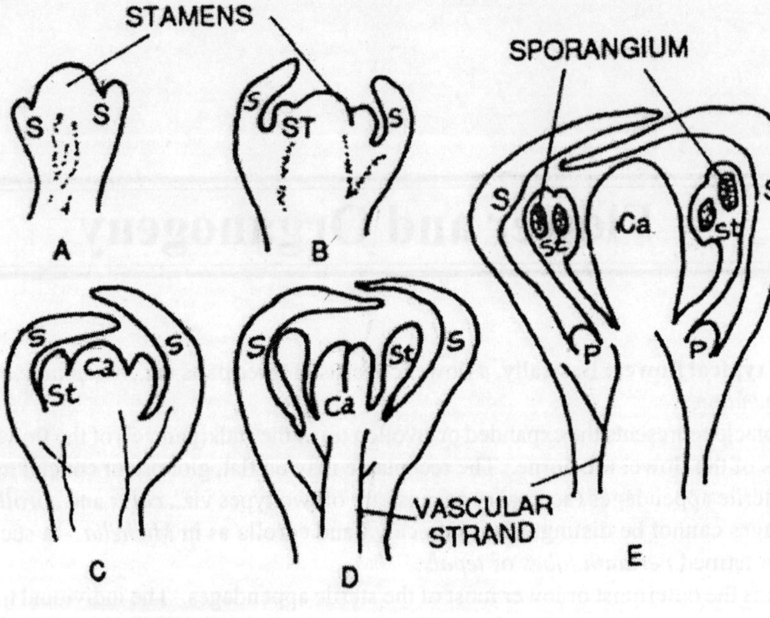

Fig. 22.1 Flower

Stages in the organogenesis of a typical flower

(S= sepal; St= stamen; P = petal ; Ca = carpel)

In the axil of the bract a conspicuous bulb like outgrowth arises and constitutes the floral primordium. By further growth, the floral primordium elongates and gives rise to two distinct regions viz., *pedicel* (stalk) and the *receptacle*. It is on the latter that the floral organs arise. In sessile flowers, the pedicel is suppressed.

The floral parts viz., K (calyx), C(corolla), A(androecium and G(gynoecium) arise in acropetalous sucession on the floral receptable. First to arise is calyx and it is followed by C.G. and A. Variations in the organogeny are met with in several members. In *Rubus* the development is K, A, G, and C while in *Rudbeckia* it is C, A, G and K.

Chapter 23

Anther and Male Gametophyte

Development of the microsporangium: A very young anther in a cross section shows a homogenous mass of cells bound by an epidermis. During further growth the anther assumes a four lobed shape. In each lobe a few of the hypodermal cells (sometimes one) become differentiated from the others by their prominent nucleus and dense cytoplasm. These are the *arhesporial cells*. In majority of the members the archesporium consists of several vertical rows of cells which appear as a plate of cells in a transverse section.

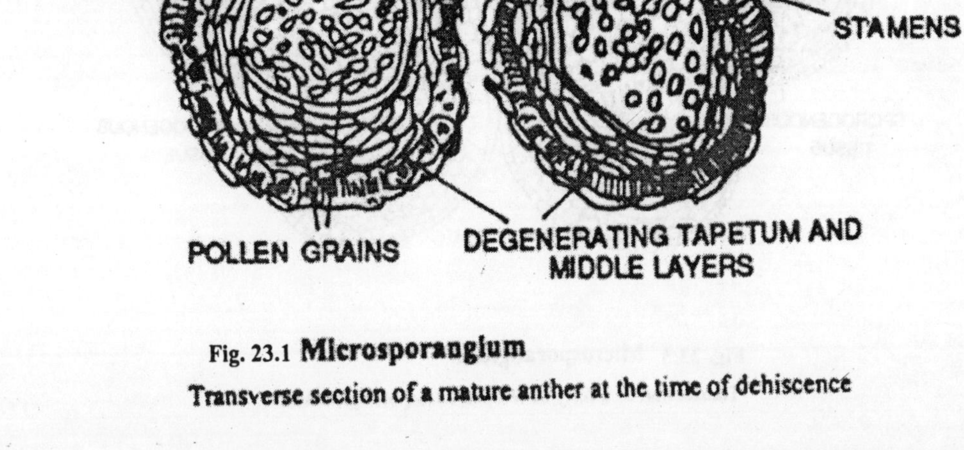

Fig. 23.1 **Microsporangium**

Transverse section of a mature anther at the time of dehiscence

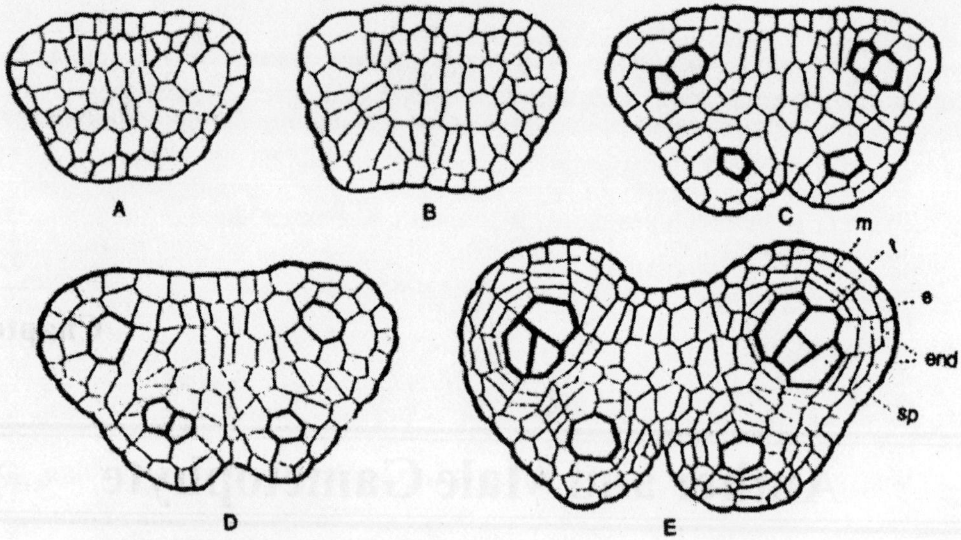

Fig. 23.2 **Microsporangium**

Developmental stages of the anther (e = Show labelling on the diagram itself (redrawn from Warning, 1873)

The archesporail cells divide periclinallay cutting off *primary parietal cells* towards the epidermis and *primary sporogenous cells* towards the interior. The primary parietal layer divides to form two *secondary parietal layers*. Of these, the inner one distinguishes itself into *tapetum* while the outer divides again periclinally to form two layers of cells. Of these, the outer one forms the *endothecium* while the inner one forms the *middle layer*. The sporogenous cells develop into microspore mother cells.

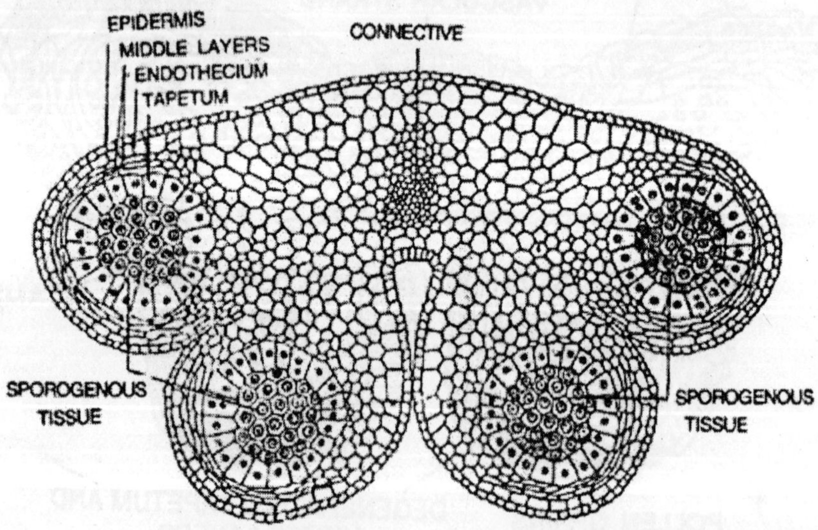

Fig. 23.3 **Microsporangium**

Transection of a fully developed tetrasporangiate anther

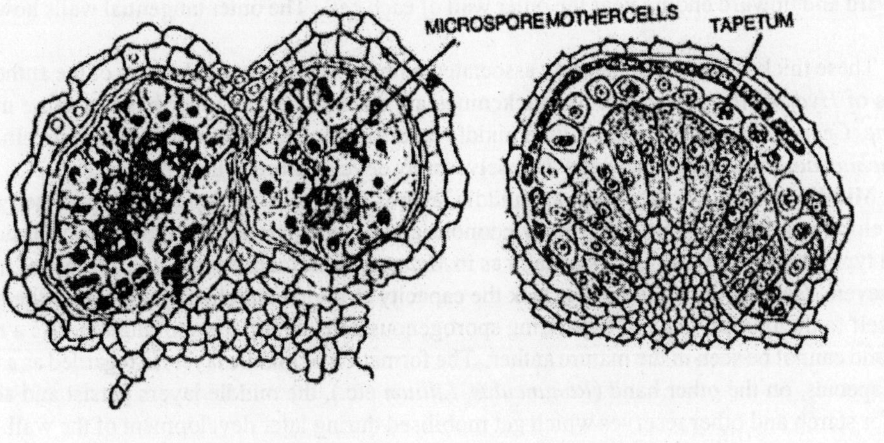

Fig. 23.4 Microsporangium. T.S. of Anther showing PMCs and tapetum

Anther Wall: A transverse section of a fully developed anther shows a group of microspore mother cells surrounded by a tapetum, a middle layer, an endothecium and an epidermis.

Epidermis: It is the outermost wall layer of the anther. During anther development, it undergoes anticlinal divisions to keep pace with the enlargement of the inner tissues. The epidermal cells become greatly stretched and flattened. In a mature anther, only the remanants of the epidermis are seen.

Endothecium: The layer of cells lying immediately below the epidermis constitute the endothecium. Endothecium attains its maximum development at the time of the anther dehiscence. Cells of this layer

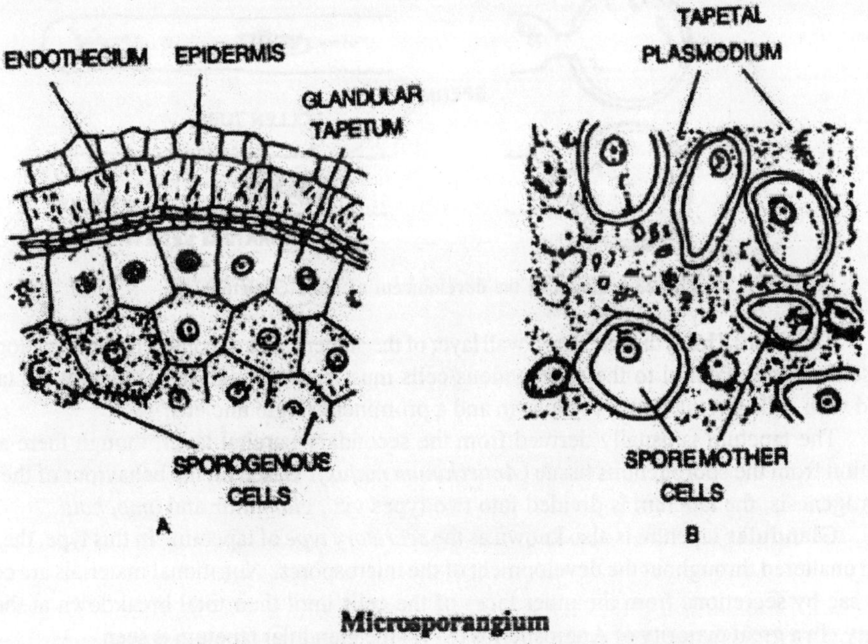

Microsporangium

Fig. 23.5 Types of Tapteum A. Glandular, B. Amoeboid

develop fibrous bands of callose which arise from the inner tangential walls (rarely from the radial walls) and run outward and upward ending near the outer wall of each cell. The outer tangential walls however remain thin.

These thickenings are apparently associated with the dehiscene mechanism of the anther. In certain members of *Hydrocharitaceae*, fibrous thickenings are absent. On the other hand in some members like *Argemone, Crinum* etc., one or more of the middle layers may acquire fibrous thickenings. In members of *Orobranchaceae*, some epidermal cells themselves may develop fibrous thickenings.

Middle layers: Usually only one middle layer is present. But, depending upon the type of anther wall development there may be one (dicotyledonous and monocotyledonous types), two (basic type) or no (reduced type), middle layers. Occassionally as in *Melastoma malabathricum*, the number of middle layers may be seven. Cells of the middle layer lack the capacity to divide anticlinally and hence the tissue cannot adjust itself to the mutiplying and expanding sporogenous cells within the sporangium. As a result, it gets crushed and cannot be seen in the mature anther. The formation of middle layers is regarded as a relic feature. In some species, on the other hand (*Ranunculus, Lillium* etc.), the middle layers persist and act as storage centres for starch and other reserves which get mobilised during later development of the wall.

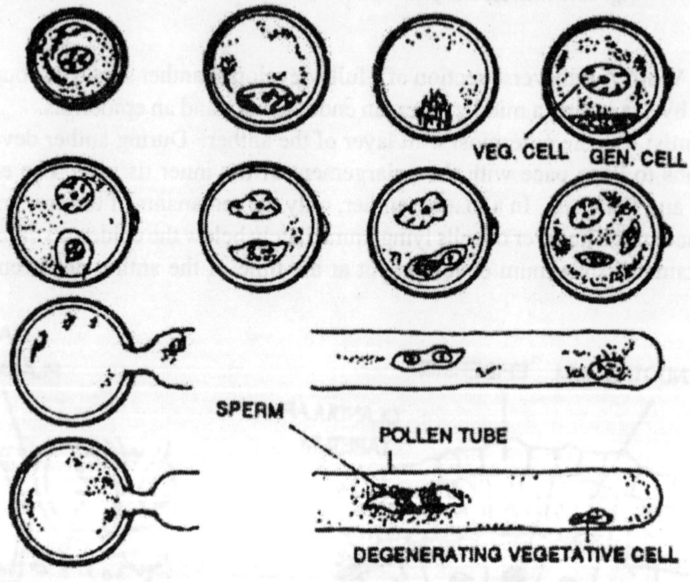

Fig. 23.6 Stages in the development of Male Gametophyte

Tapetum: This is the innermost wall layer of the anther and is of considerable physiological significance because the food material to the sporogenous cells must pass through it. Typically the tapetum is single layered with cells having dense cytoplasm and a prominent single nucleus.

The tapetum is usually derived from the secondary pareital layer, though there are reports of its derivation from the sporogenous tissue (*Antirrhinum majus*). Based on the behaviour of the cells at the time of sporogenesis, the tapetum is divided into two types viz., *glandular* and *amoeboid*.

Glandular tapetum is also known as the *secretory type* of tapetum. In this type, the cells of tapetum remain unaltered throughout the development of the microspores. Nutritional materials are contributed to the anther sac by secretions from the inner faces of the cells until their total breakdown at the time of pollen maturity. In a great majority of Angiosperms (87%) the glandular tapetum is seen.

Amoeboid tapetum is also called *invasive* or *plasmodial* type of tapetum. Here, very early during microspore development, the tapetal cells breakdown, and the protoplasts 'wander' into the anther cavity encircling the pollen mother cells. The penetration of protoplasts may take place during meiotic prophase or may be delayed until the tetrad stage. This type of tapetum is seen in *Blainvillea, Emilia, Senecio* etc.

The close association of the tapetal cytoplasm with the microspores suggests that the former must be exerting a selective influence on the transport of material into the pollen.

Sporogenous tissue: The sporogenous cells may or may not divide to increase their number. Ultimately they round off and develop into *microspore mother cells* or *pollen mother cells* (PMC).

The PMCs undergo reduction division to produce tetrads of haploid microspores. At the initiation of meiosis, the PMCs get separated from the tapetal cells. Cytokinesis or wall formation takes place at the end of meiosis and is of two types. In the *Successive type*, a dyad is formed soon after meiosis. Wall formation is by a centrifugal plate of cellulose, which starts in the centre and extends laterally. The two cells of the dyad, divide next and four spores are formed. In the *Simultaneous type* the first meiotic division is not followed by wall formation, as a result a binucleate cells is formed. Both the nuclei divide and at the end of the meiosis cytokinesis takes place by centripetal furrowing.

As stated already, after meiotic divisions microspore tetrads are formed. These are of the following types viz., *tetrahedral* (three in one plane and one in the other), *isobilateral* (all the spores in one plane), *decussate* (spores arranged crosswise), *rhomboidal* (in the form of a rhomboid), *Linear* (spores are one above the other) and 'T' *shaped* (Two spores superposed and the other two juxtaposed).

Eventually the microspores separate from the tetrad and are set free in the anther loculus.

Development of the male gametophyte: Microspores represent the first stage of the male gametophyte. After their separation from the tetrad, microspores are called pollen grains.

To start with, the pollen grain has its single nucleus situated at the centre. The first division results in the formation of two unequal cells. The larger one is the *vegetative cell* and the smaller one is the *generative cell*. The vegetative cell eventually forms the pollen tube. The generative cell is attached to the wall of the pollen grain in the beginning, but later it comes to lie freely in the cytoplasm of the vegetative cell.

Chapter 24

Ovule and Female Gametophyte

The ovule or megasporangium basically consists of two parts viz., *nucellus* and the *integument*. It is attached to the palcenta of the ovary by means of a short stalk called the funiculus. In a mature ovule the integument encloses the nucellus completely except for a small opening at one end called the *micropyle*. It is through this opening that the pollen tube gains entry into the ovule. The basal region of the ovule (region opposite the microphyle) is called the *chalaza*. Embedded in the nucellus is the female gametophyte or the embryo sac.

Mature ovules are classified into the following types, based principally on the position of the microphyle with reference to the funiulus.

Orthotropous: Here, the ovule is erect and the microphyle and the funiculus lie in a vertical line. This is seen in *Polygonaceae, Urticaceae, Piperaceae* etc.

Anatropous: The ovule takes a 180^0 turn so that the micropyle and the funiculus come to lie close to each other. This seen in almost all members of the sympetalae.

Hemianatropous: In this type, the ovule takes a 90^0 turn so that the microphyle is at right angles to the funiculus. This is seen in *Malpigiaceae, Primulaceae* etc.

Campylotropous: Here the ovule takes a turn a little less than 90^0 and is curved. This is seen in *Resedaceae, Leguminosae* etc.

Amphitropous: The curvature of the ovule is so pronounced that it also affects the embryo sac which becomes horseshoe shaped. This is seen in *Alismaceae, Butomaceae* etc.

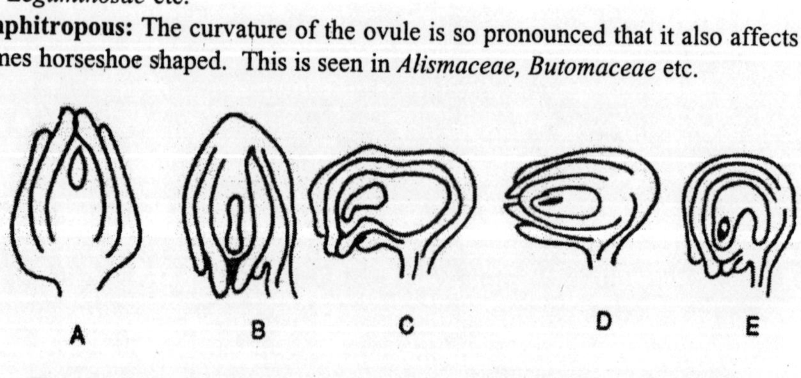

Fig. 24.1 Types of Ovules A. Orthotropous, B. Anatropous, C. Hemianatropous, D. Amphitropous, E. Campylotropous

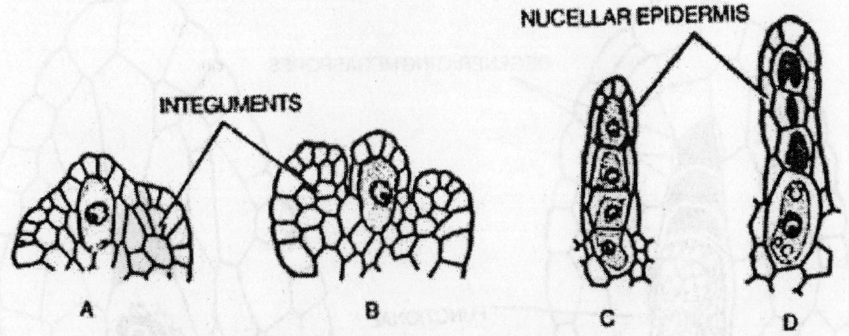

Fig. 24.2 **Female gametophyte**

Megasporogenesis in a tenuinucellate ovule (redrawn from Johri and Singh, 1959)

Circinotropous: Here the ovule first becomes anatropous. But subsequently the curvature once again lifts the micropyle to its original position so that the ovule takes a 360^0 turn. This type of ovule is seen *Plumbaginaceae* and *Cactaceae*.

Integuments: These constitute the envelopes of the ovule. In some cases as in *Loranthaceae*, the integuments are absent. Such ovules are said to be *atagmic*. Similarly ovules with one integument are called *unitegmic* and those with two, are called *bitegmic*. Rarely a third integument called *aril* is present in some members of *Ficoidaceae* (*Trianthema decandra*).

In bitegmic ovules, the micropyle may be formed only by the outer integument or only by the inner integument or by both the integuments.

Nucellus: Depending on the extent of the nucellus, the ovules are classified into two types viz., *tenuinucellate* and *crassinucellate*. In the former, the nucellus is scanty and the archesporium that arises in the hypodermal region directly functions as the megaspore mother cells (MMC), while in the latter the archesporium cuts off a few parietal cells and then functions as the MMC, so that the nucellus is bulky. A third category - *Pseudocrassinucellate* is proposed by Davis (1966), to include those ovules where the nucellar epidermis builds up the parietal tissue, while the archesporium directly functions as the MMC. But, here also the MMC becomes deep seated as in crassinucellate ovules.

MEGASPOROGENESIS

A single hypodermal cell arises in the nucellus and it differentiates itself as the *archesporial cell* by its larghe size, conspicuous nucleus and dense cytoplasm.

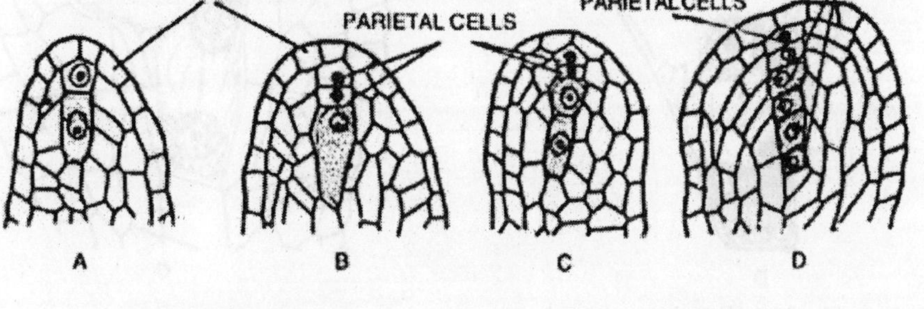

Fig. 24.3 Megasporogenesis

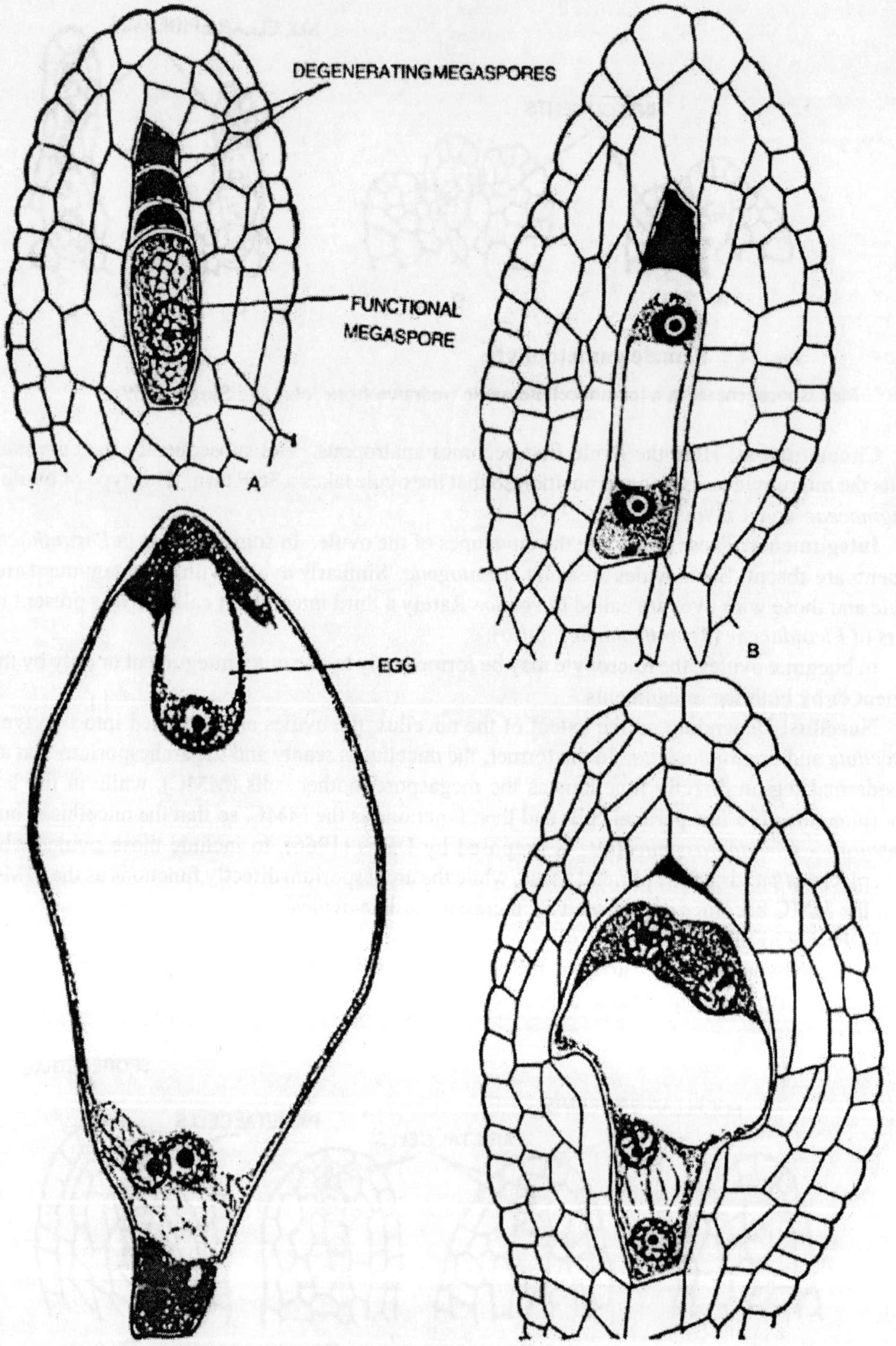

Fig. 24.4 Female gametophyte
Megasporogenesis in Hydrilla verticellata (redrawn from Maheshwari, 1950)

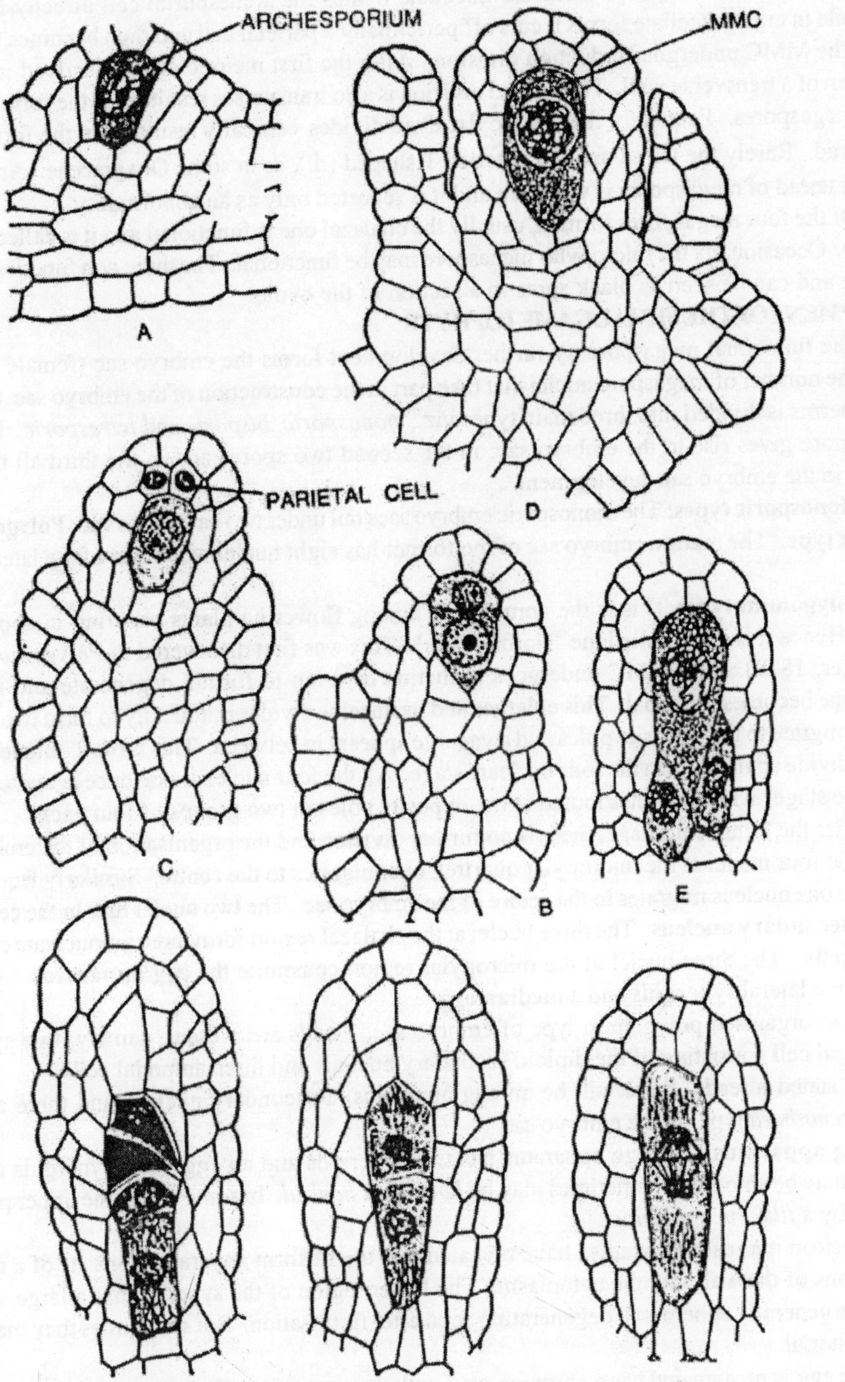

Fig. 24.5 Female gametophyte
Development of polygonum type of embryo sac in Hydrilla verticellata (redrawn from Maheshwari, 1950)

In tenuinucellate and pseudocrassinucellate ovules the archesporial cell directly functions as the MMC, while in crassinucellate forms it cuts off periclinally a parietal cell and then becomes the MMC.

The MMC undergoes reduction division. After the first meiotic division a dyad is formed by the laying down of a transverse wall. The second division is also transverses resulting in the formation of a linear tetrad of megaspores. Fequently the micropylar dyad divides vertically resulting in the formation of a 'T' shaped tetrad. Rarely the tetrad may be inverted T shaped (⊥), as in some Onagraceae. An isobilateral or tetrahedral tetrad of megaspores is very rare and it is reported only as an abnormality.

Of the four megaspores formed, usually the chalazal one is functional and it is called the functional megaspore. Occasionally the micropylar mehaspore may be functional. The three non functional megaspores degenerate and can be seen as black spots in a section of the ovule.

DEVELOPMENT OF THE FEMAL GAMETOPHYTE

The functional megaspore by further development forms the embryo sac (female gametophyte). Based on the number of megaspore nuclei that take part in the construction of the embryo sac, the embryo sac of Angiosperms is divided into three main types viz., *monosporic, bisporic* and *tetrasporic*. In the first one, only one spore gives rise to the embryo sac; in the second two spores and in the third all the four spores participate in the embryo sac development.

Monosporic types: The monosporic embryo sacs fall under two categories viz., **Polygonum type** and **Oenothera type**. The mature embryo sac of the former has eight nuclei while that of the later has only four nuclei.

Polygonum type: This is the commonest among flowering plants occuring in about 82% of the members. Hence it is often called the 'Normal type'. This was first discovered in *Polygonum divaricatum* (Strassburger, 1879). The MMC undergoes reduction division to form a degenerate and the lowermost (chalazal) one becomes funtional. This enlarges and its nucelus divides mitotically to form two nuclei. These two nuclei migrate to the opposite poles and a vacuole appears in between. This is the 2 - nucleate stage. The two nuclei divide again to form the four nucleate stage. All the four nuclei divide once again to form 8 nuclei (8 - nucleate stage) which are distributed at the opposite poles in two groups of four each.

After the 8-nucleate stage, there is no further division and the organisation of the embryo sac takes place. Of the four nuclei at the micropylar quartret, one migrates to the centre. Similarly from the chalazal quartret also one nucleus migrates to the centre of the embryo sac. The two nuclei fuse in the centre and form the diploid **secondary nucleus**. The three nuclei at the chalazal region form three uninucleate cells called the **antipodal cells**. The three nuclei at the micropylar region constitute the **egg apparatus**. The apparatus consists of two lateral **synergids** and a **median egg**.

In an organised polygonum type of embryo sac, 7 cells are present, namely, two synergids, one egge, a central cell consisting of the diploid secondary nucleus and three antpodal cells.

As stated already, there will be an egg apparatus, a secondary nucleus and three antipodal cell (except in *Oenothera* type) in the embryo sac.

Egg apparatus: The egg apparatus has two synergids and an egg. The synergids are elongated cells. They may be smooth or sometimes may be *hooked* or *beaked*. In some cases, the upper part of the cell is occupied by a *filiform apparatus*.

Electron microscopic studies have revealed that the filiform apparatus consists of a mass of inger like projictions of the wall into the cytoplasm. The lower region of the synergid has a large vacuole. The synergids are generally ephemeral, degenerating soon after fertilisation. But sometimes they may persist and become haustorial.

The egg is median and has a common wall with the synergids. The micropylar end of the cell has a large vacuole while the nucleus is situated at the lower end.

Central cell: This is the cell which consists of the secondary nucleus. Sometimes the two polar nuclei may not fuse for a long time and remain as free nuclei. It is from the secondary nucleus that the endosperm develops after fertilisation.

Antipodals: These are three in number and uninucleate. Sometimes they may be only two in which case one of them would be binucleate. The antiopodals generally degenerate soon after fertilisation. But frequently they may persist and become haustorial. Persistent antipodals are common in the tribe *Heliantheae* of *Asteraceae*.

Chapter 25

Endosperm

Endosperm is the chief nutritive tissue present in the embryo sac of Angiosperms. In Gymnosperms also there is an endosperm where the female gametophyte itself provides nutrition to the developing embryo. In Angiosperms however, the embryo sac is extremely reduced and by itself is incapable of supporting the embryo. Hence the need for endosperm arises. The origin of the Angiospermous endosperm is unique. It is formed by the fusion of three nuclei, of which two are female in origin and one is male. The resultant product i.e., primary endosperm nucleus (PEN) develops into the endosperm. Generally the endosperm is triploid, but the ploidy may vary as in *Oenothera* (diploid), *Peperomia* (polyploid) etc.

There are some exceptional Angiosperms where there is no endosperm formation. These are *Orchidaceae, Podostemaceae,* and *Trapaceae.*

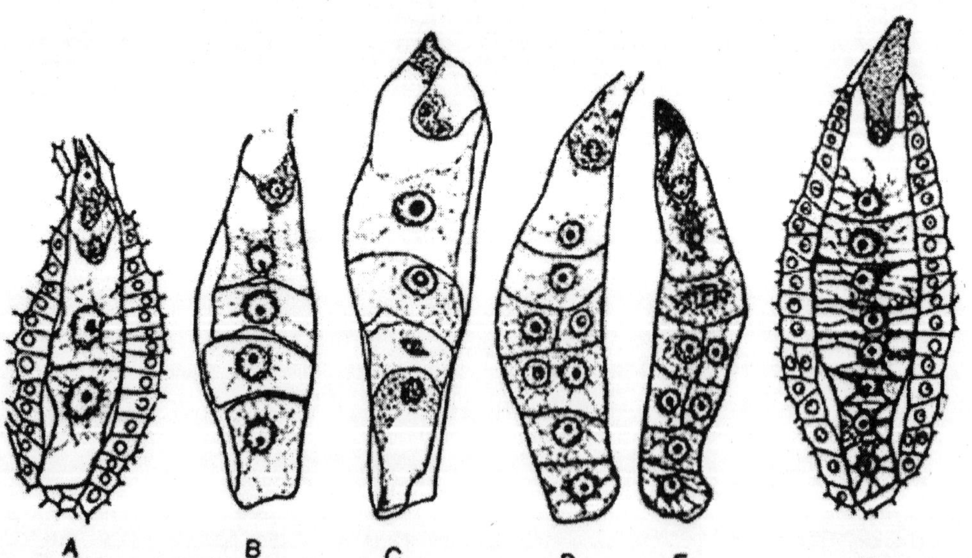

Fig. 25.1 **Endosperm**

Stages in the development of cellular endosperm in Villarsia reniformis (redrawn from Stolt,

The endosperm produced may be scanty or bulky. In the former, (exalbuminous) in the mature seeds there will be no endosperm as it is completely consumed by the developing embryo. In the latter (albuminous) the mature seed has plenty of endosperm and it is the nutritive tissue in the seed.

TYPES OF ENDOSPERM

Based on the behaviour of the PEN, the endosperm is divided into three main types. These are *Nuclear, ab initio Cellular* and *Helobial*. Of these, nuclear endosperm is found in 161 families, while cellular is seen in 72 families and Helobial is seen in 17 families.

Nuclear endosperm: The PEN undergoes a few free nuclear divisions, without any wall formation. As a result, lying in the cavity of the central cell, a few to several thousand nuclei are formed. The free nuclear stage may remain until it is consumed by the embryo (*Limnathes*) or it may become cellular at a later stage. In the latter case cell wall formation takes place centripetally. The entire endosperm may become cellular or a part of it may become cellular. In *Phaseolous*, the endosperm, only around the embryo becomes cellular, while the rest of it is free nuclear.

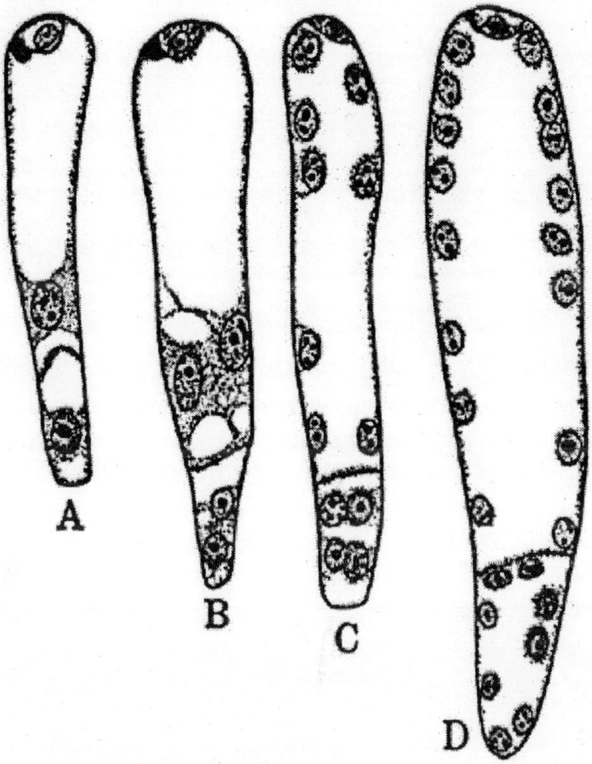

Fig. 25.2 Endosperm
Stages in the development of Helobial endosperm

ab initio Cellular: In this type, the first division of the PEN is always followed by wall formation, as a result two endosperm cells are formed. The first divisional wall may be parallel or perpendicular to the long axis of the embryo sac. Further divisions are also cellular and as a result the embryo sac gets filled by a cellular tissue.

Haustoria are very frequent in cellular endosperm also. These may arise either from the chalazal region or from the micropylar region or from both. Micropylar haustoria are seen in *Impatiens royeli* and *Hydrocera triflora* and many others. Chalazal haustoria are seen in *Magnolia obovata, Iodina rhombifolia* etc. In *Utricularia*, both chalazal and micropylar haustoria are seen.

Helobial endosperm: This is seen mostly in monocotyledons. This type of endosperm is intermediate between the nuclear and cellular types. The PEN migrates to the chalazal region of the embryo sac where it divides to form a large micropylar chamber and a small chalazal chamber. In the micropylar chamber, divisions are always free nuclear and cell wall formation, if any, starts very late. In the chalazal chamber, the nucleus remains undivided or divides only a few times. In some cases (*Phylidrum languginosum*), the chalazal chamber also becomes cellular.

Chapter 26

Embryogeny

DICOTYLEDONOUS EMBRYO

There are about 5 types of embryo development found in dicotyledons. We will study a typical type viz., *Crucifer type*. Crucifer type of embryo is seen in the family *Cruciferae*. It is worked out in great deatail in *Capsella bursa-pastoris*. The foregoing description of embryogeny is based on the study of this plant.

The first division of the zygote is transverse and it results in the formation of a basal cell *cb* and a terminal cell *ca*. *ca* divides vertically while *cb* divides transversely to form a proembryonal tetrad which is 'T' shaped. Each of the two terminal cells (derivatives of *ca*) divide by a vertical wall at right angles to the first, resulting in a *quadrant*. The quadrant cells undergo a transverse division to form the *octant*. Of these cells, lower four give rise to the *stem tip* and *cotyledons*, while the upper four give rise to the *hypocotyl*. All these eight cells, next divide periclinally to cut off 8 outer cells which form the *dermatogen*, while the inner ones give rise to the initials of the *periblem* and *plerome*.

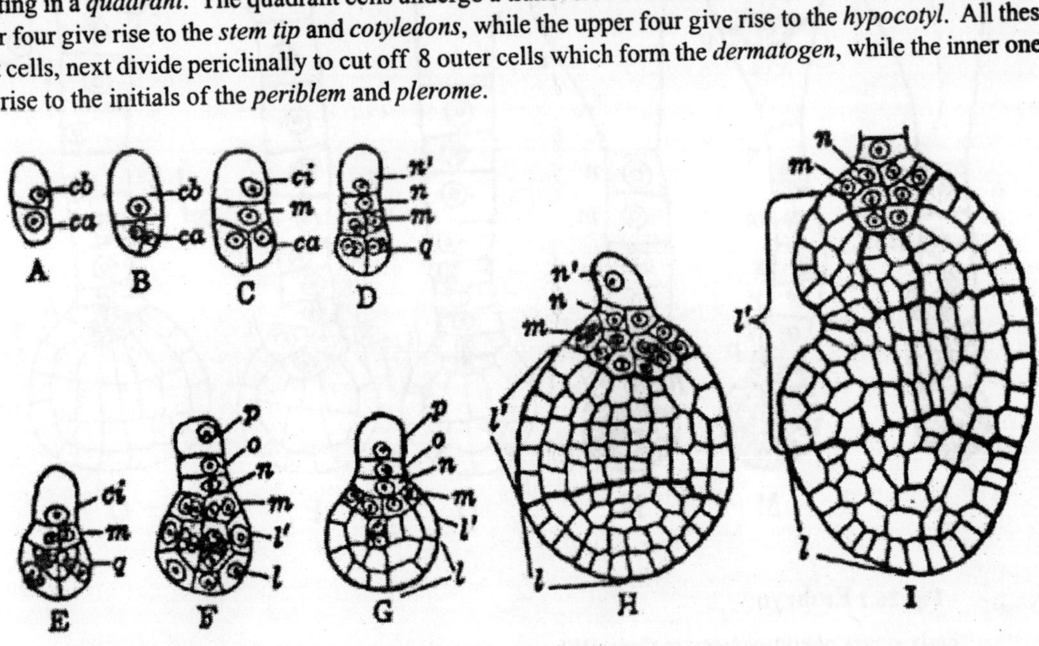

Fig. 26.1 **Embryo**

Development of embryo in Luzula forsteri (redrawn from Soueges, 1923)

Meanwhile, the two deravatives of *cb* (produced by a transverse division) namely *ci* and *m* undergo a series of transverse divisions to form a row of 6-10 *suspensor cells*. The terminal most cell of the suspensor becomes large and vesicle like and probably serves as a *haustorium*. The lowermost cell of the suspensor functions as the *hypophysis*. The hypophysis cell divides to form two tiers of four cells each. Of these, the lower tier forms the root cortex while the upper gives rise to the root cap and the root epidermis.

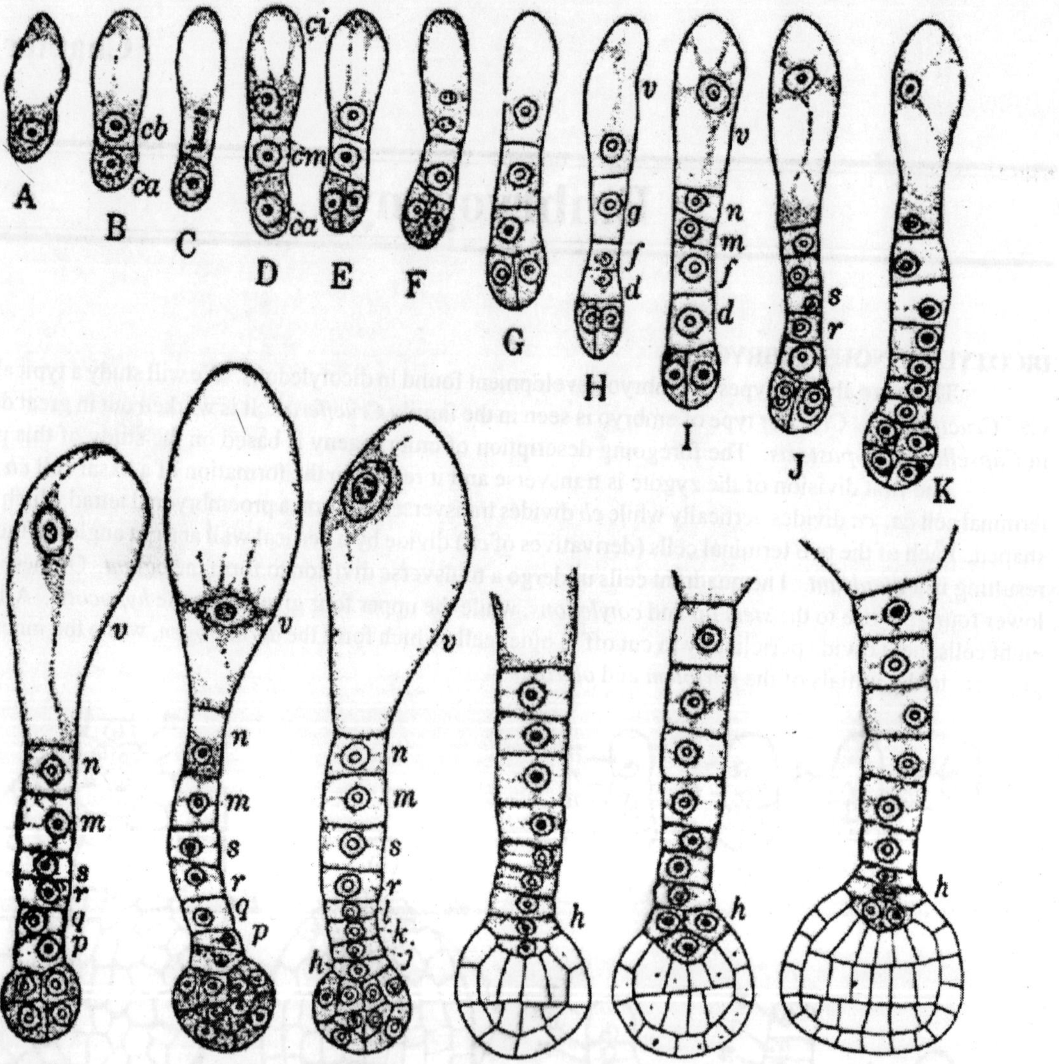

Fig. 26.2 Embryo

Early stages of embryogeny in Capsella bursa - pastoris (redrawn from Soueges, 1914, 1919)

At this stage, further growth in the embryo proper results in the formation of two cotyledons. The embryo now assumes a more or less cordate shape. The hypocotyl and cotyledons increase their size.

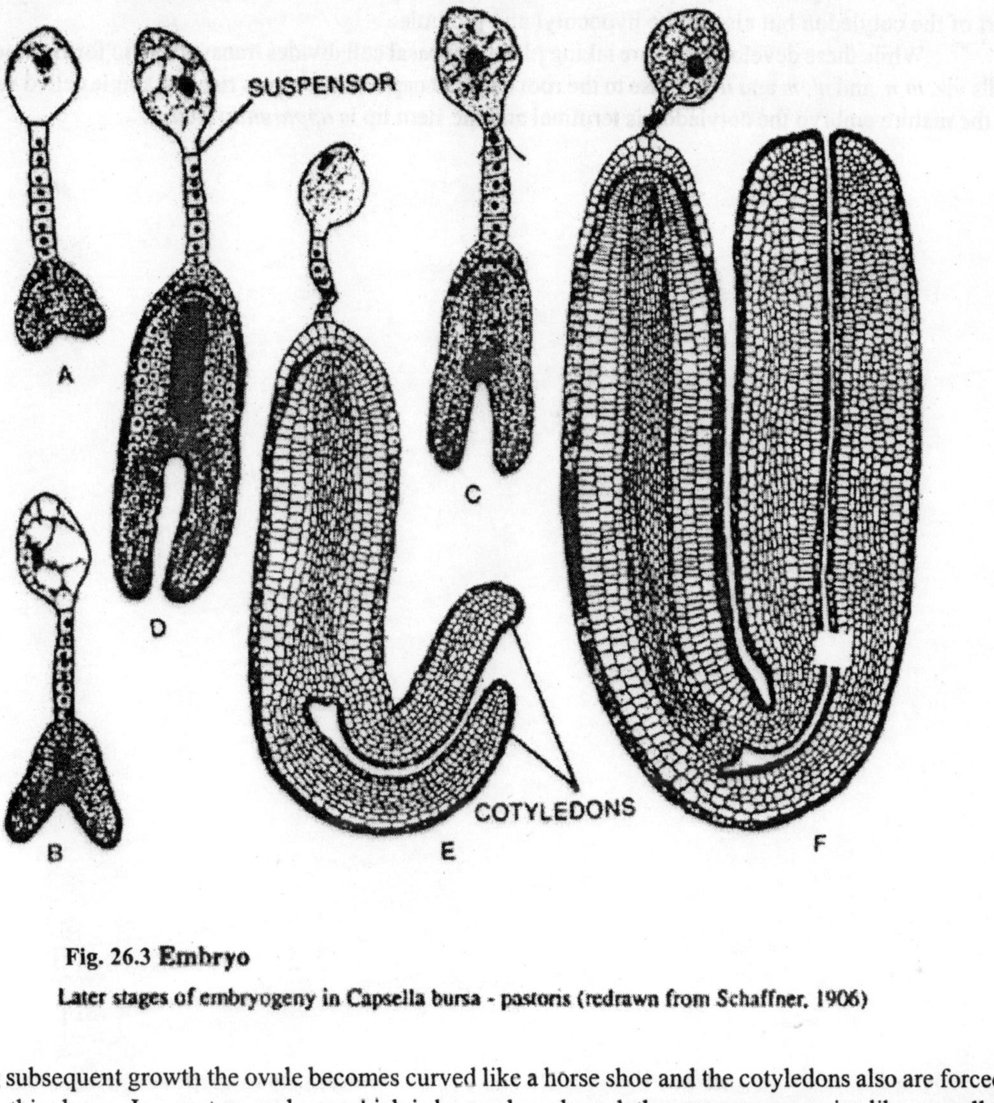

Fig. 26.3 Embryo

Later stages of embryogeny in Capsella bursa - pastoris (redrawn from Schaffner, 1906)

During subsequent growth the ovule becomes curved like a horse shoe and the cotyledons also are forced to assume this shape. In a mature embryo which is horse shoe shaped, the suspensor remains like a small tail.

MONOCOTYLEDONOUS EMBRYO

At stated already, there is not much of a difference in the development of embryo between monocotyledons and dictoyledons. In the mature embryo however, there is difference between the two. It is generally held that in a dicotyledonous embryo, stem tip is terminal and cotyedons are lateral, while in a monocotyledonous embryo it is *vice versa*. Some of the works of Swamy and his co-workers (Lakshmanan, 1972) however have established that even in monocotyledons, stem tip is terminal and later it is pushed towards a side due to rapid growth of the cotyledon. The developement of a typical monocotyledonous (*Luzula forsteri*) embryo proceeds as follows.

The zygote divides transversely to form a basal cell and a terminal cell. This is next followed by the formation of a T shaped proembryo. Quadrants are formed in the terminal cells by a vertical division at right angles to the previous plane of division. Periclinal divisions in the quadrant separate the dermatogen from the central cells. These cells divide transversely to form two tiers l, l^{l}.

The tier *l*, gives rise to a part of the single cotyledon while the tier l^{l} gives not only to the remaining part of the cotyledon but also to the hypocotyl and plumule.

While these developments are taking place, the basal cell divides transversely to form a row of three cells viz. *m*,*n*, and n^{l}, *m* and *n* give rise to the root and root cap while n^{l} gives rise to a single celled suspensor. In the mature embryo the cotyledon is terminal and the stem tip is ***apparently*** lateral.